Gaofeng Zheng

Electrohydrodynamic Direct-Writing

Also of interest

Nanoelectronics.
From Device Physics and Fabrication Technology to Advanced Transistor Concepts
Knoch, 2024
ISBN 978-3-11-105424-7, e-ISBN (PDF) 978-3-11-105442-1,
e-ISBN (EPUB) 978-3-11-105501-5

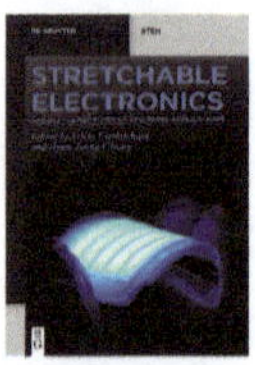

Stretchable Electronics.
The Next Generation of Emerging Applications
Carmichael, Chung (Eds.), 2024
ISBN 978-3-11-075718-7, e-ISBN (PDF) 978-3-11-075728-6,
e-ISBN (EPUB) 978-3-11-075731-6

Surface-Functionalized Nanomaterials.
Environmental, Energy Storage, Energy Conversion Applications
Rasheed, Verma (Eds.), 2025
ISBN 978-3-11-138164-0, e-ISBN (PDF) 978-3-11-138180-0,
e-ISBN (EPUB) 978-3-11-138189-3

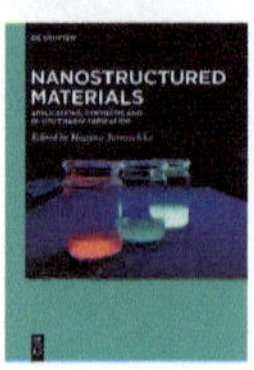

Nanostructured Materials.
Applications, Synthesis and In-Situ Characterization
Terraschke (Ed.), 2023
ISBN 978-3-11-045829-9, e-ISBN (PDF) 978-3-11-045909-8,
e-ISBN (EPUB) 978-3-11-045833-6

Ethics in Nanotechnology.
Emerging Technologies Aspects
Van de Voorde, Jeswani (Eds.), 2021
ISBN 978-3-11-070181-4, e-ISBN (PDF) 978-3-11-070188-3,
e-ISBN (EPUB) 978-3-11-070197-5

Gaofeng Zheng

Electrohydrodynamic Direct-Writing

Micro- and Nanofabrication

Translated by Shumin Guo, Shufan Li, Jiaxin Jiang

DE GRUYTER

厦门大学出版社 XIAMEN UNIVERSITY PRESS | 国家一级出版社 全国百佳图书出版单位

Authors

Gaofeng Zheng
Pen-Tung Sah Institute of Micro-Nano Science and Technology
Xiamen University
361102 Xiamen
People's Republic of China

Xiang Wang
School of Mechanical and Automotive Engineering
Xiamen University of Technology
361024 Xiamen
People's Republic of China

Jiaxin Jiang
School of Mechanical and Automotive Engineering
Xiamen University of Technology
361024 Xiamen
People's Republic of China

Wenwang Li
School of Mechanical and Automotive Engineering
Xiamen University of Technology
361024 Xiamen
People's Republic of China

Translators

Shumin Guo
School of Mathematical Sciences
Xiamen University
361005 Xiamen
People's Republic of China

Shufan Li
School of Opto-electronic and Communication Engineering
Xiamen University of Technology
361024 Xiamen
People's Republic of China

Jiaxin Jiang
School of Mechanical and Automotive Engineering
Xiamen University of Technology
361024 Xiamen
People's Republic of China

ISBN 978-3-11-079756-5
e-ISBN (PDF) 978-3-11-079787-9
e-ISBN (EPUB) 978-3-11-079795-4

Library of Congress Control Number: 2025942108

Bibliographic information published by the Deutsche Nationalbibliothek
The Deutsche Nationalbibliothek lists this publication in the Deutsche Nationalbibliografie; detailed bibliographic data are available on the Internet at http://dnb.dnb.de.

Cover image: PhonlamaiPhoto/iStock/Getty Images Plus
Typesetting: Integra Software Services Pvt. Ltd.
Printing and binding: CPI books GmbH, Leck

www.degruyterbrill.com
Questions about General Product Safety Regulation:
productsafety@degruyterbrill.com

MIX
Papier | Fördert gute Waldnutzung
FSC® C083411

Preface

Precision electrohydrodynamic direct-writing (EDW) micro-nanoprinting technology is an emerging additive manufacturing technology, which is remarkably and widely applied to integrated precision manufacturing in micro-nanodimensions. As a typical external-driven micro-nano jet printing technology, it possesses the properties of easy operation, compact size, and material and process compatibility, opening up new applications in various fields such as flexible electronics, biomedicine, and precision sensing. Research on EDW technology further promotes the theoretical study of microfluid dynamics, multi-field coupling, interface rheology, and other aspects, hence providing an effective and accurate manufacturing method for flexible micro nanosystems.

This book provides the concepts of application requirements of EDW technology in integrated manufacturing of micro-nanosystems and elaborates the research achievements of jet dynamic behaviors, jet electrical characteristics, construction of the control system, precise positioning and deposition of micro-nanostructures with external field control strategy, formation of three-dimensional structures, integrated development and application of functional devices, and so on. This book especially focuses on the discussion of the control strategy and integrated application of precision EDW micro/nano jet printing manufacturing technology.

In organization, this book serves as a technical textbook, featuring a logical blend of introduction and applications, along with the abundant and well-illustrated figures. Our intent has been to provide a systematic track to study the EDW technology, as well as assist professionals in other fields to expand their understanding of the technology.

The preliminary research work was supported by fundings, including National High Technology Research and Development Program of China (863 Program) (no. 2007AA04Z308), National Natural Science Foundation of China (NSFC) Key Program (no. 51035002), NSFC Project (no. 50675184; no. 51305373; no. 5227557), Natural Science Foundation of Fujian Province (no. 2013J05083), and Natural Science Foundation of Guangdong Province (2018A030313522). More than 10 doctoral and master's students and many excellent undergraduates participated in the research work at different stages. Our research team has achieved great results in precision EDW micro-nano jet printing manufacturing technology that are reflected in more than 100 academic papers in peer-reviewed journals, and drawn extensive attention from peers. We also own more than 50 national invention patents. Given the important role of precision EDW micro-nano jet printing manufacturing technology in the integrated manufacturing and application of the micro-nano systems, we take this opportunity to organize a comprehensive summary and analysis of the research work and achievements, to attract more scientific and technological professionals to contribute in the community.

Professor Gaofeng Zheng participated in the writing of Chapters 2, 3, 4, and 5. Professor Xiang Wang participated in the writing of Chapters 3 and 6. Dr. Jiaxin Jiang

https://doi.org/10.1515/9783110797879-202

participated in the preparation of Chapters 1, 4, and 5. Professor Wenwang Li participated in the compilation of Chapters 1 and 4, and reviewed the whole book.

Many of our students contributed to this content in various ways over the years. We particularly appreciate the outstanding contributions of those students who did great research work during their postgraduate study: Lingling Sun, Kai Zhang, Zhaojie Yu, Yihong Lin, and Guangqi He. We are grateful for the assistance given to us in the course of the research work and proofreading of this book: Huatan Chen, Guoyi Kang, Zungui Shao, Biao Qu, Junyu Chen, Xinyi Zhang, Fangqin Shao, Peiyuan Liu, Qingfeng Wang, and Hao Peng. Our thanks also go to the legions of unnamed students who contributed in the research work and provided invaluable ideas and outcomes. Finally, we extend our sincere appreciation to Professor Liwei Lin at the University of California, Berkeley, Professor Sun Daoheng at Xiamen University, and other well-known scholars home and abroad for their years of guidance throughout the research work and the course of preparation of this book.

Any thoughts/comments are welcome and appreciated.

Contents

Chapter 1
Introduction to electrohydrodynamic printing

1.1 Micro-nano inkjet printing technology

The fabrication technology for the micro-nanodevices is maturing and showing a trend of diversification during recent development. Device flexibility, miniaturization, and green manufacturing have become the mainstream of the electronic industry development. The new generation of electronic manufacturing industry, particularly represented by flexible electronic circuits and organic micro-nanodevices, is showing the broad market prospect. Therefore, there emerge the special needs for the manufacturing process, equipment, and materials of flexible electronics and micro-nanoelectronics in the next-generation electronics manufacturing technology and industry.

At present, the micro-nanomanufacturing is mainly concentrated on the traditional semiconductor industry represented by silicon devices, whose manufacturing and integration are based on photolithography process. Although photolithography achieves high control precision, the material consumption, environmental pollution, complicated process, high cost, and the demanding environmental requirements become primary issues obstructing it from satisfying the needs of new-generation flexible and green manufacturing. In addition, the photolithography technology is restricted from most materials with biological and chemical activities due to its material compatibility, which also limits the applications of the traditional micro-nanoprocessing technology to the emerging fields of chemical and biological devices.

Inkjet printing technology is an additive manufacturing process, which pumps jets of various functional ink materials and deposits them to form structural devices. Inkjet printing technology now employs a variety of organic, inorganic, and even biological materials and can also fulfill the preparation of large-area devices on non-silicon substrates (especially flexible substrates). It has been used in the manufacturing of organic transistors, flexible capacitors, flexible displays, smart wearable devices, electronic skins, and other devices [1–4], as shown in Figure 1.1. The inkjet printing technology considerably simplifies the process flow from the traditional manufacturing, improves the production efficiency, and minimizes material waste and environmental pollution, all realized with the cost-effective equipment. It is widely recognized by academia and industry as one of the best technologies for the preparation of flexible electronics and micro-nanoelectronics.

Inkjet printing technology is classified into continuous inkjet printing and drop-on-demand inkjet printing according to the jet generation mechanism [5]. The schematic diagram of the continuous inkjet printing technology is shown in Figure 1.2(a). The solution ejects from the spinneret in the form of a continuous jet under pressure and is split into small droplets under the drive of Rayleigh instability. The deposition of the charged droplets is controlled by the electric or magnetic field. Piezoelectric driving

https://doi.org/10.1515/9783110797879-001

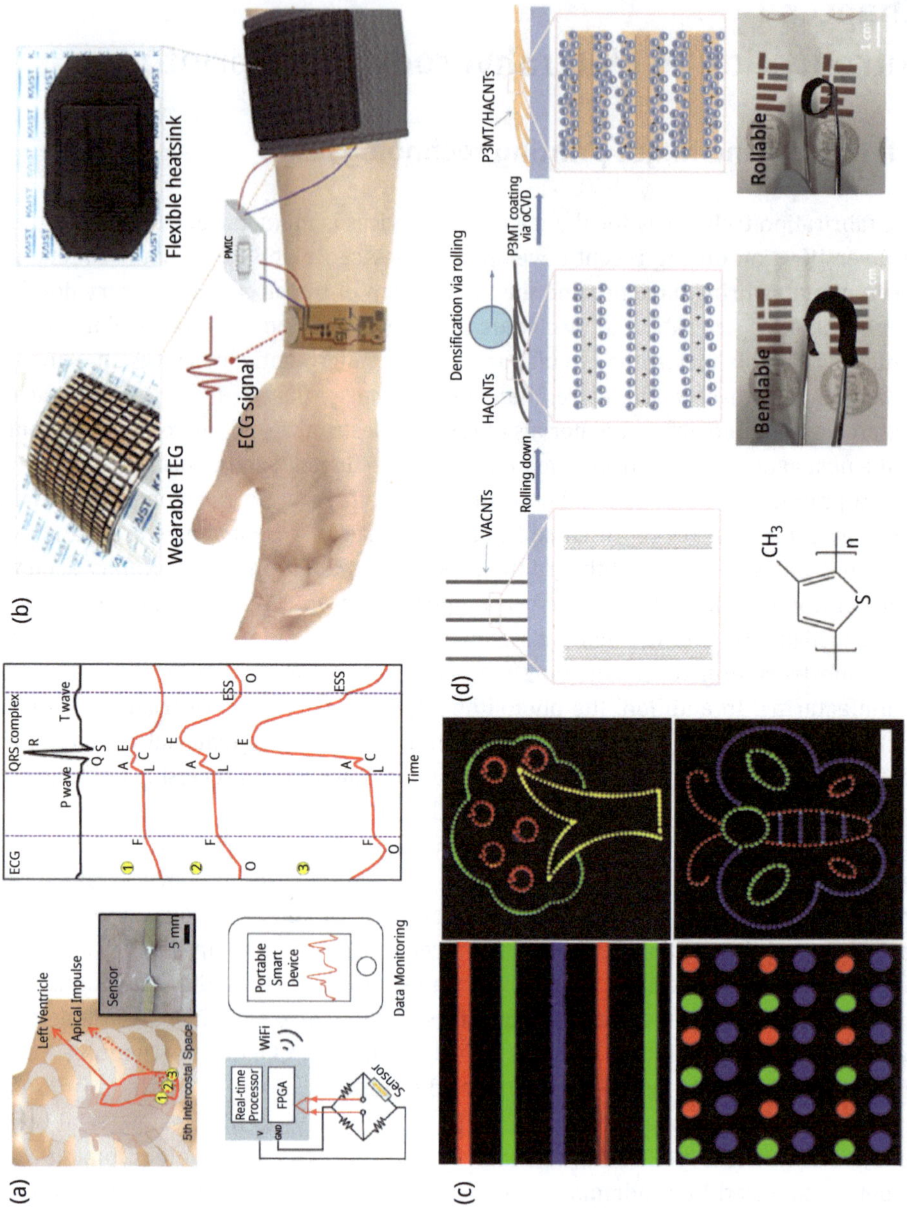

Figure 1.1: Flexible electronic devices: (a) electronic skins [1]; (b) wearable devices [2]; (c) flexible displays [3]; and (d) flexible capacitors [4].

leads the process. Continuous inkjet printing technology achieves high printing quality and efficiency and is mostly used for high-speed printing. In drop-on-demand inkjet printing, droplets are produced in real time on demand as the name suggests, and the deposition of droplets is controlled mechanically by moving the spinneret. Based on different jet-driving methods, drop-on-demand inkjet printing is divided into thermal inkjet (bubble inkjet) printing (Figure 1.2(b)), piezoelectric inkjet printing (Figure 1.2(c)), and electrostatic inkjet printing [6]. Thermal inkjet printing heats up and vaporizes a small amount of solution to produce bubbles and thus squeezes the solution from nozzle. For piezoelectric inkjet printing and electrostatic inkjet printing, the rapid deformation of piezoelectric ceramics or electrostatic drive directly act on the solution, causing it to undergo deformation and ejection.

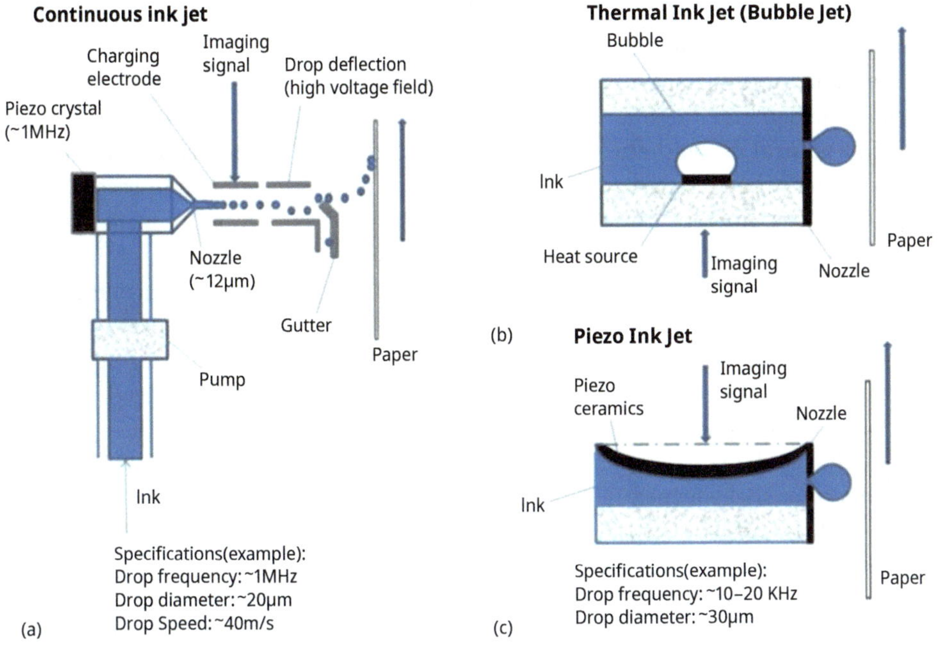

Figure 1.2: Inkjet printing technology [5]: (a) continuous inkjet printing technology; (b) thermal inkjet printing; and (c) piezo inkjet printing.

The electrohydrodynamic (EHD) printing [7] deforms and ejects the solution into fluid flows by the external electric field applied between spinneret and substrate. This is different from the traditional jet printing which forces ejection by the interior pressure in the liquid chamber. EHD printing overcomes the limitation of the traditional jet printing that is only applicable to low viscosity ink and realizes direct printing of high viscosity (up to 15,000 mPa s) solution.

In addition, the jet/droplet originated from tip tip of the Taylor cone achieves a finer dimension than the minimum line width of 20 μm in the traditional jet printing technology, no longer being restricted by the inner diameter of the spinneret. This feature further promotes the application of printing technology in micro-nanosystem manufacturing. By controlling the solution properties and printing process parameters, EHD printing can fulfill the requirements of different manufacturing in terms of continuous fiber structure, discrete micro-nanodroplets, and so on.

1.2 Overview of electrospinning technology

Throughout the development of micro-nanotechnology since the 1980s, electrospinning technology has attracted the attention in various fields owing to its advantages, including simple equipment, easy operation, low manufacturing cost, small fiber diameter, and large surface-area-to-volume ratio, and has become one of the most important methods for the micro-nanostructure manufacturing. So far, more than 100 kinds of natural or synthetic organic materials, ceramic materials, and biomedical materials have been successfully used to produce micro-nanofibers. In recent years, research focus on electrospinning technology has been progressing from spinning mechanism, equipment, and process control to more extensive research including mass production, precise control of electrospinning process, preparation, and application of functional nanofibers.

Figure 1.3 shows the schematic diagram of a typical electrospinning equipment containing the syringe pump, spinneret, high-voltage power supply, and collector. The precision syringe pump delivers electrospinning polymer solution to the spinneret, and the high-voltage power supply activates the jet ejection process [8]. The solution at spinneret orifice shrinks to form the suspended hemispherical droplet under the liquid surface tension as in Figure 1.4(a)–(d). The high-voltage electric field applied between the spinneret and the collector accumulates charges on the surface of polymer droplet, inducing the droplet deformation. When the electric field on droplet and the liquid surface tension reach an equilibrium, the suspended droplet forms Taylor cone at the spinneret, as shown in Figure 1.4(e). Once the electric field force increases to overcome the solution surface tension, the charged liquid jet will eject from the tip of the Taylor cone [9], as shown in Figure 1.4(f).

1.2.1 Instability analysis of the electrospinning jet

The fluid instability and the jet ejection behavior under high electric field provide the theoretical foundations of electrospinning technology. Study of fluid instability under high electric field dates back to the 1880s. In 1882, Rayleigh [10] studied the instability of a droplet under high electric field and concluded as *Rayleigh instability*: when a

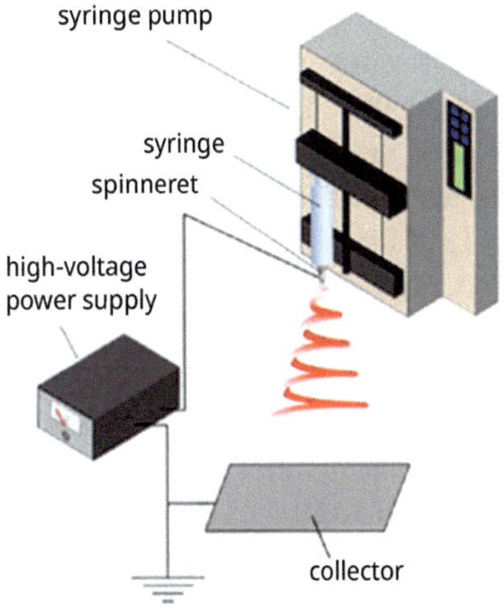

Figure 1.3: Schematic diagram of a typical electrospinning setup [8].

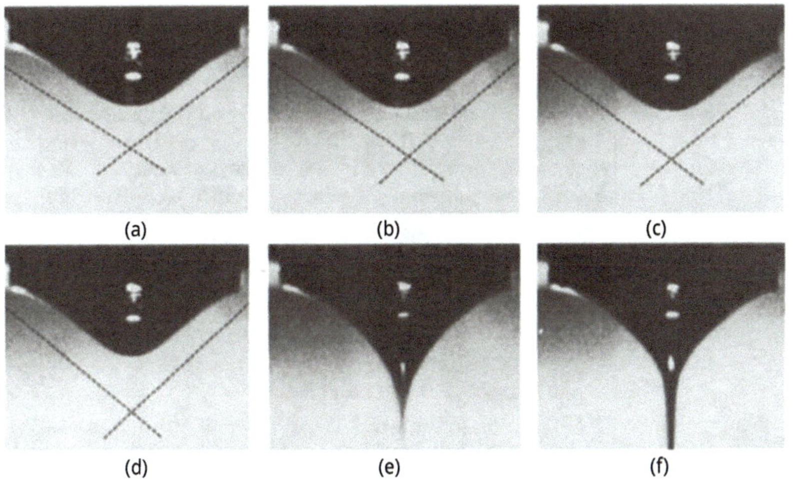

Figure 1.4: Formation of electrospinning jet under high applied voltage [9].

droplet carries a charge exceeding a certain theoretical threshold, it will split and form a micro-jet on the surface. In the 1910s, Zeleny [11, 12] pointed out that the shape of a Newtonian fluid droplet would become unstable when the electric field strength exceeds the critical value, and a conical tip would appear at the end of the droplet. In the 1930s, Formhals [13, 14] developed a production equipment for polymer fibers using electric field force and held a series of patents. This is considered as the start of practical application of electrospinning technology. In the 1960s, Taylor [15, 16] studied the water jet under high electric field. He reported that as the voltage increases, the charged liquid gradually forms a conical droplet at the end of the capillary tube and eventually generates a jet at tip of the cone known as "Taylor cone". He concluded that the cone presents a half-cone angle of about 49.3°. Thereafter, Yarin et al. [17] experimentally studied the elongation behavior of suspended droplets under the action of strong electric field and found that the surface of the droplets became unstable and the cone angle of the Taylor cone was 33.35° instead of 49.3°. They pointed out that the cone angle of a Newtonian fluid is independent of the fluid nature. For viscoelastic fluid, the half-cone angle depends on the viscoelastic force and the surface tension.

The early research work on fluid instability and jet generation laid the foundation for EHD printing technology, but it was not until the past two decades with the rise of micro-nanotechnology that the field of EHD printing saw significant development. EHD printing technology has recaptured attention, the peer attention for its simple device and low cost, and gradually becomes a main approach for micro-nanodroplets and fiber manufacturing. Meanwhile the community gradually shifts the research focus from the equipment and process control to the profound investigation of ejection process model and internal mechanism. EHD printing involves a variety of ejection modes which was carefully classified by Jaworek and Krupa [18, 19] as shown in Figure 1.5. Lots of researchers, such as Hartman et al. [20–22], Reneker et al. [23],

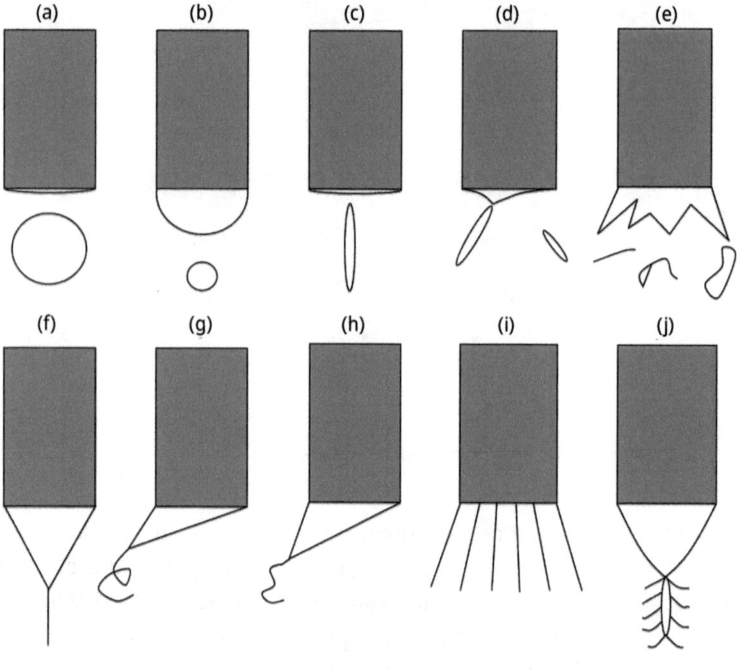

Figure 1.5: Ejection modes of electrohydrodynamic printing jets: (a) dripping; (b) micro-dripping; (c) spindle; (d) multi-spindle; (e) ramified-meniscus; (f) cone-jet; (g) spiral jet; (h) oscillating jet; (i) multi-jet; and (j) ramified jet.

Yarin et al. [17, 24], Hohman et al. [25, 26], and Feng et al. [27, 28] systematically studied the physical characteristics of EHD printing and have made important contributions to the electrospinning technology.

The charged jet in EHD printing experiences an unstable state after a stable straight trajectory. Hartman et al. [22] analyzed the jet instability phenomenon in electrospray based on the perturbation theory and brought in a number of simplifications to model the jet process. They described the jet surface in harmonic perturbation as given as follows:

$$r_s = r_{\text{jet}} + \alpha_0 e^{(wt+jm\theta-jkz)}, \quad k = 2\pi/\lambda \tag{1.1}$$

where r_s is the jet radius, r_{jet} is the jet radius of the unperturbed jet, α_0 is the perturbation amplitude at $t=0$, w is the perturbation growth rate, m is the perturbation coefficient, k is the perturbation wavenumber, and λ is the perturbation wavelength. According to the value of the perturbation coefficient m, three modes of jet perturbation can be distinguished, as shown in Figure 1.6: (1) when $m=0$, the jet splits into droplets, which is an axisymmetric instability; (2) when $m=1$, the jet whips under external forces like charges or air resistance, that is called bending instability; and (3) when $m=2$, the jet ramifies when it is highly charged.

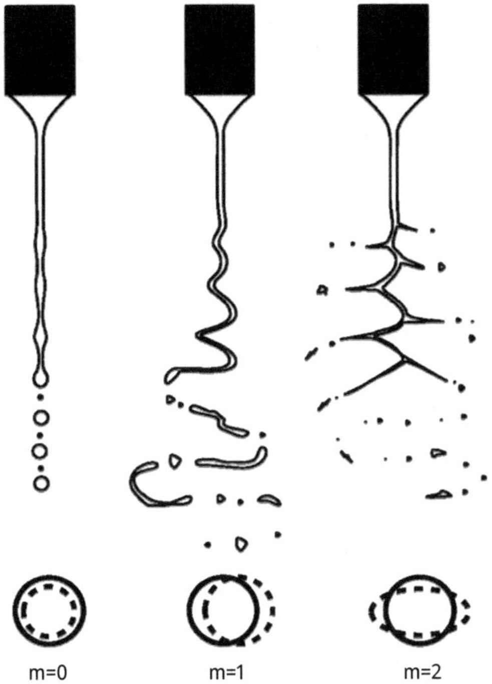

Figure 1.6: Jet breakup modes [22].

Reneker and Yarin [29] analyzed the motion behavior of the jet and summarized the typical path of the charged jet as three stages, as illustrated in Figure 1.7: the jet starts with a straight segment (red) followed by a spiral whipping motion of increasing diameter (yellow) that is caused by the horizontal disturbance. Meanwhile, the spinning jet is elongated and refined by the electric force. Due to the electric repulsion, the path of jet again became unstable and a succession of three or more new smaller diameter electrical bending instabilities develop (green). The new electrical bending instability forms a smaller coil nested onto the larger coil. The bending instability spiral motion diameter decreases periodically as well as the jet diameter, until the jet finally solidifies as disordered micro-nanofibers.

Shin et al. [25, 30, 31] studied and identified the jet instability into three types: (1) the classical Plateau-Rayleigh instability; (2) the electric field-induced axisymmetric varicose instability; (3) non-axisymmetric bending instability (whipping instability). Plateau-Rayleigh instability is mainly induced by the surface tension and viscoelastic forces [32]; axisymmetric varicose instability is mainly driven by charge repulsion between adjacent regions of the jet [23, 33] and non-axisymmetric bending instability is an unstable whipping of jet caused by other electrical stresses and disturbances [24]. Among them, Plateau-Rayleigh instability and axisymmetric instability are responsible for the jet stretching, and they are considered the most likely mechanism of beaded fiber formation along the jet for their symmetry [34]; the non-axisymmetric bending instability greatly affects the further stretching and refinement of the jet. The solvent volatilization and jet refinement into micro-nanofibers are mainly completed in this stage [29].

Carroll et al. [33] closely studied the electrospinning process and concluded that the axisymmetric instability of highly conducting jet is caused by the charge repulsion between adjacent regions of the jet, instead of the instability of solution surface fluctuation. Increasing the electric field intensity and the viscoelastic force of the polymer was conducive to improving the jet stability. Reneker et al. [23] simulated and analyzed the influence of the electric field force on the jet stability. They found that it is the lateral perturbation induced by the repulsive force between adjacent jet charges that mainly causes jet instability. In contrast, the axial viscoelastic force and liquid surface tension improve the stability of jet. They developed a localized approximation to calculate the bending electric field force acting on an electrified polymer jet and formulated the dynamics of unstable motion of the spinning jet. The results confirmed that the charge repulsion force is the dominant factor driving the unstable motion of the jet [24].

1.2.2 Charge transfer during electrospinning process

Electrospinning utilizes the electric field force to stretch the viscoelastic solution to generate a jet, thus achieving the preparation of nanofibers. The charge transfer and trans-

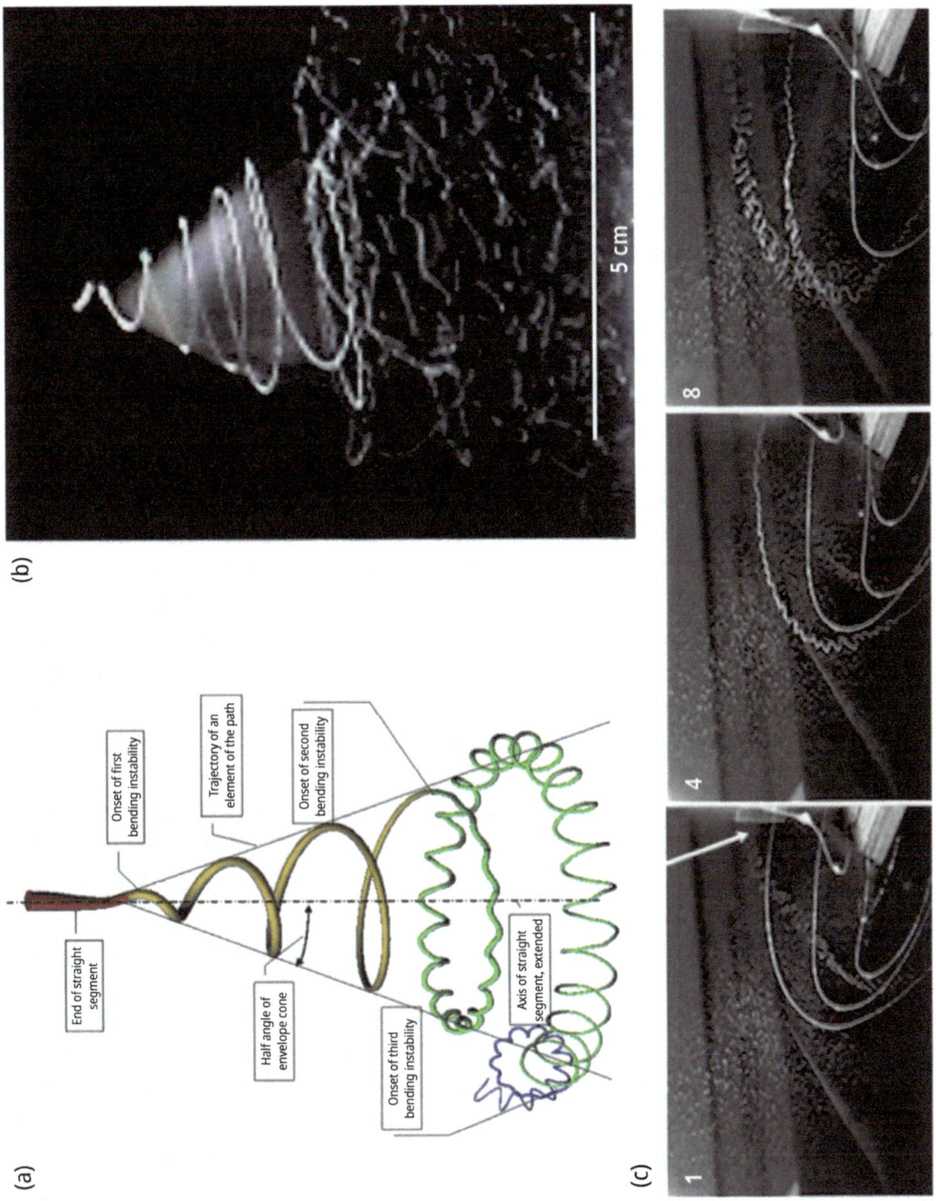

Figure 1.7: Motion state of electrospinning jet [29]: (a) schematic illustration of the instantaneous position of the jet; (b) stereographic and stroboscopic picture of the unstable jet; and (c) evolution of the electrical bending instabilities.

mission in polymer solution and jet directly reflect the electric mechanism of electrospinning as well as directly affect the spinning jet flow motion and nanofiber formation. The change in spinning current represents a macroscopic vision of the charge transfer and transmission in electrospinning process. It also provides an effective way for online real-time monitoring of electrospinning. The spinning current can be categorized into three portions: free electron flow current, Ohmic bulk conduction current, and convention current [28]. The research groups of Reznik et al. [35], Hohman et al. [25, 26], and Yarin et al. [24], established simulation models to predict the magnitude of spinning current. The research results showed that Ohmic bulk conduction current and convention current played the dominant role in the rheological and motion behavior of the spinning jet, while the free electron flow has minor effects on the jet that can be negligible. Mora [36] analyzed the charge transfer in Taylor cone in electrospinning and electrospraying processes. He summarized that the solution conductivity was the main factor affecting the forming of the Taylor cone, which directly affected the size of the Taylor cone apex angle during the jet ejection.

The movement of the charge in the electrospinning solution under the external electric field drives the surrounding solution to move together. The charge migration behavior is different from that in conductors or insulators. Melcher and Taylor [37, 38] developed a leaky dielectric model to describe the unique electrical properties of the polymer solution in the electrospinning process: the charge migration in electrospinning exhibits both features of that in conductors and insulators. The leaky dielectric model is developed on the presupposition that there are two independent charge migrations in the electrospinning: assumes that the electrospinning process involves two independent charge migration behaviors: Ohmic bulk charge conduction and surface charge advection that can be directly superimposed. This model has been widely recognized in the community and had extensive application in studies of electrospinning Taylor cones, jet deformation, and so on.

Monitoring the spinning current is an effective method to control the jet ejection state and spinning process accurately. Kim et al. [39] studied the impact of spinning current on the nanofiber morphology and optimized the current for fabricating uniform nanofibers. Samatham et al. [40] developed an electrospinning current examination system. It used the electric current as a control variable in the electrospinning process to control the quality of fibers. Electric current and the morphology of fibers corresponding to four types of jet regimes were analyzed. A relation between the electric current, type of jet, and morphology of the fibers was established to monitor the electrospinning status. Yang et al. [41] also analyzed the jet current's impact on jet ejection status and nanofiber diameter and reported that uniform diameter nanofibers can be well fabricated when the jet current has small change rate and there is a Taylor cone at the nozzle tip. Bhattacharjee et al. [42] also verified this finding by other way that they scaled the spinning current as a formula of specific electric field strength, solution flow rate, and conductivity. It summarized that increasing electric field gives rise to undulations on jet surface and secondary jets and thus severely affects the nanofiber quality. Munir et al. [43, 44] further investigated the impact of spinning current on nanofiber diameter uniformity and developed a constant-current electrospinning system which observed the real-time electrospinning current as closed-loop feedback variable. This system successfully fabricated the thinner and more uniform nanofibers.

1.3 Controlled deposition of electrospun nanofibers

The irregular and disordered deposition of electrospun nanofibers is a major obstacle to their application in research and industry. There is a critical need for controllable nanofiber deposition especially in the fields of manufacturing tissue structure, flexible electronics, micro-nanodevice systems, and so on. The key point is to overcome the electrospinning instability so that one can achieve controllable deposition of nanofibers.

1.3.1 Positioning deposition of electrospun nanofiber

1.3.1.1 Positioning deposition by collector inducing

The collector-induced method is a creative electrospinning strategy that directs the charged jets to deposit nanofibers in specific morphology and location by modifying the collector electric field distribution with auxiliary electronics. Li et al. [45] designed a collector composed of two conductive strips separated by a variable insulating gap (Figure 1.8(a)) to generate nanofibers in uniaxially aligned arrays over large area and could be further stacked. This successful case opened new doors for electrified collector controlling the electrospun nanofiber deposition. Theron et al. [46] designed a disk collector with tapered edge (Figure 1.8(b)) winding to collect the nanofibers. The electric field concentrated between pendant drop and disk edge takes control of continuous production of nanofibers with diameters ranging from 80 to 400 nm in parallel arrays and controllable intervals.

Salim et al. [47] used a gold-coated polydimethylsiloxane mask on the selective deposition of electrospun nanofibers. The variable grid voltage applied on the gold film generates a repulsive force on charged nanofibers, guiding them to deposit on target position. Ding et al. [48] electrospun the nanofibers on an elastomeric substrate with gold-coated pyramidal protrusions. The electric field distribution closed to flat-tip pyramidal structures forces nanofibers to deposit on the pyramids, as shown in Figure 1.9. The effective positioning deposition of nanofibers on micro-collectors would greatly enhance the applications of nanofibers in micro-nanosystems.

Lee et al. [49] used a pair of metal cylindrical side-wall and sharp-pin ground electrodes as the constraint and collector electrodes to guide deposition while restraining the disordered instability movement of nanofibers, as shown in Figure 1.10. Produced fiber diameter is less than 200 μm. Zhang et al. [50] introduced a conductive needle

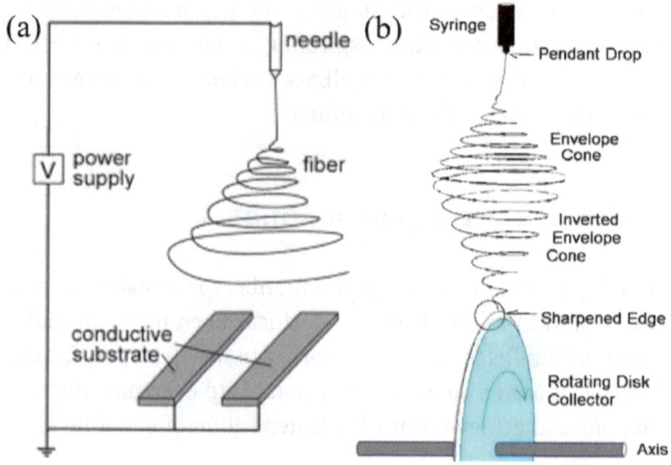

Figure 1.8: Positioning deposition by collector inducing: (a) collector composed of two conductive substrates separated by a void gap [45] and (b) rotating disk collector [46].

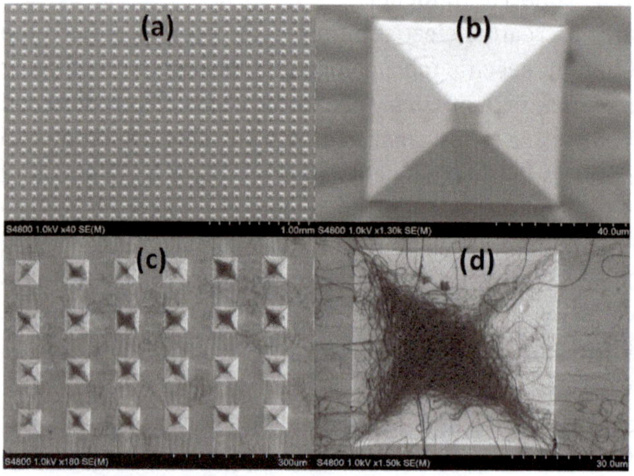

Figure 1.9: Positioning deposition of electrospun nanofibers on the substrate with flat-tip pyramidal protrusions: (a) elastomeric substrate with gold-coated pyramidal protrusions; (b) a single pyramidal protrusion before spinning; (c) patterned fibers on substrate; and (d) a single pyramid with deposited fibers on top [48].

matrix to collect the electrospun nanofibers in arrangement of charged needle array and stack to fabricate three-dimensional structures.

The collector patterned with electroplated or lithography electrodes can configure the electrostatic field distribution in fine space on collector and thus guides and restrains the deposition of charged spinning jet. Zhang et al. [51] used a metal wire

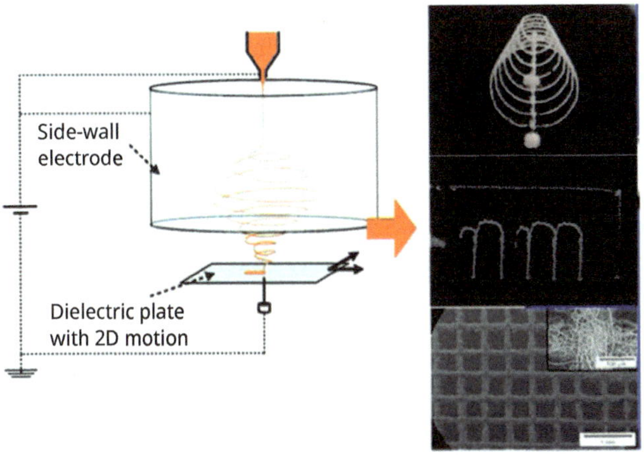

Figure 1.10: Sharp-pin electrode-induced fiber positioning deposition [49].

template as the collector, providing a simple approach to deposit electrospun nanofibers at specific location over a large area. Lee et al. [52] electroplated platinum line of 20 μm width on the glass plate to focus electric field and the charged jet. The nanofibers are strongly attached and deposited on the stationary, straight-line platinum ground, and forms the nanowalls of thickness 180 nm by precisely stacked fibers, as shown in Figure 1.11.

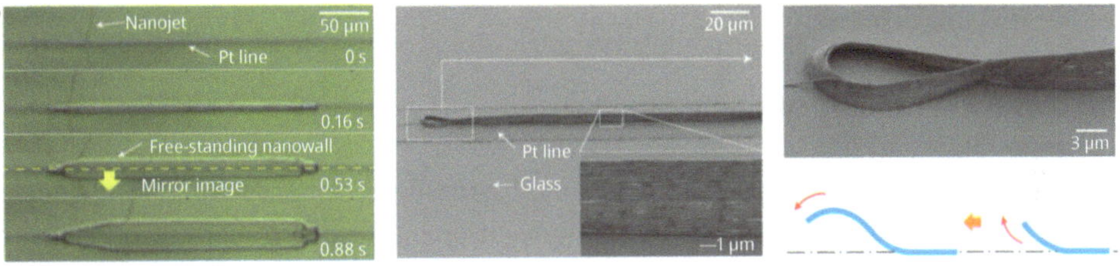

Figure 1.11: Collector-induced controlled nanofiber stacking deposition [52].

1.3.1.2 Positioning deposition based on guiding electrode

Guiding electrode is an effective assist to configure the electric field between the spinneret and the collector in the electrospinning system and therefore constrain the spinning jet generation, spatial motion, and deposition process. It directly guides the deposition of the electrospun nanofibers. Deitzel et al. [53] employed the multiple ring electrodes concentrically around the spinneret as shown in Figure 1.12(a). They applied biased voltages to the spinneret and rings (where both are positive in polarity and ring voltage is lower than that of spinneret and jet) and a negative high voltage to the collector. This electrospinning apparatus forms the electric field toward the electrically grounded collector so that it effectively controls the nanofiber deposition as well as dampens the jet instability. Neubert et al. [54] also installed an electrode of positive high voltage coaxially below the spinneret. It controls the electrospun nanofibers deposition onto collector by the strength of electric field, flow rate of solution, and the distance between spinneret tip, guiding electrode, and collector, as shown in Figure 1.12(b). Yang et al. [55] used an insulating tube instead to reconfigure the electric field by the accumulated charges in tube. The electric field distribution restricts the instability of electrospinning jet and charged nanofibers and therefore improves the deposition precision.

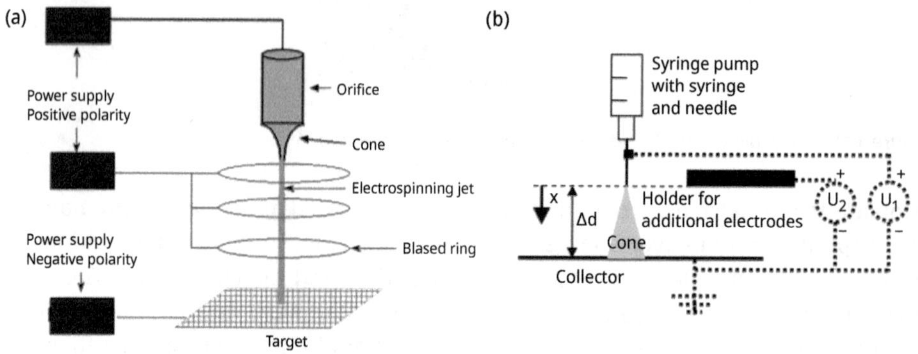

Figure 1.12: Guiding electrode induced-positioning deposition: (a) multiring electrode [53] and (b) auxiliary electric field guiding [54].

1.3.2 Precision control of single nanofiber

The development and extensive application of nanofibers in micro-nanoelectronic devices, micro-actuators, high-sensitivity microsensors, and so on, raise an urgent need for precise positioning and morphology control of single electrospun nanofiber.

1.3.2.1 Scanning tip electrospinning

The research group of Craighead [56, 57] developed a scanning tip electrospinning method for controlled deposition using a microfabricated scanning tip as an electrospinning source. They deposited the polymer solution on scanning tip by dipping in solution (Figure 1.13(a) and (b)) or dispensing by a pipette. The tip acquired a solution droplet of about 20 μL and was placed 2 cm from the wafer electrode. A potential of 4–6 kV was applied between the tip and wafer substrate to form jet ejection. Well-ordered nanofibers of diameters ranging from 100 to 1,800 nm could be fabricated on the wafer whose high-speed rotation is driven by an optical chopper motor. Figure 1.13(c) shows the schematic diagram of polymeric nanofiber fabrication process.

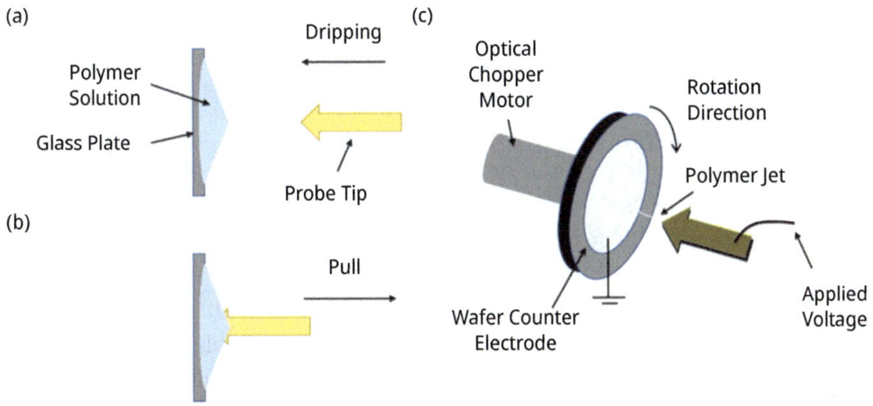

Figure 1.13: Polymeric nanofiber fabrication process [57]: (a) scanning tip was dipped in a polymeric solution; (b) scanning tip was taken out of the polymeric solution and a liquid droplet was formed on the scanning tip; and (c) ordered nanofiber is deposited on the wafer counter electrode.

1.3.2.2 AFM-assisted electrospinning

Atomic force microscope (AFM)-assisted electrospinning adopts AFM as the electrospinning platform, and the AFM tip as the spinneret. The fiber collector replaces the AFM detection platform. A low voltage is applied between the spinneret and the substrate to initiate the spinning jet ejection. Wu et al. [58] presented the AFM-based electrospinning in their work shown in Figure 1.14. They used the AFM tip as the spinneret and dipped it in polymeric solution. The ultra-short distance electrospinning can be realized when the conditions are set: distance between the tip and collector ranges from 5 to 25 μm, and a voltage of 8–12 V is applied in between, and the electric

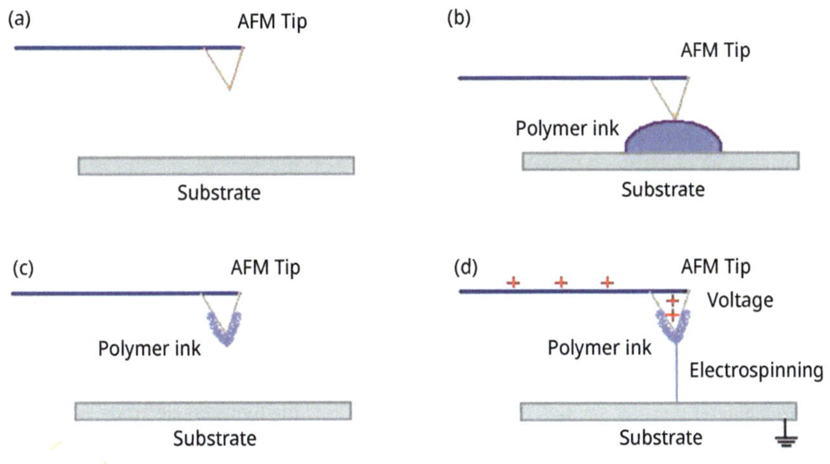

Figure 1.14: Schematic illustration of AFM-assisted electrospinning [58].

field strength is controlled at 0.8 V/μm. Single well-ordered nanofiber is obtained by driving the collector in a predesigned routine at a speed of 5 μm/s.

1.3.2.3 Near-field electrospinning

In 2006, Sun et al. [59] reported the near-field electrospinning (NFES) technology, which addresses the issue that traditional electrospinning generates spiral whipping in nature and has random deposition of nanofibers. It has been developed to deposit solid nanofibers in a direct, continuous, and controllable manner. NFES uses tungsten electrode tip as the nozzle and shortens electrode-to-collector distance within 0.5–3.0 mm. Orderly nanofibers are constructed in position-controllable deposition as the stable liquid jet is formed immediately outside the nozzle under the assistance of intensified electric field strength.

Figure 1.15 shows the schematic illustration of NFES and the Taylor cone and jet status at the nozzle. A tungsten electrode probe with a tip diameter of 25 μm is used as the spinneret, reducing the diameter of the Taylor cone and spinning jet, that allows a faster evaporation of solvent and thus constructs solid nanofibers within a minimal spinning distance. The voltage on electrode reduced to minimum 600 V settles a lower requirement for the insulation property of electrospinning equipment and also reduces distraction of unstable factors in working process. These conditions contribute to a controllable deposition of nanofibers. The controllability of nanofibers is demonstrated by moving the collector using an *x–y* stage. The nanofibers are constructed at fixed position or in predetermined trajectory according to *x–y* stage motion. NFES process shows good visibility that it can monitor the operations in real time, including probe positioning, solution dipping, nanofiber direct writing, and posi-

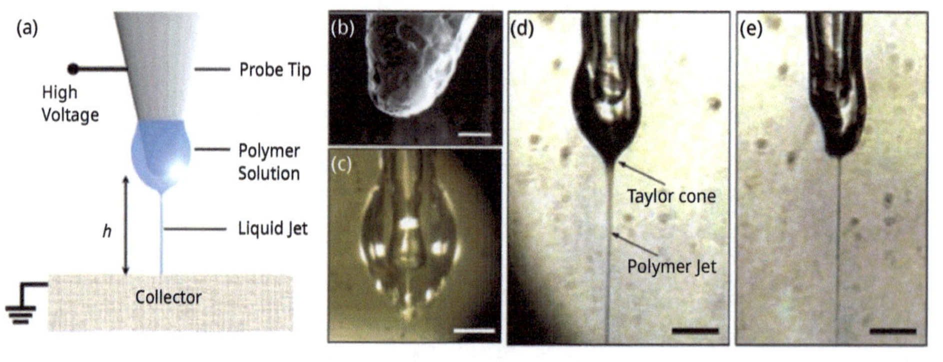

Figure 1.15: Near-field electrospinning [59]: (a) schematic diagram of the NFES using tungsten electrode probe; (b) SEM image of the tip region of the tungsten probe, scale bar 10 μm; (c) an optical photo showing a 50 μm diameter polymer solution droplet collected on tip of the tungsten electrode, scale bar 20 μm; (d) a polymer jet is ejected from the apex of a Taylor cone under applied electrical field and observed under an optical microscope. Scale bar 25 μm; and (e) the size of the polymer droplet decreases as the polymer jet continues to electrospinning. Scale bar 25 μm.

tioning, under microscope. It greatly improves the controllability of nanofiber deposition. Figure 1.16 shows the NFES nanofibers.

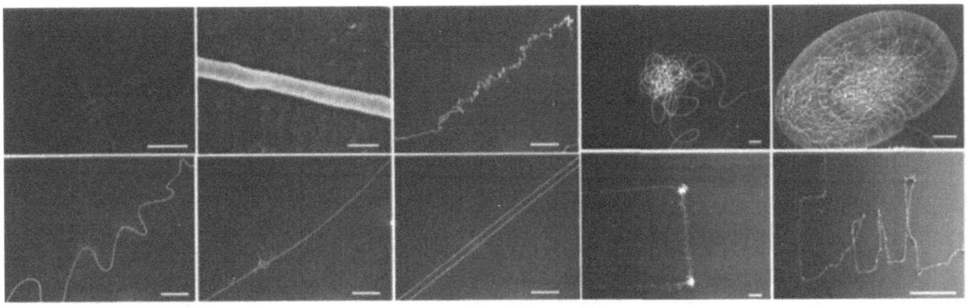

Figure 1.16: Aligned nanofibers and patterns prepared by near-field electrospinning [59].

Compared to scanning tip electrospinning and AFM-assisted electrospinning, the NFES technology possesses simpler equipment and higher visibility and thus is easier to operate. NFES can realize precise positioning and pattern construction of nanofibers on large planar substrates with a single two-dimensional stage.

1.3.2.4 Continuous electrohydrodynamic direct-writing (EDW)

The amount of polymeric solution limits the total length of the directly written structure in a single NFES process. Continuous consumption of polymer droplets also affects size uniformity. To address these issues, a hollow needle is used to replace the solid probe in conventional NFES device. A precision syringe pump delivers stable and constant polymeric liquid to the hollow needle. The EDW system is then able to function continuously and stably for a long time.

The larger curvature radius of the liquid droplet from hollow needle nozzle than that from solid probe tip makes thicker jet ejected from the hollow needle. Larger jet diameter and thus smaller surface-area-to-volume ratio lowers the solvent evaporation rate in jet. It might collect the jet in liquid film form if it has not fully solidified before reaching the collector. To obtain solid nanofibers, Chang et al. [60] developed a continuous EDW process using tungsten probe-induced jet ejection. This technique uses a tungsten probe with 1-µm tip diameter to mechanically poke inside the droplet meniscus that aids electrical force in overcoming surface tension to initiate the jet ejection at subcritical voltage. The continuous EDW of nanofibers is shown as a schematic diagram in Figure 1.17.

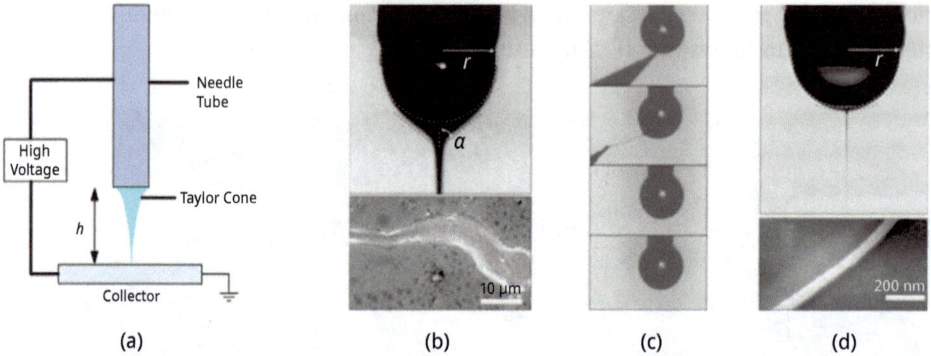

Figure 1.17: Continuous electrohydrodynamic direct-writing (EDW): (a) schematic diagram of EDW apparatus; (b) the syringe needle, droplet, and cone of conventional NFES setup; (c) mechanical drawing of a fiber from polymer droplet by a tungsten probe tip poking inside the polymer meniscus; and (d) the syringe needle, droplet, and cone of continuous EDW [60].

1.4 Applications of electrohydrodynamic printing technology

EHD printing enables the rapid production of ordered micro-nanostructures. Thin-film or three-dimensional structures can be fabricated by directly ejecting the solution or solid fiber on substrates. Zhu et al. [61] achieved a high-resolution dot matrix of 5 μm with full-color and large scalability (Figure 1.18(a)). Lee et al. [62] produced fiber coil structures and else spiral-type inductors by EDW (Figure 1.18(b)). Luo et al. [63] realized the construction of 3D grid structures by consistent and spatially controlled fiber-by-fiber stacking (Figure 1.18(c)).

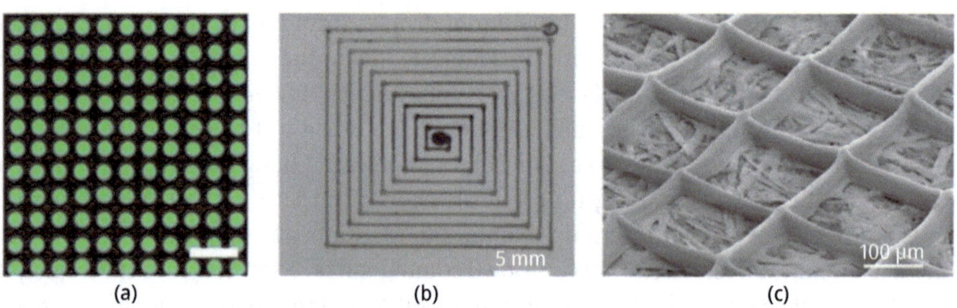

Figure 1.18: Microstructures prepared by EHD printing: (a) high-resolution dot matrix [61]; (b) 2D square-shape pattern [62]; and (c) 3D grid fiber structure [63].

EHD printing technology is also used in the manipulation and assembly of micro-nanoparticles by the ejection and deposition of solutions containing these particles. The particles are retained on substrates and self-assemble during the solvent evaporation. For example, Chen et al. [64] developed the electrohydrodynamic drop-and-place particle deployment (Figure 1.19(a)); Poon et al. [65] and Korkut et al. [66], respectively, achieved the self-assembly and ordered arrangement of polystyrene microspheres (Figure 1.19(b) and (c)). In addition, Onses et al. [67] achieved hierarchical patterns of three-dimensional block-copolymer films with repeatable complex geometries and self-assembly ability by EHD printing.

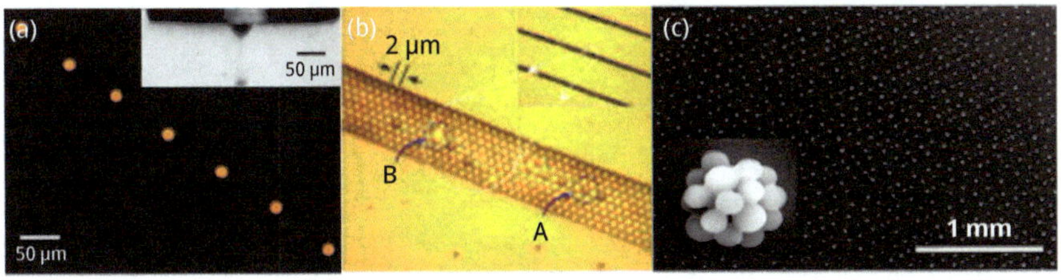

Figure 1.19: Manipulation of micro-nanoparticles by EHD printing technology: (a) uniformly arrangement of micro-nanoparticles [64]; (b) self-assembly of polystyrene microspheres [65]; and (c) patterned aligned polystyrene microspheres [66].

Patterned EHD printing of functional materials possesses an important position in processing technology in micro-nanomanufacturing and flexible manufacturing. It is now commonly used for fabrication of functional structures in the device. Fuh et al. [68] produced a self-powered active sensor with concentric circular topography of piezoelectric fibers, which could mechanically deform in arbitrary direction and be possible to collect the mechanical energy in any deformation (Figure 1.20(a)). Chang et al. [69] made a high-performance piezoelectric nanogenerator with electrospun PVDF nanofibers with in situ mechanical stretching and electric poling characteristics (Figure 1.20(b)). Jeong et al. [70] directly micropatterned lines of 3-hexylthiophene without polymer binder by EHD jet printing to form (OFETs) (Figure 1.20(c)). Liu et al. [71] mixed zinc acetate with polyvinyl alcohol to prepare a spinning solution and used it to fabricate the polycrystalline nanowire-based high-performance photodetectors on a flexible substrate (Figure 1.20(d)). Park et al. [72] demonstrated the controlled EHD printing fabrication of submicrometer-sized droplet of various forms of DNA with less denaturation as well as the aptamer-based biosensor (Figure 1.20(e)).

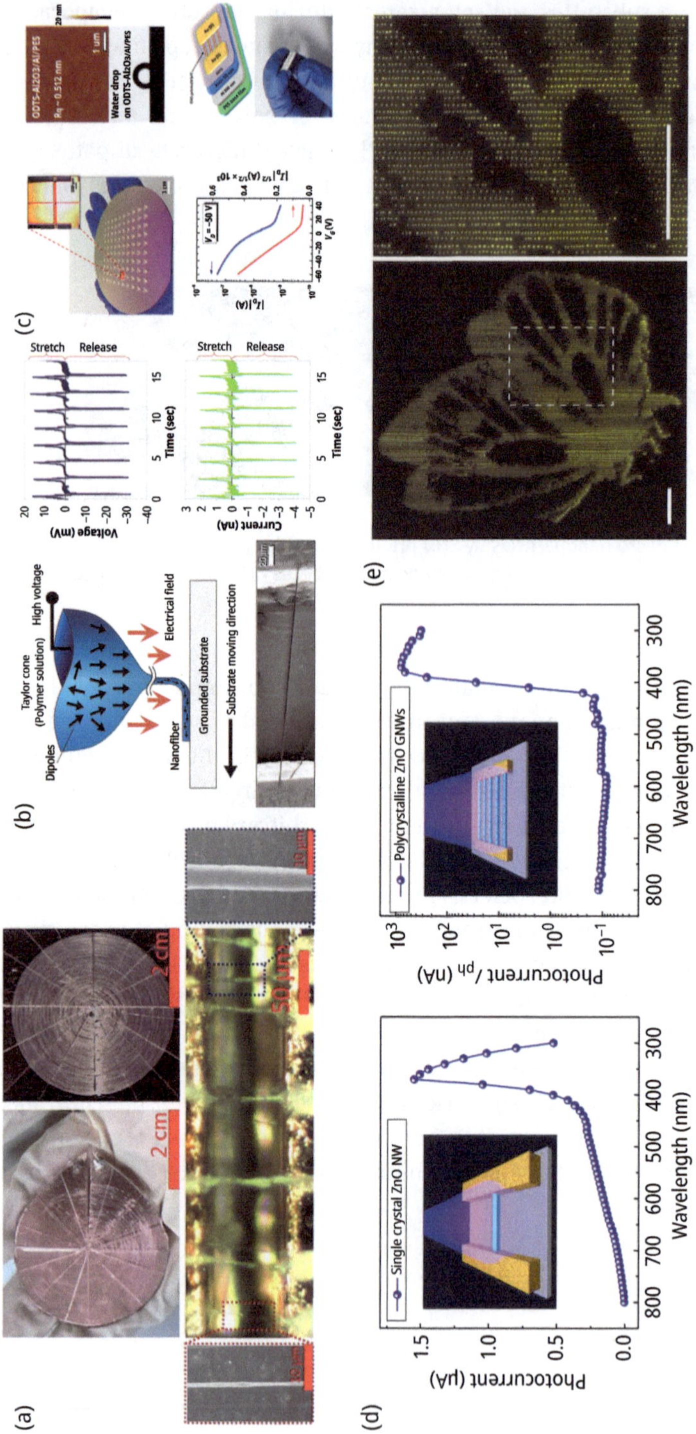

Figure 1.20: Applications of electrohydrodynamic direct-writing technology: (a) self-powered active sensor with concentric circular topography [68]; (b) high-performance piezoelectric nanogenerator [69]; (c) organic field-effect transistors [70]; (d) polycrystalline nanowire-based high-performance photodetector [71]; and (e) dot matrix pattern formed with ink of ssDNA [72].

References

[1] You I, Kim B, Park J. et al. Stretchable e-skin apexcardiogram sensor. Adv Mater 2016, 28(30), 6359–6364.

[2] Kim C S, Yang H M, Lee J. et al. Self-powered wearable electrocardiography using a wearable thermoelectric power generator. ACS Energy Lett 2018, 3(3), 501–507.

[3] Zhu M H, Duan Y Q, Liu N. et al. Electrohydrodynamically printed high-resolution full-color hybrid perovskites. Adv Funct Mater 2019, 29(35), 1903294.

[4] Zhou Y, Wang X, Acauan L. et al. Ultrahigh-areal-capacitance flexible supercapacitor electrodes enabled by conformal p3mt on horizontally aligned carbon-nanotube arrays. Adv Mater 2019, 31(30), 1901916.

[5] Weng B, Shepherd R L, Crowley K, Killard A J, Wallace G G. Printing conducting polymers. Analyst 2010, 135(11), 2779–2789.

[6] Dong H, Carr W W, Morris J F. An experimental study of drop-on-demand drop formation. Phys Fluids 2006, 18(7), 072102.

[7] Park J U, Hardy M, Kang S J. et al. High-resolution electrohydrodynamic jet printing. Nat Mater 2007, 6(10), 782–789.

[8] Xue J J, Xie J W, Liu W Y, Xia Y N. Electrospun nanofibers: new Concepts, materials, and applications. Accounts Chem Res 2017, 50(8), 1976–1987.

[9] Larrondo L, Manley R S J. Electrostatic fiber spinning from polymer melts. I. Experimental observations on fiber formation and properties. J Polym Sci Pol Phys 1981, 19(6), 909–920.

[10] Rayleigh L. On the equilibrium of liquid conducting masses charged with electricity. Philos Mag 1882, 14(87), 184–186.

[11] Zeleny J. The electrical discharge from liquid points, and a hydrostatic method of measuring the electric intensity at their surfaces. Phys Rev 1914, 3(2), 69–91.

[12] Zeleny J. Instability of electrified liquid surfaces. Phys Rev 1917, 10(1), 1–6.

[13] Formhals A. Process and apparatus for preparing artificial threads. 1934: U.S., 1975504 [P].

[14] Formhals A. Method and apparatus for the production of fibers. 1938: U.S., 2123992 [P].

[15] Taylor G. Disintegration of water drops in an electric field. Proc R Soc London 1964, 280(1382), 383–397.

[16] Taylor G. Electrically driven jets. Proc R Soc Lond A 1969, 313(1515), 453–475.

[17] Yarin A L, Koombhongse S, Reneker D H. Taylor cone and jetting from liquid droplets in electrospinning of nanofibers. J Appl Phys 2001, 90(9), 4836–4846.

[18] Jaworek A, Krupa A. Classification of the modes of EHD spraying. J Aerosol Sci 1999, 30(7), 975.

[19] Jaworek A, Krupa A. Jet and drops formation in electrohydrodynamic spraying of liquids: a systematic approach. Exp Fluids 1999, 27(1), 43–52.

[20] Hartman R P A, Borra J P, Brunner D J, Marijnissen J C M, Scarlett B. The evolution of electrohydrodynamic sprays produced in the cone-jet mode: a physical model. J Electrostat 1999, 47(3), 143–170.

[21] Hartman R P A, Brunner D J, Camelot D M A, Marijnissen J C M, Scarlett B. Electrohydrodynamic atomization in the cone-jet mode physical modeling of the liquid cone and jet. J Aerosol Sci 1999, 30(7), 823–849.

[22] Hartman R P A, Brunner D J, Camelot D M A, Marijnissen J C M, Scarlett B. Jet break-up in electrohydrodynamic atomization in the cone-jet mode. J Aerosol Sci 2000, 31(1), 65–95.

[23] Reneker D H, Yarin A L, Fong H, Koombhongse S. Bending instability of electrically charged liquid jets of polymer solutions in electrospinning. J Appl Phys 2000, 87(9), 4531–4547.

[24] Yarin A L, Koombhongse S, Reneker D H. Bending instability in electrospinning of nanofibers. J Appl Phys 2001, 89(5), 3018–3026.

[25] Hohman M M, Shin M, Rutledge G, Brenner M P. Electrospinning and electrically forced jets. I Stability Theory Phys Fluids 2001, 13(8), 2201–2020.
[26] Hohman M M, Shin M, Rutledge G, Brenner M P. Electrospinning and electrically forced jets. II. Appl Phys Fluids 2001, 13(8), 2221–2236.
[27] Feng J J. The stretching of an electrified non-Newtonian jet: a model for electrospinning. Phys Fluids 2002, 14(11), 3912–3926.
[28] Feng J J. Stretching of a straight electrically charged viscoelastic jet. J Non-Newton Fluid Mech 2003, 116(1), 55–70.
[29] Reneker D H, Yarin A L. Electrospinning jets and polymer nanofibers. Polymer 2008, 49(10), 2387–2425.
[30] Shin Y M, Hohman M M, Brenner M P, Rutledge G C. Electrospinning: a whipping fluid jet generates submicron polymer fibers. Appl Phys Lett 2001, 78(8), 1149–1151.
[31] Shin Y M, Hohman M M, Brenner M P, Rutledge G C. Experimental characterization of electrospinning: the electrically forced jet and instabilities. Polymer 2001, 42(25), 9955–9967.
[32] Yu J H, Fridrikh S V, Rutledge G C. The role of elasticity in the formation of electrospun fibers. Polymer 2006, 47(13), 4789–4797.
[33] Carroll C P, Joo Y L. Axisymmetric instabilities in electrospinning of highly conducting, viscoelastic polymer solutions. Phys Fluids Fluids 2009, 21(10), 103101.
[34] Zuo W W, Zhu M F, Yang W, Yu H, Chen Y M, Zhang Y. Experimental study on relationship between jet instability and formation of beaded fibers during electrospinning. Polym Eng Sci 2005, 45(5), 704–709.
[35] Reznik S N, Yarin A L, Theron A, Zussman E. Transient and steady shapes of droplets attached to a surface in a strong electric field. J Fluid Mech 2004, 516, 349–377.
[36] de la Mora, J F. The fluid dynamics of Taylor cones. Annu Rev Fluid Mech 2007, 39, 217–243.
[37] Melcher J R, Taylor T G I. Electrohydrodynamics: a review of the role of interfacial shear stresses. Annu Rev Fluid Mech 1969, 1(1), 111–146.
[38] Saville D A. Electrohydrodynamics: the Taylor-Melcher leaky dielectric model. Annu Rev Fluid Mech 1997, 29(1), 27–64.
[39] Kim S J, Shin K M, Kim S I. The effect of electric current on the processing of nanofibers formed from poly (2-acrylamido-2-methyl-1-propane sulfonic acid). Scr Mater 2004, 51(1), 31–35.
[40] Samatham R, Kim K J. Electric current as a control variable in the electrospinning process. Polym Eng Sci 2006, 46(7), 954–959.
[41] Yang Y, Jia Z D, Li Q, Guan Z C. Controlling the electrospinning process by jet current and Taylor cone. In Proceedings of the conference on electrical insulation and dielectric phenomena 2005, 453–456.
[42] Bhattacharjee P K, Schneider T M, Brenner M P, McKinley G H, Rutledge G C. On the measured current in electrospinning. J Appl Phys 2010, 107(4), 044306.
[43] Munir M M, Iskandar F, Khairurrijal, Okuyama K. High performance electrospinning system for fabricating highly uniform polymer nanofibers. Rev Sci Instrum 2009, 80(2), 026106.
[44] Munir M M, Iskandar F, Khairurrijal, Okuyama K. A constant-current electrospinning system for production of high quality nanofibers. Rev Sci Instrum 2008, 79(9), 093904.
[45] Li D, Wang Y L, Xia Y N. Electrospinning nanofibers as uniaxially aligned arrays and layer-by-layer stacked films. Adv Mater 2004, 16(4), 361–366.
[46] Theron A, Zussman E, Yarin A L. Electrostatic field-assisted alignment of electrospun nanofibres. Nanotechnology 2001, 12(3), 384–390.
[47] Salim A, Son C, Ziaie B. Selective nanofiber deposition via electrodynamic focusing. Nanotechnology 2008, 19(37), 375303.
[48] Ding Z W, Salim A, Ziaie B. Selective nanofiber deposition through field-enhanced electrospinning. Langmuir 2009, 25(17), 9648–9652.

[49] Lee J, Lee S Y, Jang J, Jeong Y H, Cho D W. Fabrication of patterned nanofibrous mats using direct-write electrospinning. Langmuir 2012, 28(18), 7267–7275.

[50] Zhang K, Wang X F, Jing D Z, Yang Y, Zhu M F. Bionic electrospun ultrafine fibrous poly (L-lactic acid) scaffolds with a multi-scale structure. Biomed Mater 2009, 4(3), 035004.

[51] Zhang D M, Chang J. Patterning of electrospun fibers using electroconductive templates. Adv Mater 2007, 19(21), 3664–3667.

[52] Lee M, Kim H Y. Toward nanoscale three-dimensional printing: nanowalls built of electrospun nanofibers. Langmuir 2014, 30(5), 1210–1214.

[53] Deitzel J M, Kleinmeyer J D, Hirvonen J K, Tan N C B. Controlled deposition of electrospun poly (ethylene oxide) fibers. Polymer 2001, 42(19), 8163–8170.

[54] Neubert S, Eblenkamp M, Pliszka D, Sundarrajan S, Ramakrishna S, Wintermantel E. Precise deposition of electrospun nanofibers and electrospraying of nanoparticles as enabling techniques for biomedical engineering applications [C]. In Proceedings of the micro- and nanosystems in medicine, active implants, biosensors 2009, 124–127.

[55] Yang Y, Jia Z, Hou L, Li Q, Guan Z. Controlled deposition of electrospinning jet by electric field distribution from an insulating material surrounding the Barrel of the polymer solution. IEEE T Dielect El In 2008, 15(1), 269–276.

[56] Czaplewski D, Kameoka J, Craighead H G. Nonlithographic approach to nanostructure fabrication using a scanned electrospinning source. J VAC SCI Technol B 2003, 21(6), 2994–2997.

[57] Kameoka J, Orth R, Yang Y N. et al. A scanning tip electrospinning source for deposition of oriented nanofibres. Nanotechnology 2003, 14(10), 1124–1129.

[58] Wu Y, Johannes M S, Clark R L. AFM-based voltage assisted nanoelectrospinning. Mater Lett 2008, 62(4–5), 699–702.

[59] Sun D H, Chang C, Li S, Lin L W. Near-field electrospinning. Nano Lett 2006, 6(4), 839–842.

[60] Chang C, Limkrailassiri K, Lin L. Continuous near-field electrospinning for large area deposition of orderly nanofiber patterns. Appl Phys Lett 2008, 93(12), 214–216.

[61] Zhu M H, Duan Y Q, Liu N. et al. Electrohydrodynamically printed high-resolution full-color hybrid perovskites. Adv Funct Mater 2019, 29(35), 1903294.

[62] Lee D Y, Shin Y S, Park S E, Yu T U, Hwang J. Electrohydrodynamic printing of silver nanoparticles by using a focused nanocolloid jet. Appl Phys Lett 2007, 90(8), 3.

[63] Luo G X, Teh K S, Liu Y M, Zang X N, Wen Z Y, Lin L W. Direct-write, self-aligned electrospinning on paper for controllable fabrication of three-dimensional structures. ACS Appl Mater Inter 2015, 7(50), 27765–27770.

[64] Chen C H, Saville D A, Aksay I A. Electrohydrodynamic 'drop-and-place' particle deployment. Appl Phys Lett 2006, 88(15), 154104.

[65] Poon H F, Saville D A, Aksay I A. Linear colloidal crystal arrays by electrohydrodynamic printing. Appl Phys Lett 2008, 93(13), 133114.

[66] Korkut S, Saville D A, Aksay I A. Collodial cluster arrays by electrohydrodynamic printing. Langmuir 2008, 24(21), 12196–12201.

[67] Onses M S, Song C, Williamson L. et al. Hierarchical patterns of three-dimensional block-copolymer films formed by electrohydrodynamic jet printing and self-assembly. Nat Nanotechnol 2013, 8(9), 667–675.

[68] Fuh Y K, Huang Z M, Wang B S, Li S C. Self-powered active sensor with concentric topography of piezoelectric fibers. Nanoscale Res Lett 2017, 12, 8.

[69] Chang C E, Tran V H, Wang J B, Fuh Y-K, Lin L W. Direct-write piezoelectric polymeric nanogenerator with high energy conversion efficiency. Nano Lett 2010, 10(2), 726–731.

[70] Jeong Y J, Lee H, Lee B S. et al. Directly drawn poly(3-hexylthiophene) field-effect transistors by electrohydrodynamic jet printing: Improving performance with surface modification. ACS Appl Mater Inter 2014, 6(13), 10736–10743.
[71] Liu X, Gu L L, Zhang Q P, Wu J, Long Y, Fan Z. All-printable band-edge modulated ZnO nanowire photodetectors with ultra-high detectivity. Nat Commun 2014, 5, 4007.
[72] Park J U, Lee J H, Paik U, Lu Y, Rogers J A. Nanoscale patterns of oligonucleotides formed by electrohydrodynamic jet printing with applications in biosensing and nanomaterials assembly. Nano Lett 2008, 8(12), 4210–4216.

Chapter 2
Hydrodynamic behavior of EDW jet

2.1 Study on the motion behavior of EDW jet

2.1.1 Principle of electrohydrodynamic direct-writing

Electrohydrodynamic direct-writing (EDW) technology fabricates the micro-nanostructures by external pulling force. It addresses the drawbacks of traditional pressure-induced jet printing technology including large droplet formation and thus low resolution. The principle of the very fine jet formation is shown in Figure 2.1. A spherical suspending droplet is formed at the spinneret due to the surface tension when no electric field exists. After applying an increasing external electric field at the spinneret, the charges distributed on solution surface stretch and deform the large droplet to a cone shape under the electric field. The solution finally overcomes the surface tension and viscoelastic force to eject the cone tip at a critical electric field, which further stretches solution volume to form micro-nanostructures and deposits them on the collector. The electrospinning jet diameter is much smaller than that of the spinneret, restricting the resolution dependence on the spinneret diameter comparing to the traditional inkjet printing technology. It grants the EDW great advantages in high-precision and low-cost fabrication of micro-nanostructures.

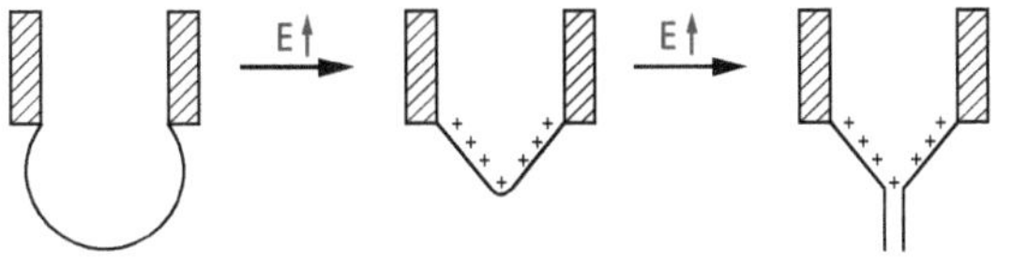

Figure 2.1: Schematic diagram of EDW technology.

In the traditional electrospinning process, the jet exhibits unstable behaviors such as whipping and bending after an initial period of straight, stable motion. These instabilities result in the deposition of disordered nanofibers on the collector after solvent evaporation. EDW technology leverages the straight, stable phase of the traditional electrospinning process.

By reducing the spinneret-to-collector distance, the jet deposits on the collector before any unstable motion occurs, thus enabling precise control over the jet's deposition position.

https://doi.org/10.1515/9783110797879-002

After ejection, the jet undergoes solvent evaporation and eventually deposits as either discrete droplets or continuous fibers on the collector. Depending on application requirements, liquid membranes or fibers can be prepared, as shown in Figure 2.2. If most of the solvent evaporates before the jet reaches the collector, nanofibers can be directly written onto the collector (Figure 2.2(a)). Conversely, if the jet has not completely solidified before deposition, liquid film (Figure 2.2(b)) or discrete droplets (Figure 2.2(c)) will be deposited. By controlling the motion trajectory of the collector during the EDW process, patterned deposition of nanofibers, micron films, and discrete liquid droplets can be achieved, as shown in Figure 2.2(d)–(f).

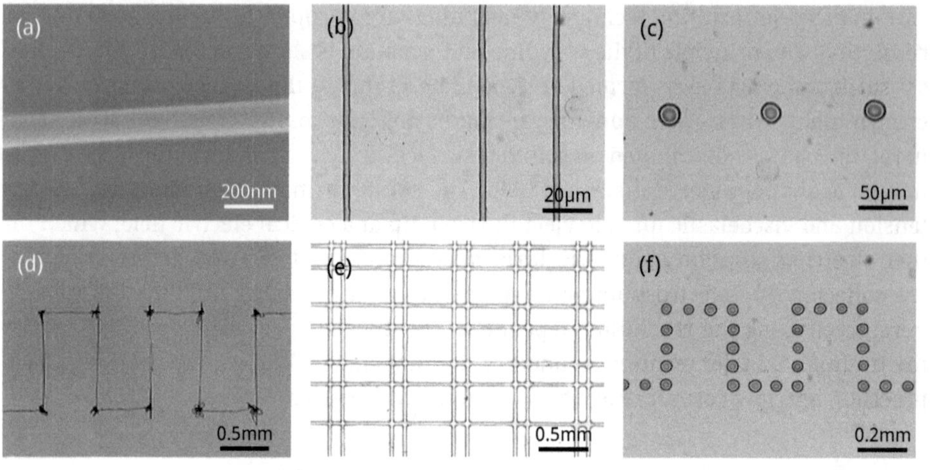

Figure 2.2: Electrohydrodynamic direct-written micro/nanostructures: (a) nanofiber; (b) micron thin films; (c) discrete droplets; and (d)–(f) patterned structures.

2.1.2 Experimental system for electrohydrodynamic direct-writing

The EDW experimental system primarily comprises a custom-built precision EDW jet-printing setup, which includes modules for the spinneret, power supply, solution delivery, jet detection, and motion control. The specific system composition is shown in Figure 2.3. During the experiment, the spinneret is positioned above the collector, which is secured to the motion platform and moves along with it. The anode of the power supply is connected to the spinneret, while the cathode is connected to the collector and grounded. Jet morphology and rheological process are monitored by camera. The patterned deposition is achieved by controlling the collector's moving speed and trajectory.

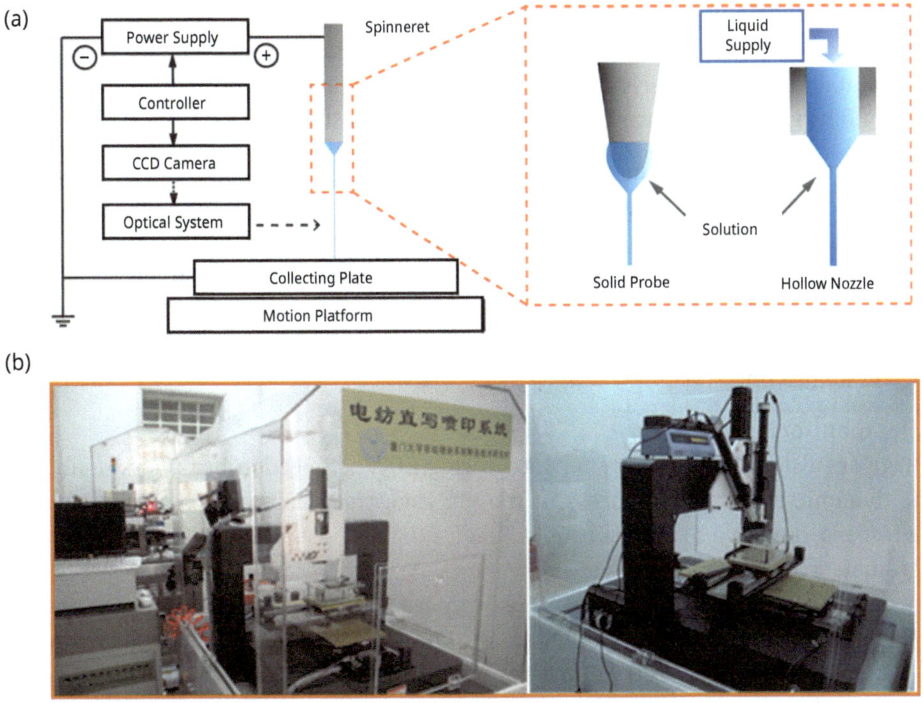

Figure 2.3: EDW experimental system: (a) schematic diagram of module connection and (b) motion platform system.

2.1.2.1 EDW setup

The equipment

The main equipment discussed here is utilized in the EDW experiments in the subsequent sessions.

1. High-voltage power supply

The high-voltage DC source (DW-P603-1ACF1, 0–40 kV, Tianjin Dongwen high voltage Co., Ltd.) is linked for data communication and remote control by host computer through an RS232 interface

2. Spinneret

Two commonly used types of spinnerets are hollow spinneret and the solid probe spinneret. The hollow spinneret features a simple structure and can be continuously supplied with liquid by a precision syringe pump, making it suitable for extended use in EDW. The solid probe, on the other hand, uses a liquid dipping method, allowing for direct writing for approximately 10 s per dipping. The small curvature radius of the solid probe enhances the electric field strength at the probe tip and reducing the required voltage.

3. Precision syringe pump

A precision syringe pump (Harvard 11025 Pico Plus, Harvard, USA) is used to deliver the solution to the spinneret by a stepped flow mechanism. It provides a highly controlled flow rate of liquid ranging from 1.28 pL/min to 88.28 mL/min.

4. Motion control platform

A precision two-dimensional motion platform controls the collector's motion trajectory. The Parker ACR8020 eight-axis motion control card serves as the core controller, managing the X-axis linear motor (Parker TR10-2) and Y-axis linear motor (Parker TR7-1). Advantech IPC-610H industrial control computer operates as the host system. The platform offers moving speed ranging from 0 to 2 m/s, with a positioning accuracy of up to 5 μm.

5. Jet monitoring system

Image acquisition is necessary for monitoring EDW jet ejection behaviors to analyze the dynamic rheological process of the jet. Image acquisition equipment includes high-speed camera and charge coupled device (CCD). The specific product models and parameters are listed as follows:

(1) High-speed camera: GX-1 high-speed camera (NAC Image Technology Inc., Japan) is used, providing a maximum resolution of 1,280 pixels × 1,024 pixels, a maximum frame rate of 200,000 frames per second (FPS), and a minimum exposure time of 2 μs.

(2) CCD: uEye UI-2250G CCD camera (IDS, German) is used, offering a resolution of 1,600 pixels × 1,200 pixels and a frame rate ranging from 12 to 30 FPS.

6. Sample testing and characterization facilities

The solution properties and the morphology and physical characteristics of the samples prepared by EDW are analyzed. The testing instruments and primary characterization methods used are listed as follows:

(1) Sample morphology

① Optical microscope: this comprises a tool microscope (Mitutoyo, Japan) paired with a digital video camera (Sony SSC-DC80, Japan). The microscope offers an optical magnification of 20–2,000×.

② Scanning electron microscope (SEM): Two SEMs are used for analysis. The first is a field-emission SEM (LEO1530, Germany) with a magnification range of 20–900,000× and a resolution of 1 nm. The second is an environmental SEM (XL30, FEI, the Netherlands) with a magnification range of 6–100,000× and a resolution of 3.5 nm.

(2) Electrical properties

① Digital multimeter: A 6.5-digit digital multimeter (Agilent 34410, Agilent Technologies Co., Ltd, USA) is used to confirm the electrical properties of the samples. Its performance parameters include a voltage range of 0–750 V, a current range of 0–3 A, and a resistance range of 0–1 GΩ.

② Semiconductor characteristic measurement system: A semiconductor parameter analyzer (Keithley 4200-SCS), produced by Keithley Corporation (USA), is used. It features a voltage measurement range of 1 μV to 200 V and a current measurement range is 100 fA to 100 mA.

(3) Solution properties

① Conductivity meter: A DDS-307A conductivity meter (Shanghai INESA Scientific Instrument Co., Ltd.) is used, featuring a measurement range of 0–100 mS/cm. The electronic unit has a basic error of ±1.0% of full scale (FS), and the instrument's overall rate is ±1.5% FS.

② Rotary viscometer: The rotary viscometer NDJ-8S (Shanghai Fangrui Instrument Co., Ltd.) is used with the measurement range of 10 mPa·s-2,000 Pa·s, and the accuracy of ±2%.

(4) Mass

Precision electronic balance: Sartorius BS 124 Electronic Balance (Sartorius, Germany) is used, featuring a weighing range of 0–120 g, resolution of 0.1 mg, and repeatability of 0.1 mg.

2.1.2.2 Primary experimental materials for electrohydrodynamic direct-writing

1. Polyethylene oxide(PEO) is supplied by Changchun Dadi Fine Chemical Co., Ltd, China. The molecular formula is $(CH_2 - CH_2)_n$, having a molecular weight of 300,000 g/mol, melting point of 66–70 °C, a thermal decomposition temperature of 423–425 °C, and an apparent density of 0.15–0.3 g/cm^3. Unless stated otherwise, all solutions used in the experiments presented in this book are PEO solutions.
2. Polyvinyl pyrrolidone (PVP) is supplied by Acros Organics. The molecular formula is $(C_6H_9NO)_n$. The molecular weight is 1,300,000 g/mol (K90). The melting point is 110–180 °C and the apparent density is 1.2 g/cm^3.
3. Polyimide (PI) is supplied by Hangzhou Surmount Technology Co., Ltd. The PI solution is made of 30% w/w with volatile solvent.
4. Deionized water is supplied by Xiamen University, Pen-Tung Sah Institute of Micro-nano Science and Technology. The conductivity is approximately 0.056 μS/cm.
5. Other commonly used chemicals, including glycerol ($C_3H_8O_3$) and zinc acetate ($Zn(CH_3COO)_2 \cdot H_2O$), are supplied by Sinopharm Chemical Reagent Co., Ltd. The purity standard is certified analytical reagent (AR).

2.1.2.3 Experimental procedure for EDW

The solute is solved in solvent and stirred thoroughly with a magnetic stirrer at room temperature for 24 h. The solution is then left to stand for another 24 h for later use.

The main steps of EDW experiment are as follows:

1. Extract the appropriate amount of solution using a disposable syringe and place it on the precision syringe pump.

2. Start the precision syringe pump and deliver the solution to the spinneret with adjustable solution supply rate.
3. Once a stable liquid supply is achieved, turn on the high-voltage power supply to induce solution ejection and deposition on the collector.
4. Monitor parameters and capture images during the ejection process, while recording the jet behavior.
5. Analyze the morphology of the electrospun structures using optical microscopy, SEM, and other analytical instruments.

2.1.3 Ejection behavior of EDW jet

In traditional electrospinning or electrospraying, nanofibers or droplets mainly form during the unstable phase of the jet, with minimal influence from the initial dynamic jet formation. The spinneret-to-collector distance is relatively large, so the fibers or droplets deposited on the collector have little impact on the overall process. As a result, the initial jet formation and its stability control are less prioritized in these methods. However, the EDW jet is deposited on the collector almost immediately after its generation, so its formation and ejection behavior directly affect the quality of the resulting patterns. In EDW, stable jet ejection is essential for precise pattern formation, making it necessary to investigate the jet's formation process and develop control strategies to regulate its stability.

2.1.3.1 Jet formation process

1. Jet formation process with solid probe spinnerets

By mounting a high-speed camera onto the microscope platform, the droplet deformation at the probe tip and the jet formation process under a strong electric field can be clearly recorded, as shown in Figure 2.4. The moment when the DC high voltage is first applied is marked as 0 s. In Figure 2.4(a), the polymer solution at the probe tip retracts into a hemisphere due to surface tension. As the electric field is applied, free charges accumulate on the solution surface, causing the deformation of the polymer solution into a Taylor cone (Figure 2.4(b) and (c)). As the charge density on the surface increases, the electric field force acting on the polymer solution increases. Once the electric field force is strong enough to overcome surface tension, a jet is ejected from the tip of the Taylor cone as shown in Figure 2.4(d) at 0.983 s. The jet continues to extend under the influence of the electric field, making its diameter gradually decrease (Figure 2.4(e)–(h)). Before jet formation, a high density of charges accumulates on the solution surface at the spinneret, increasing the electric field force on the polymer solution. This results in the jet carrying away a larger volume of polymer solution as it ejects from the Taylor cone. Surface tension causes the solution to shrink to a bead-like structure at the leading edge of the jet, as highlighted by the box in Figure 2.4(f) and (g).

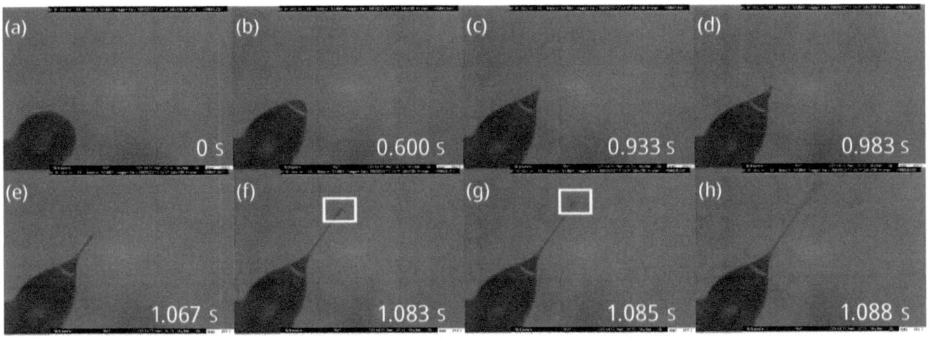

Figure 2.4: Taylor cone and jet formation process in EDW with solid probe (a PEO solution concentration of 16%, a spinneret-to-collector distance of 3 mm, an applied voltage of 2.4 kV, and a spinneret diameter of 40 μm).

The axial stress generated by the electric field force plays a key role in maintaining the jet's stability, while the repulsive forces between like charges on the jet cause chaotic movement. Within the confines of the electric field force, the jet maintains a brief, stable linear motion, with the oscillation-free length of the straight jet being approximately 100 μm, as shown in Figure 2.4(f). As the jet continues to eject, it begins to exhibit oscillatory behavior due to factors such as charge repulsion, as shown in Figure 2.4(g) and (h) (the high-frequency oscillations result in a ghosting effect in the high-speed camera images). The front of the jet advances under the electric field force, its length increases, while its diameter decreases, and the amplitude of oscillations increases. These oscillations alter the jet's trajectory and affect its deposition position.

2. Jet formation with hollow spinneret

When using a hollow spinneret, a continuous liquid supply is maintained via a syringe pump. Once the jet is formed, it self-regulates and tends toward stability during the spinning process, ensuring a prolonged stable state. The formation process of EDW jet with hollow spinneret is shown in Figure 2.5. Initially, the solution at the spinneret forms a pendent spherical droplet due to surface tension. Upon the application of voltage, the droplet begins to deform under the influence of the electric field force, eventually taking on a conical shape and ejecting. During the initial ejection phase, the lower part of the droplet has a higher charge density, causing the jet to carry more solution, resulting in the formation of a small droplet at the jet front. The volume of the droplet increases with the solution viscosity. Once the jet exits the Taylor cone, it accelerates toward the collector under the electric field, with its diameter decreases as the stretching continues. As shown in Figure 2.5, the diameter of the initial jet gradually decreases after deposition (from 16.0 to 16.4 ms), while the morphology of the Taylor cone also changes and eventually stabilizes at around 17.6 ms.

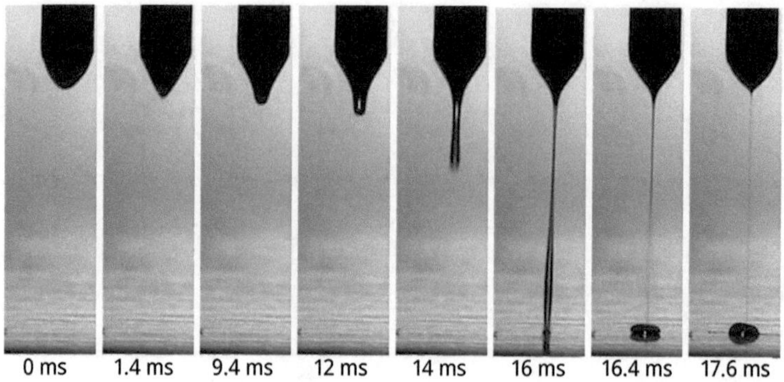

Figure 2.5: Ejection process of EDW jet with hollow spinneret (glycerin solution, the applied voltage, the inner diameter of the spinneret, the liquid supply rate, and spinneret-to-collector distance are 3.2 kV, 0.6 mm, 800 μL/h, and 6 mm, respectively).

The initial jet diameter and the droplet volume at the jet front can affect the size of the initial deposition spot. To gain a better understanding of the jet formation process and achieve the control over the jet behavior, we analyze the effects of applied voltage and solution supply rate on the initial jet state. The increasing applied voltage enhances the electric field force on the solution surface, resulting in a reduction of both the initial jet diameter and the droplet volume at the jet front, as shown in Figure 2.6 (a). Figure 2.6(b) illustrates the morphology of the initial jet at different solution supply rates. When the supply rate increases from 400 to 1,200 μL/h, no significant change is observed in the initial jet diameter or the droplet size at the jet front.

Figure 2.6: Initial jet morphology at different voltages and liquid supply rates: (a) glycerin solution with a feeding rate of 400 μL/h, an inner spinneret diameter of 0.6 mm, and a spinneret-to-collector distance of 6 mm and (b) glycerin solution with an applied voltage of 3.5 kV, an inner spinneret diameter of 0.6 mm, and a spinneret-to-collector distance of 6 mm.

3. Motion state of the jet front

To further investigate the movement behavior of the charged jet, the entire EDW jet processes are recorded using a high-speed camera. Figure 2.7 shows the generation and movement process of the charged jet.

Under the influence of the electric field, the droplet forms a Taylor cone, and a jet is ejected from the cone tip, as shown in Figure 2.7(a)–(f). The charged jet accelerates toward the collector due to the electric field. However, the jet exhibits significant oscillatory motion caused by factors such as electric field forces and charge repulsion, as shown in Figure 2.7(g)–(k). The oscillation amplitude increases with the jet length. In Figure 2.7(j), the jet length reaches about 300 μm, and with a distance of 450 μm from the collector surface, the oscillation amplitude is 180 μm. As the distance between the jet front and the collector surface decreases below 200 μm, the oscillation amplitude begins to diminish. In Figure 2.7(k), with the distance between the jet front and the collector reduced to 100 μm, the oscillation amplitude is only 26 μm. Ultimately, driven by the electric field force, the jet deposits at the intersection of the extension line of the spinneret axis and the surface of the collector, as indicated by the small circle in Figure 2.7(l).

When the jet front reaches the deposition point on the collector, it is constrained and effectively suppressing large-amplitude oscillations. The resulting straight-line jet between the spinneret and the collector can be used for the controllable preparation of single nanofibers, as shown in Figure 2.7(l).

It takes 5.6 ms for the jet front to travel 0.75 mm, resulting in an average velocity of 0.13 m/s (Figure 2.7(g)–(l)). By tracking the trajectory of the jet front, the variation in jet velocity over time and distance can be obtained, as shown in Figure 2.8. The jet is accelerated by the electric field force, with the increasing velocity in the jet progress. The jet front reaches a maximum speed of 0.50 m/s. Before the jet front deposits on the collector, the electric field stretches the jet to thin, reducing the constraint on the bead-like structure at the jet front, which results in greater acceleration. When the jet further approaches the collector, its moving speed rises rapidly, making control difficult as the front experiences minimal restriction, moves faster, and oscillates more widely. Once the jet starts to deposit on the collector, the speed decreases due to the obstruction of the deposited portion.

The variation in the maximum velocity of the jet front as a function of applied voltage and solution concentration is shown in Figure 2.9. As the solution concentration increases, the jet front velocity decreases, while it increases with higher applied voltage. The increase in voltage amplifies the electric field force acting on the jet, thus boosting the velocity of the jet front. Conversely, a higher solution concentration increases the viscoelastic forces acting on the jet, which reduces its velocity.

4. Critical voltage for jet ejection

Once the applied voltage surpasses a certain threshold, known as “critical voltage” or “starting voltage”, the electric field force acting on the polymer solution becomes

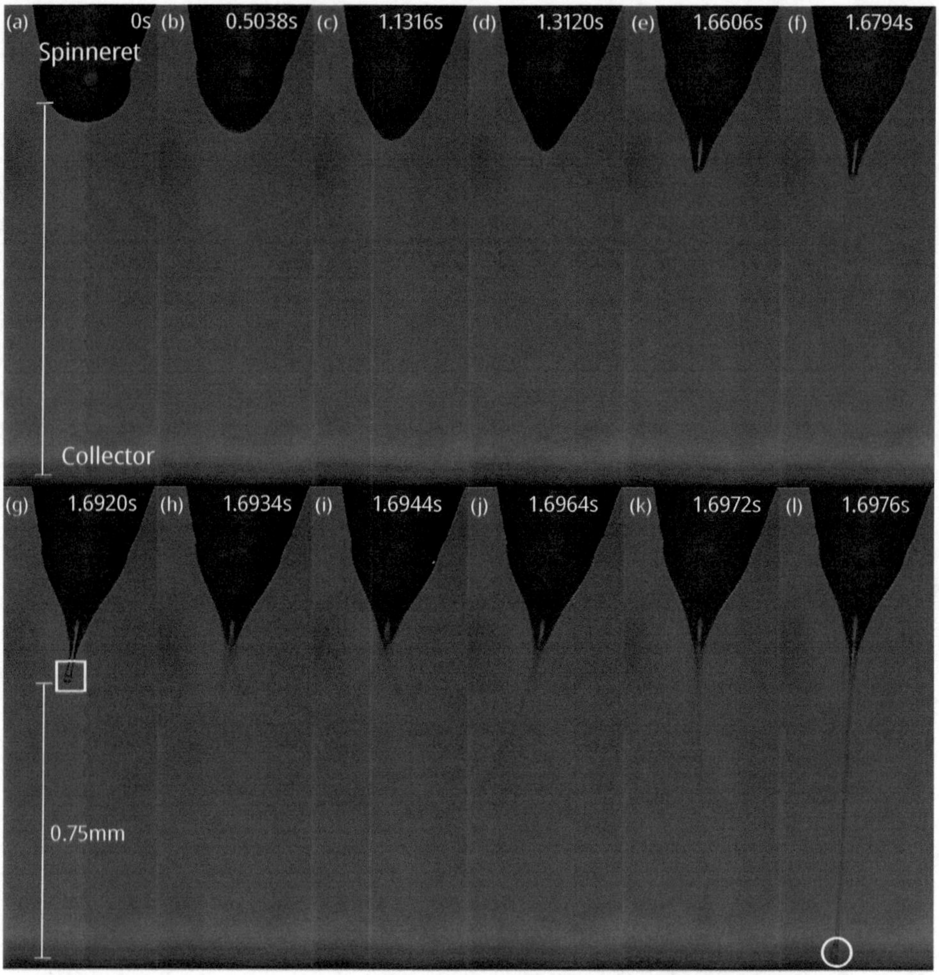

Figure 2.7: Motion dynamic of EDW jet (silicon substrate, a shooting speed of 2,500 FPS, a spinneret-to-collector distance of 1 mm, an applied voltage of 1.7 kV, and solution concentration of 20%).

strong enough to overcome the surface tension of the solution, causing a jet to eject from the tip of the Taylor cone. The effects of the spinneret diameter, spinneret-to-collector distance, and solution concentration on the critical voltage of EDW are analyzed, and the corresponding curves are shown in Figures 2.10–2.12, respectively.

The applied voltage in the electrospinning process primarily influences the jet ejection by affecting both the electric field strength at the spinneret and the charge density accumulated on the surface of the solution [1–3]. For a given applied voltage, the electric field strength at the nozzle increases as the spinneret diameter decreases. Additionally, a smaller spinneret diameter leads to a higher accumulate of charges on the surface of solution droplet at the spinneret. Therefore, the critical voltage required for

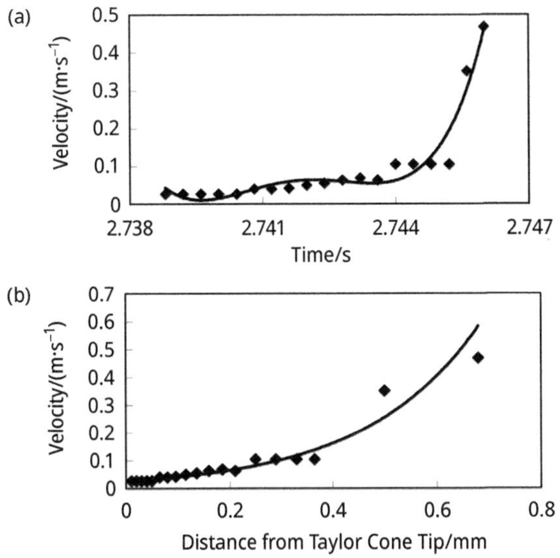

Figure 2.8: Velocity of the jet front: (a) variation of velocity over ejection time. The horizontal axis indicates the time elapsed after the application of DC high voltage and (b) variation of jet velocity with travel distance. The horizontal axis indicates the jet travel distance from the Taylor cone apex (silicon substrate, with a nozzle-to-collector distance of 1 mm, an applied voltage of 1.7 kV, a solution concentration of 20%, and a spinneret diameter of 40 μm).

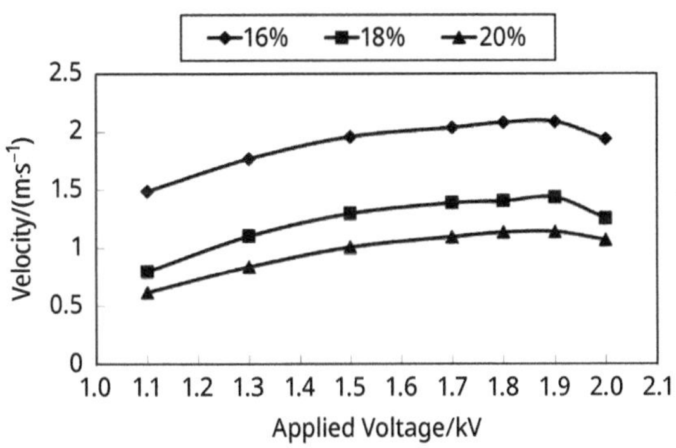

Figure 2.9: Maximum velocity of jet front (silicon substrate with a spinneret-to-collector distance of 1 mm and a spinneret diameter of 40 μm).

EDW increases with the spinneret diameter, as shown in Figure 2.10. Increasing the spinneret-to-collector distance weakens the space electric field strength, thereby increasing the critical voltage required for EDW, as shown in Figure 2.11. Furthermore, increasing the solution concentration rises the viscosity of the solution, which impedes the jet ejection, causing the critical voltage *t* increase with the solution concentration, as shown in Figure 2.12.

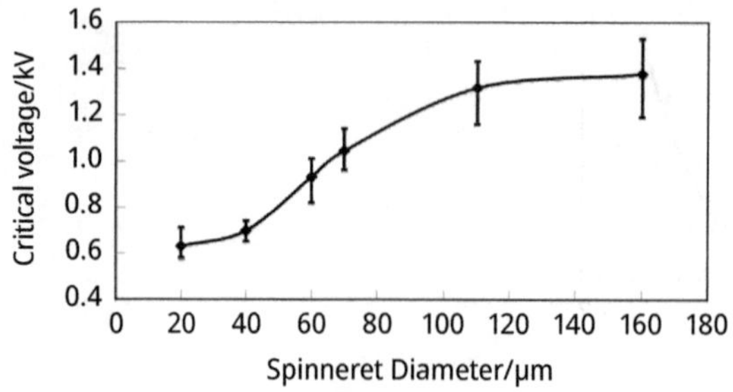

Figure 2.10: Plot of critical voltage versus spinneret diameter (silicon substrate with a spinneret-to-collector distance of 0.5 mm and solution concentration of 10%).

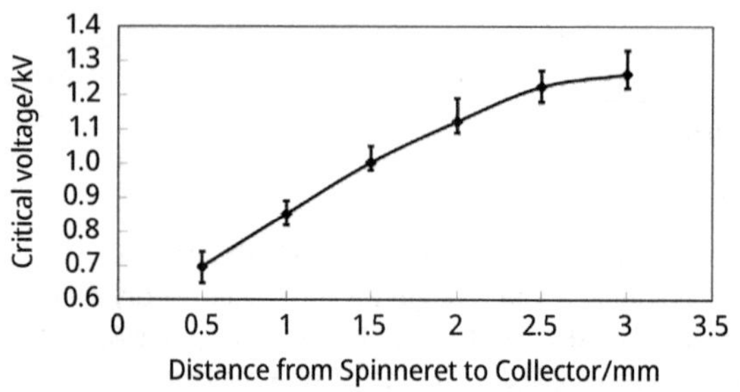

Figure 2.11: Plot of critical voltage versus spinneret-to-collector distance (silicon substrate with a spinneret diameter of 40 μm and solution concentration of 10%).

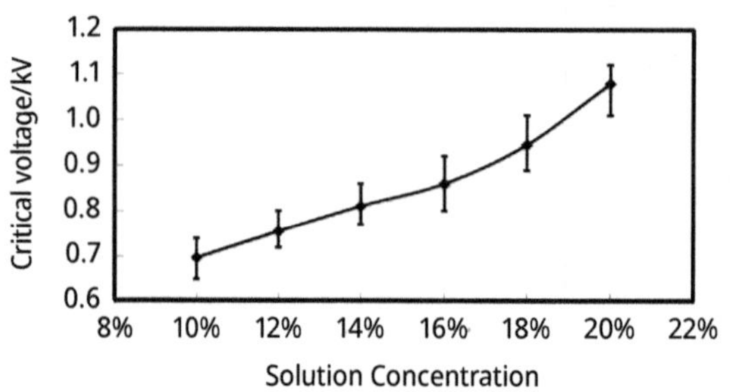

Figure 2.12: Plot of critical voltage versus solution concentration (silicon substrate, with a spinneret diameter of 40 μm and spinneret-to-collector distance of 0.5 mm).

5. Jet ejection delay time

After applying DC high voltage, charges migrate and accumulate to the polymer solution surface at the spinneret tip. The electric field force generated by these charges causes the polymer solution to deform and eject. The jet ejection progress, as shown in Figure 2.4, reveals a time delay between the voltage application and the solution ejection, which is referred to as the ejection delay time in this study. This delay primarily reflects the viscous characteristics of the viscoelastic fluid during the electrohydrodynamic rheological process [4, 5].

From the perspective of charge migration, a certain amount of time is needed after applying voltage to allow sufficient charge accumulation on the solution surface at the spinneret, providing necessary electric field force for jet generation. In terms of the polymer solution rheology, the electrospinning solution exhibits relaxation properties in response to external forces, which manifest as a time delay between the application of the external force and the solution deformation.

In the electrospinning, the jet formation and the ejection delay time τ_d primarily depend on the applied voltage U, the fluid characteristic time τ_H, and the charge relaxation time of the fluid under the action of strong electric field τ_c [6, 7]. The charge relaxation time can be given as [8]

$$\tau_c = \varepsilon/(4\pi\kappa) \tag{2.1}$$

The characteristic hydrodynamic time is given as [9]

$$\tau_H = \mu a_0/\gamma \tag{2.2}$$

where ε is the relative permittivity of the solution, κ is the electric conductivity, μ is the viscosity, a_0 is the characteristic size of the liquid droplet, and γ is the surface tension coefficient.

The leaky dielectric model [10, 11] is the most widely used theoretical framework for describing the behaviors of polymer solutions during the electrospinning process. According to this model, under the influence of a strong electric field, the conductivity of the polymer solution is intermediate between that of a conductor and an insulator. Research by Reneker et al. [8] shows that when $\tau_c \ll \tau_H$, the fluid exhibits the characteristics of a leaky dielectric. This behavior is a necessary condition for electrostatic ejection of polymer solutions, while the ejection delay time becomes a fundamental characteristic of the solution's leaky dielectric rheology.

To better understand the rheological properties of the spinning jet, the effects of applied voltage and solution concentration on the jet ejection delay time are analyzed. The variation in ejection delay time as a function of applied voltage is shown in Figure 2.13. Rising the applied voltage intensifies the space electric field intensity, accelerating the charge migration speed during the jet ejection process, thus shortening the ejection delay time. In contrast, as the solution concentration increases, the viscosity of the solution also increases, which further hinders the deformation process of the polymer solution. As a result, the jet ejection delay time increases with the solution concentration.

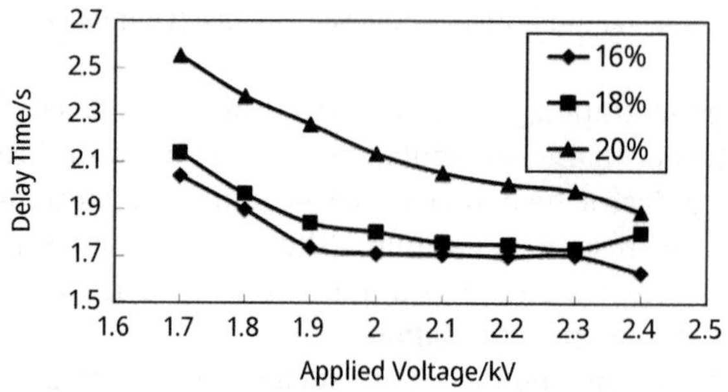

Figure 2.13: Variation in jet ejection delay time in EDW (silicon substrate with a spinneret-to-collector distance of 1 mm and spinneret diameter of 40 μm).

2.1.3.2 Jet ejection mode

The EDW process is affected by a variety of factors, and the jet behaviors are greatly affected by the solution properties, experimental setup, and control parameters. These include:

(1) Solution characteristics: Viscosity, density, conductivity, dielectric constant, surface tension, and other related properties.
(2) Experimental setup: Factors such as the inner diameter of the spinneret, spinneret-to-collector distance, and other physical configurations of the setup.
(3) Control parameters: applied voltage, solution supply rate, and other operational parameters that influence the process.

For a given EDW system, the applied voltage and solution supply rate are crucial parameters that jointly determine the rheological behaviors and ejection modes of the solution. For the following experiment, a glycerol solution is used with a spinneret inner diameter of 0.6 mm and a spinneret-to-collector distance of 6 mm.

At a fixed solution supply rate, different jet ejection modes can be achieved by adjusting the applied voltage, as shown in Figure 2.14. These modes include dripping, pulse jet, stable cone jet, and multi-jet modes. When the voltage is too low, the electric field force is insufficient to induce the formation of a Taylor cone or spray from the hanging droplet at the spinneret. In this case, the solution is ejected under the influence of gravity, resulting in a **dripping mode**, as shown in Figure 2.14(a). As the applied voltage increases, the electric field force acting on the hanging droplet intensifies, leading to a dripping mode that has reduced droplet size and an increased dripping frequency. When the electric field force reaches a certain threshold, the system transitions to **pulse jet mode** (Figure 2.14(b)), where the electric field force is still not sufficient to maintain a stable ejection state. Figure 2.14(c) shows the jet ejection morphology in the **cone-jet mode**, in which the jet behavior is stable and controllable, making it an ideal mode to print micro-nanostructures. As the voltage continues to increase, no distinct Taylor cone is formed. The jet is ejected directly from the inside of the spinneret, moving along the edge of the spinneret, resulting in **unstable deflection** as shown in Figure 2.14(d). This instability arises when the solution supply rate is lower than the jet ejection speed, preventing the solution from maintaining a stable Taylor cone shape. Due to the small curvature radius at the spinneret, as the voltage increases further, the high electric field strength at the spinneret tip induces multiple jets to eject, as shown in Figure 2.14(e), potentially leading to **air breakdown discharge**.

Generally, the occurrence of unstable deflection and multiple jets can lead to significant jet positioning errors and line width inconsistencies, which should be avoided in practical applications.

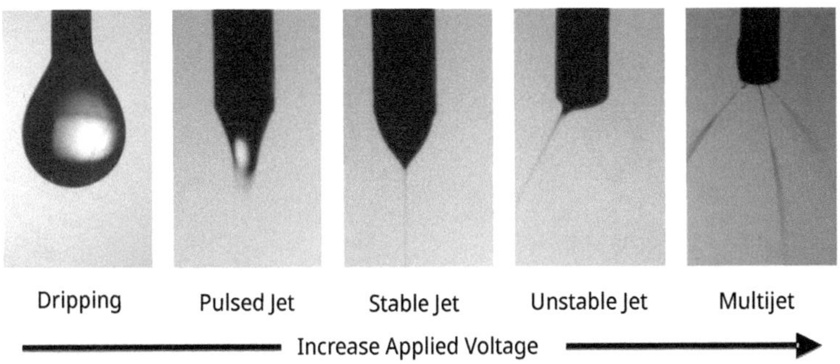

Figure 2.14: Schematic diagram of jet ejection modes.

The ejection behavior in the steady cone-jet mode is further investigated. Figure 2.15 illustrates the changes in jet morphology under progressively increasing voltage levels. As the voltage increases, the volume of Taylor cone decreases, but the cone angle increases. Notably, under stable jet ejection conditions, lowering the applied voltage help to reduce the surface charge, the jet speed, and the unstable disturbance. This reduction enhances control over the jet's positioning and deposition, thereby improving the precision of micro-nanoscale printing [12].

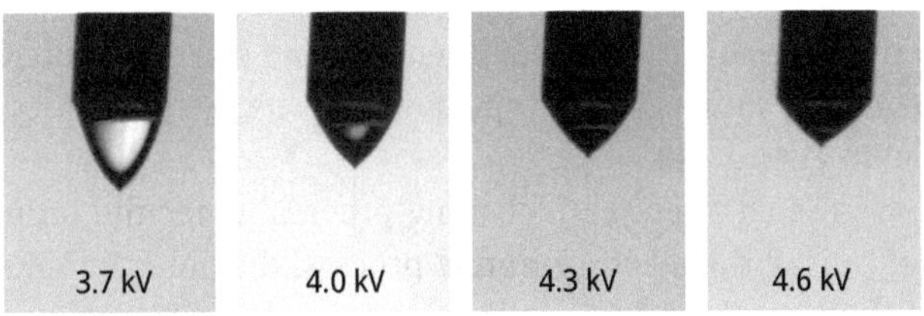

Figure 2.15: Jet morphology at different applied voltages (glycerin solution, spinneret inner diameter is 0.6 mm, liquid supply rate is 400 μL/h, and spinneret-to-collector distance is 6 mm).

Figure 2.16 shows the jet morphology at different solution supply rates. In the stable ejection state, both the Taylor cone volume at the spinneret and the jet diameter increases as the solution supply rate rises. In addition, the solution supply rate directly influences the jet deposition speed, which is critical for precise micro-nanostructure fabrication. Excessive jet diameter and deposition speed can compromise the accuracy of the printed structures. When the solution supply rate is lower than the solution jetting speed, a **pulse jet mode** emerges. Conversely, if the solution supply rate exceeds the ejection speed, the ejection process becomes unstable, causing the solution volume at the spinneret to continuously accumulate, eventually leading to a pulse-like dripping mode.

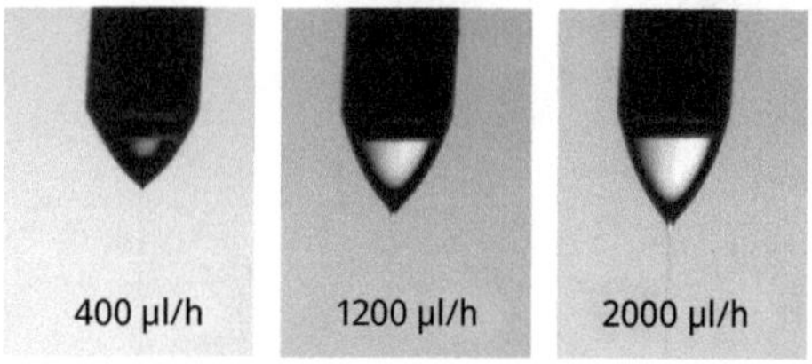

Figure 2.16: Jet morphology at different solution feeding rates (glycerin solution is used, spinneret inner diameter is 0.6 mm, applied voltage is 4.3 kV, and the spinneret-to-collector distance is 6 mm).

In summary, achieving a stable cone-jet ejection in the EDW process requires balancing the solution supply rate and applied voltage. Experimental investigations conducted at different applied voltage and solution supply rate levels have yielded distinct jet ejection patterns, as illustrated in Figure 2.17. These findings provide a valuable experimental foundation for controlling jet ejection modes and morphology in electrospinning applications.

2.1.4 Deposition behaviors of EDW jet

2.1.4.1 Jet spiral deposition

As the jet front deposits onto the collector, it undergoes bending and spiraling due to the combined effects of electric field force, charge repulsion, and its intrinsic viscoelastic properties, as illustrated in Figure 2.18.

Figure 2.18(a) and (b) shows the charged jet depositing process under the influence of the electric field force. During the electrospinning process, the polymer solution behaves as a viscoelastic fluid. As the solvent evaporates, the liquid viscosity of the jet gradually decreases while the solid elasticity increases. The deposited nanofibers exert a viscoelastic force on the incoming jet, causing it to bend under the combined action of the electric field force (Figure 2.18(c)). Due to this viscoelastic force, the jet undergoes a coiling process (Figure 2.18(d) and (e)). Consequently, the jet/nanofiber deposit in a spiral pattern on the collector (Figure 2.18(f)–(h)).

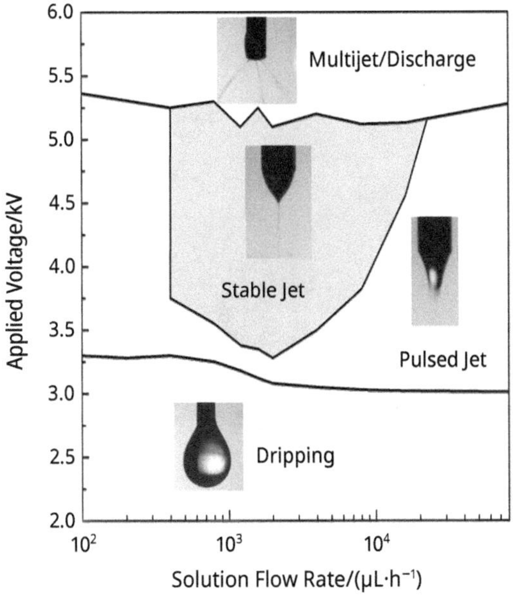

Solution Flow Rate/ (μL·h^{-1})	Applied Voltage/kV			
	Dripping	Pulsed Jet	Stable Jet	Multijet/ Discharge
1.0 X 10^2	<3.30	3.30~5.30	--	>5.30
2.0 X 10^2	<3.28	3.30~5.27	--	>5.27
4.0 X 10^2	<3.30	3.30~3.75	3.75~5.25	>5.25
8.0 X 10^2	<3.25	3.25~3.55	3.55~5.30	>5.30
1.2 X 10^3	<3.18	3.18~3.38	3.38~5.10	>5.10
1.6 X 10^3	<3.12	3.12~3.35	3.35~5.25	>5.25
2.0 X 10^3	<3.08	3.08~3.28	3.28~5.10	>5.10
4.0 X 10^3	<3.05	3.05~3.50	3.50~5.20	>5.20
8.0 X 10^3	<3.03	3.03~3.82	3.82~5.12	>5.12
1.6 X 10^4	<3.02	3.02~4.56	4.56~5.13	>5.13
2.0 X 10^4	<3.02	3.02~4.96	4.96~5.30	>5.30
2.3 X 10^4	<3.02	3.02~5.16	--	>5.16
4.0 X 10^4	<3.01	3.01~5.20	--	>5.20
8.0 X 10^4	<3.01	3.01~5.27	--	>5.27

Figure 2.17: Jet ejection modes at different applied voltages and solution flow rates (glycerin solution, spinneret inner diameter is 0.6 mm, and spinneret-to-collector distance is 6 mm).

The residual charge on the collector introduces a repulsive force that influences the deposition of the subsequent jet. Once the charged jet lands on the collector, the accumulated charge gradually dissipates through the collector, a process that requires a certain relaxation time. The time required for complete dissipation of residual charge primarily depends on the charge relaxation time of the collector [13]. As deposition continues, the surface charge density of the deposited nanofibers decreases.

However, uneven distribution of the residual charge induces the spiraling effect around the initial deposition point of the jet front, as shown in Figure 2.18(i)–(p), where the small arrow indicates the deposition position of the charged jet. The figure demonstrates that the jet rotates clockwise around the previously deposited nanofibers. When the collector remains stationary, the nanofibers accumulate layer by layer, forming a multilayer structure with increasing height over time.

2.1.4.2 Multi-ring nanofiber structure

As the charged nanofibers accumulate on the collector, the charge gradually dissipates, leading to a decrease in charge density along the direction indicated by the arrow in Figure 2.19.

Consider a point A on the deposited nanofiber, as shown in Figure 2.19. Before reaching point A, most or all of the surface charge on the nanofiber has been transferred

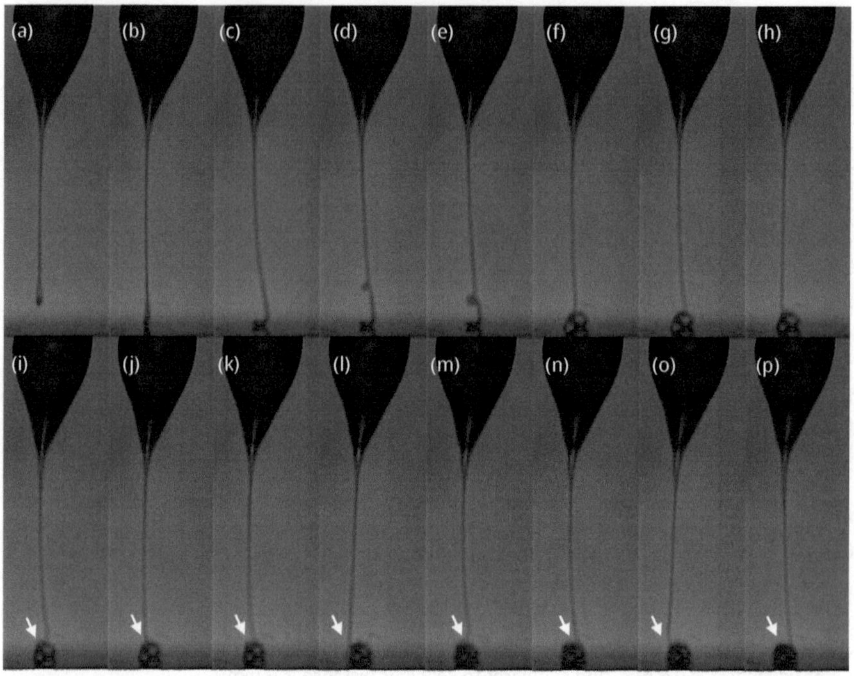

Figure 2.18: Jet deposition behavior and spiral deposition of nanofibers on substrate (the collector is kept still, spinneret-to-collector distance is 1 mm, applied voltage is 1.9 kV, solution concentration is 18%, and spinneret diameter is 40 μm, silicon substrate).

away, resulting in little to no repulsive effect on the deposition of subsequent fibers. However, beyond point A, the surface charge density remains relatively high, generating a repulsive force. Under the combined influence of charge repulsion and the nanofiber's inherent viscoelasticity, the jet undergoes a spiral whipping motion.

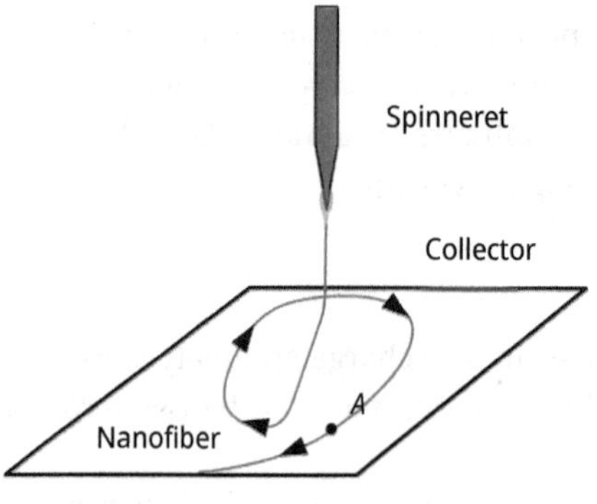

Figure 2.19: Charge density on the charged nanofibers decreases along the arrow direction.

In the experiment, two types of collectors – semiconductor and insulator – are used for EDW of nanofibers. A monocrystalline silicon wafer (500-μm thick, purchased from CETC 46th Research Institute, Tianjin) serves as the semiconductor collector. For the insulator collector, a 2-μm-thick silicon dioxide layer is grown on a monocrystalline silicon wafer using thermal oxidation method. Due to the guiding effect of the jet's helical motion, nanofibers are spirally deposited on the stationary collector, forming a multi-coil structure as shown in Figure 2.20, with the deposition time of 3 s.

When the nanofibers deposit on the silicon dioxide substrate, the carried charge cannot dissipate quickly, producing a strong repulsive force on the deposition of subsequent jet. The charge repulsion dominates the jet deposition process, amplifying the jet whipping amplitude and expanding the nanofiber deposition area, ultimately forming a spider-web-like structure. As shown in Figure 2.20(a), the outer diameter of the spider-web-like structure is about 728 μm, with adjacent nanofiber circles spaced 10–20 μm apart.

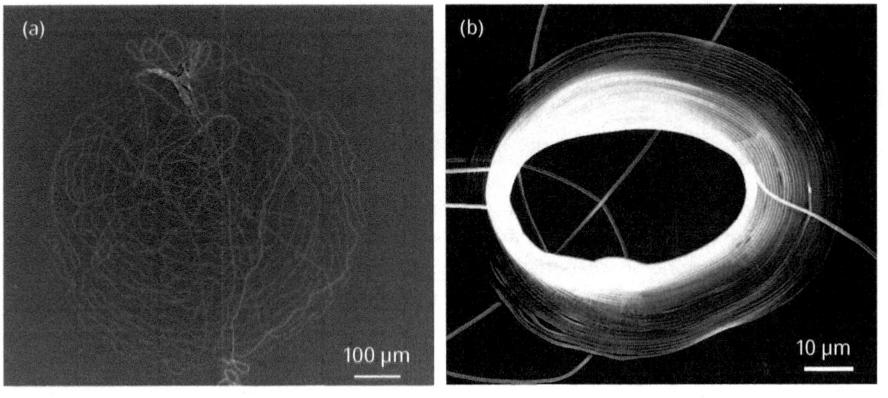

Figure 2.20: Multi-loop structures deposited on stationary collector: (a) spider-web-like structure on a silicon dioxide substrate and (b) hat-shaped structure on a silicon substrate (solution concentration is 20%, nozzle diameter is 40 μm, nozzle-to-collector distance is 2 mm, and applied voltage is 1.7 kV).

The charge carried by nanofibers deposited on silicon substrate immediately dissipate, resulting in weaker repulsive force on subsequent charged jet. As a result, viscoelastic force becomes the dominant factor, constraining the jet to spiral within a small range and stack layer upon layer, as shown in Figure 2.18. As deposition time increases, the number and the height of the deposited layers grow, while the diameter of the nanofiber loops gradually decreases, forming a three-dimensional hollow hat-shaped structure with a wide base and a narrow top as shown in Figure 2.20(b). This structure consists of a total of 48 layers of nanofiber loops, with a bottom diameter of 76 μm and a top opening diameter of 47 μm – significantly smaller than the outer diameter of the spider-web-like structure on the silica substrate.

When the collector moves at a low speed of 0.01 m/s, an array of hat-shaped structure, as shown in Figure 2.21(a), is formed on the silicon substrate. A single hat-shaped structure, depicted in Figure 2.21(b), has the inner and outer diameters of 40 and 48 μm, respectively. A magnified view of the hat-shaped three-dimensional structure in Figure 2.21(c) reveals a total of 11 layers of nanofiber ring layers within the structure.

As the collector moves, the helical whipping of the charged jet weakens, leading to sparsely deposited nanofibers due to the shorter deposition time in each area. Consequently, both the diameter and the number of layers in the hat-shaped structure formed on the moving collector (Figure 2.21) are smaller compared to those on the stationary collector (Figure 2.20(b)).

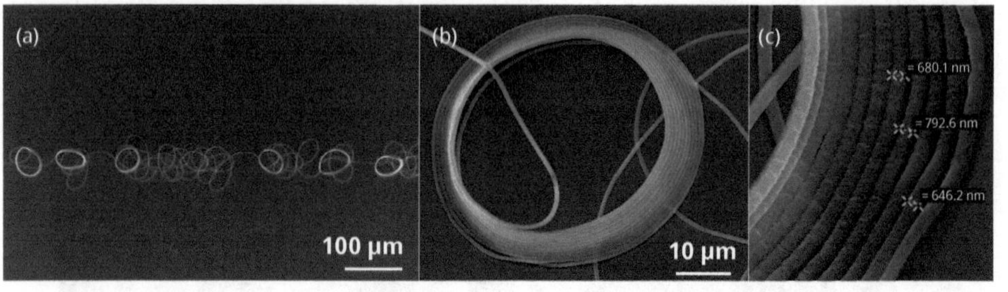

Figure 2.21: Hat-shaped structures deposited on a moving silicon substrate: (a) array of hat-shaped structures on the deposition track; (b) a single hat-shaped structure; and (c) local amplification of the hat-shaped structure (the structures are conducted at solution concentration of 20%, spinneret diameter of 40 μm, spinneret-to-collector distance of 2 mm, applied voltage of 1.7 kV, and collector motion speed of 0.01 m/s, silicon substrate).

The EDW process shortens the distance between the spinneret and the collector, enhancing the influence of residual charges on the deposition behavior of the charged jet. The combined effect of the repulsive force from residual charge on the collector and the viscoelastic force of the solution flow induces the jet's spiral whipping motion. The formation of spider-web-like and hat-shaped structures on the collector visually represents this spiral whipping behavior.

2.1.4.3 Nanofiber spiral deposition

1. Influence of collector conductivity on nanofiber spiral deposition

This section investigates the influence of collector conductivity on the spiral deposition of electrospun nanofibers. To achieve better control over nanofiber deposition, two collectors with different conductivities – silicon dioxide and silicon – are used. Figure 2.22 shows the nanofibers obtained on silica and silicon collectors at a low collector speed.

On the silicon dioxide substrate, the deposited nanofibers experience a stronger repulsive force due to retained surface charges. As a result, the jet exhibits dominant whipping behavior perpendicular to the moving direction of the collector, leading to a wavy fiber structure, as shown in Figure 2.22(a). In contrast, on the silicon substrate, where the charge dissipation is more efficient, the repulsive force is weaker, and viscoelastic forces become the primary influence. This causes the charged jet to whip in a circular manner, forming a circular nanofiber structure with an average diameter of 48.56 μm, as shown in Figure 2.22(b).

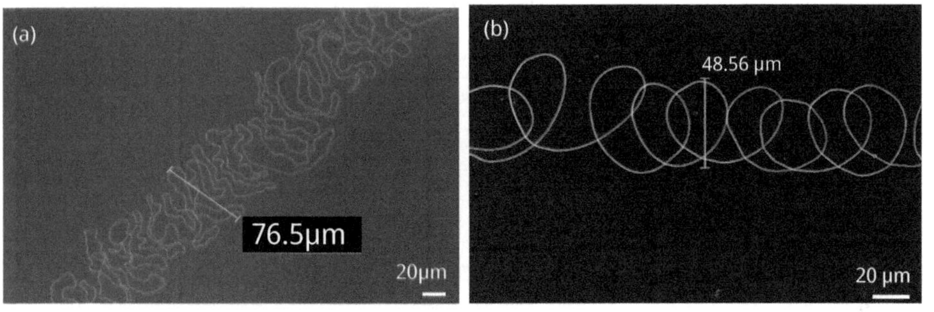

Figure 2.22: Nanofibers deposited on (a) silicon dioxide substrate and (b) silicon substrate (collector motion speed is 0.08 m/s, spinneret diameter is 40 μm, spinneret-to-collector distance is 1 mm, applied voltage is 1.7 kV, and solution concentration is 20%).

According to Figure 2.22, the distribution width and spacing of the ribbon-shaped nanofibers deposited on the silicon dioxide substrate are relatively large. In Figure 2.22(a), the distribution width of the nanofibers on the silicon dioxide substrate is 76.5 μm, while in Figure 2.22(b), the distribution width of the nanofibers on the silicon substrate is equal to the diameter of the circular structure, which is 48.56 μm.

To eliminate the discrepancies in the deposition results caused by variations in process parameters and to accurately analyze the effects of collector conductivity, a patterned silicon dioxide layer (thermal oxidation, thickness of 2 μm) is prepared on a silicon substrate, as shown in Figure 2.23. This configuration enables simultaneous EDW deposition of nanofibers on both silicon and silicon dioxide under identical experimental conditions, ensuring a more precise comparison.

The patterned nanofiber depositions on the collector at a low collector speed (0.03 m/s) are shown in Figure 2.24. The distribution width of the ribbon-shaped nanofibers in silicon dioxide region is significantly wider than that in the silicon region. Specifically, the line width in the silica region ranges from 30 to 200 μm, whereas in the silicon region, it is only 5–60 μm.

The deposition of a single nanofiber on the collector with a patterned structure for different collector speeds is shown in Figure 2.25. At a moving speed of 0.10 m/s, the nanofibers exhibit a wavy structure in both the silicon and silicon dioxide regions,

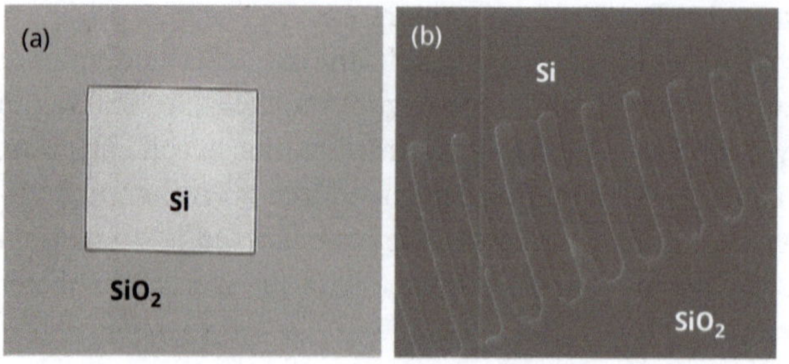

Figure 2.23: Silicon substrates with patterned layers of silicon dioxide: (a) squared layer and (b) serrated layer.

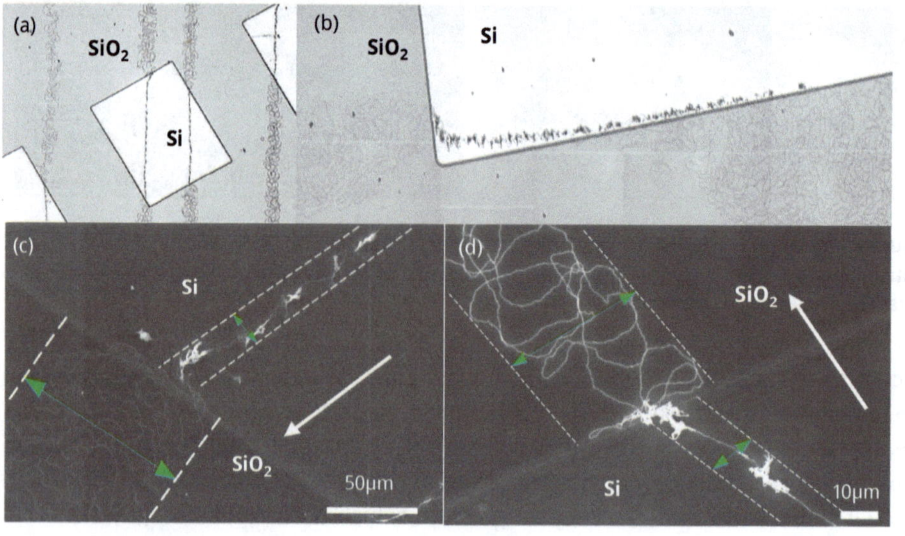

Figure 2.24: Nanofibers deposited on a collector with patterned structures: (a, b) optical microscopic images with magnification of 20 and 100 times, respectively; (c, d) SEM images (solution concentration is 20%, spinneret diameter is 70 μm, spinneret-to-collector distance is 1 mm, collector speed is 0.03 m/s, and applied voltage is 1.46 kV; single arrow indicates the direction of collector movement).

as shown in Figure 2.25(a). As the collector speed increases to 0.13 m/s, the wavy morphology gradually diminishes, transitioning toward a linear structure, as shown in Figure 2.25(b). At 0.20 m/s, nanofibers in the silicon region have already formed a linear structure, while those in the silicon dioxide region remain wavy, as shown in Figure 2.25 (c). Even at 0.26 m/s, nanofibers in the silica area still exhibit a wavy morphology, as shown in Figure 2.25(d). When the collector speed increases to 0.43 m/s, nanofibers in both the silicon and silicon dioxide regions become straight, as shown in Figure 2.25(e). Additionally, Figure 2.25(f) presents the straight nanofibers directly written onto a collector with a serrated pattern at a collector speed of 0.38 m/s.

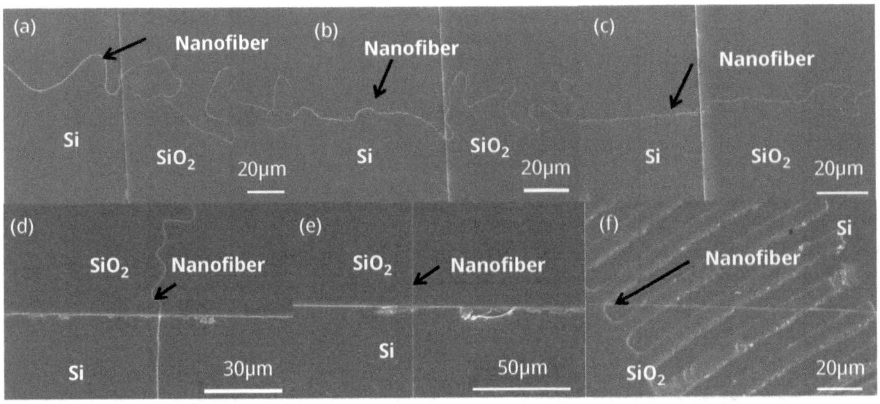

Figure 2.25: Morphology of nanofibers deposited on the collector with patterned structure at different collector speeds: (a) 0.10; (b) 0.13; (c) 0.20; (d) 0.26; (e) 0.43; and (f) 0.38 mm/s (solution concentration is 18%, spinneret diameter is 70 μm, spinneret-to-collector distance is 1 mm, and applied voltage is 1.46 kV).

2. Influence of the collector moving speed on nanofiber spiral deposition

The nanofiber deposition patterns at different collector moving speeds are shown in Figure 2.26. As the collector speed increases, the influence of jet whipping on the deposition process gradually diminishes. At low collector speeds, the strong whipping effect leads to nanofiber structures with coiled fibers stacking on each other, as shown in Figure 2.26(a). As the collector moving speed increases, the degree of overlap between the circular nanofiber structures gradually reduces, forming an independent nanofiber helix as shown in Figure 2.26(b). With a further increase in collector speed, whipping behavior weakens further, causing a decrease in both the adjacent spirals' spacing and the spiral diameter, as shown in Figure 2.26(c). When the collector speed matches the jet deposition speed, the effect of jet spiral whipping is minimized, generating the linear nanofibers as shown in Figure 2.26(d). This specific collector speed, at which spiral whipping is effectively suppressed, is defined as "matching speed". To investigate the matching speed under different experimental conditions, we analyze its dependence on applied voltage, solution concentration, and spinneret-to-collector distance, as presented in Figures 2.27–2.29.

The matching speed is primarily determined by the deposition speed and the extent of jet whipping. An increase in applied voltage enhances the electric field force acting on the charged jet, thereby accelerating the jet deposition speed and increasing the number of deposited nanofibers per unit time. In addition, a higher voltage or solution concentration results in a greater charge density on the jet surface, intensifying the spiral whipping. Consequently, the matching speed increases with both applied voltage and solution concentration, as shown in Figures 2.27 and 2.28. Furthermore, as the spinneret-to-collector distance increases, the jet deposition speed rises, making the jet more susceptible to external disturbances, which promotes spiral whipping. Therefore, the matching speed also increases with spinneret-to-collector distance, as shown in Figure 2.29.

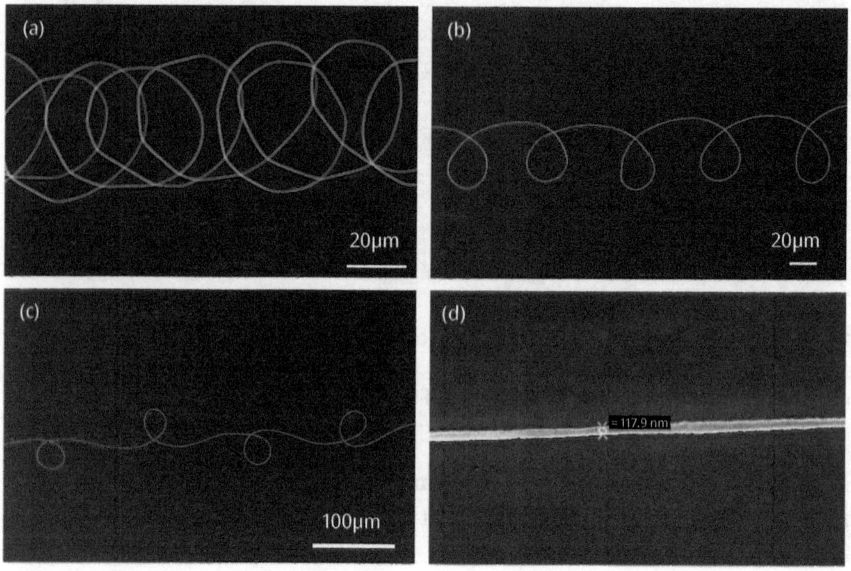

Figure 2.26: Nanofibers deposited at different collector speeds. The collector speeds are (a) 0.06; (b) 0.20; (c) 0.25; and (d) 0.40 m/s (spinneret diameter is 40 μm, spinneret-to-collector distance is 1 mm, applied voltage is 1.7 kV, and solution concentration is 18%, silicon substrate).

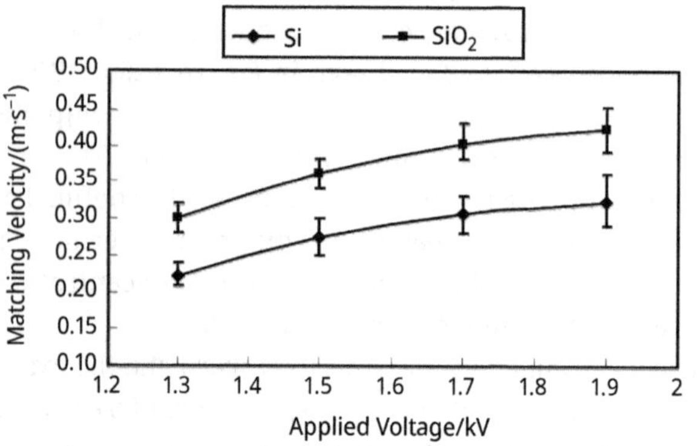

Figure 2.27: Relationship between matching velocity and applied voltage (spinneret-to-collector distance is 1 mm, solution concentration is 18%, and the spinneret diameter is 40 μm).

Comparing cases for the two collectors, the matching speed of the silicon dioxide substrate is higher than that of the silicon substrate. As the silicon dioxide is low in conductivity, the charge migration is slow, so the residual charge density on the collector

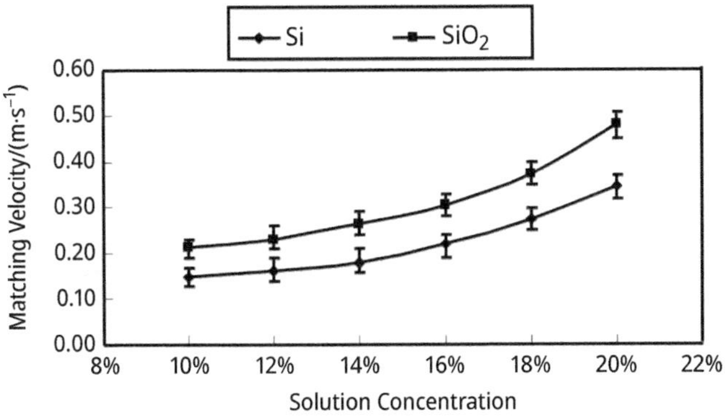

Figure 2.28: Relationship between matching velocity and solution concentration (the spinneret-to-collector distance is 1 mm, the applied voltage is 1.4 kV, and the spinneret diameter is 40 μm.).

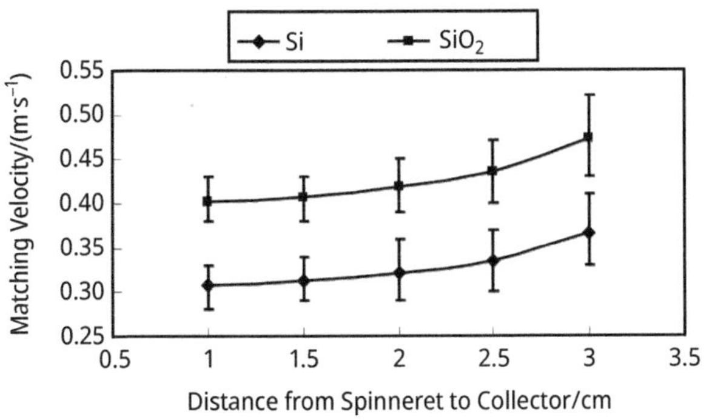

Figure 2.29: Relationship between matching velocity and spinneret-to-collector distance (solution concentration 18%, applied voltage 1.7 kV, and nozzle diameter 40 μm).

is high, and the charge repulsion force to the subsequent charged jet is strong, thus exacerbating the whipping behavior in the deposition process. Investigation to the changes of the collector matching speed has laid an experimental foundation for the control of nanofiber deposition.

2.1.5 Electrohydrodynamic direct-writing micro-nanostructures

2.1.5.1 Electrohydrodynamic direct-writing nanofibers

Nanofibers deposited on silicon and silicon dioxide substrates are shown in Figure 2.30. As the silicon substrate has high conductivity, the carried charges can be conducted away in time after the charged jet or nanofibers deposits on the collector, resulting in less accumulated charge on the collector and less repulsive force for the subsequently deposited nanofibers. The charged jet generates spiral whipping along the motion trajectory of the collector, inducing the nanofibers to spirally deposit on the silicon substrate to form the looped structure as shown in Figure 2.30(a) and (b), with a diameter of 46.53 μm for a single-looped nanofiber structure, which is approximately equal to the width of the fiber deposition area on the collector.

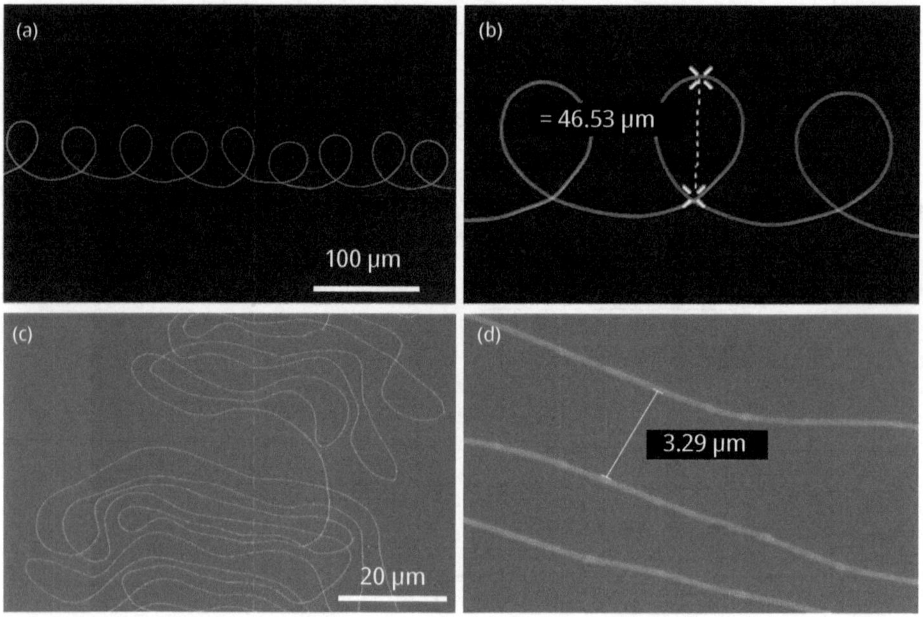

Figure 2.30: Electrohydrodynamic direct-written nanofibers: (a) SEM image of nanofiber deposited on the silicon substrate; (b) magnification of nanofiber segment on the silicon substrate; (c) SEM image of nanofibers deposited on the silicon dioxide substrate; and (d) magnification of nanofiber segment on silicon dioxide substrate (solution concentration is 18%, spinneret-to-collector distance is 1 mm, collector speed is 0.02 m/s, applied voltage is 1.46 kV, and spinneret diameter is 70 μm).

As the silicon dioxide substrate is of low conductivity and slow charge transfer rate, the deposition of the charged nanofibers on the silica substrate results in more charge accumulation on the collector, which generates a larger charge repulsion force on the subsequent deposition of nanofibers, resulting in a wavy structure as shown in Figure 2.30(c). The nanofibers deposited on the silicon dioxide substrate show a distribution width of about 63 μm, and the distance between two adjacent nanofibers is 3.29 μm (Figure 2.30 (d)). The width of the nanofiber distribution area on the silicon dioxide substrate is greater than that on the silicon substrate.

Using silicon and silicon dioxide substrates as the collectors respectively, the variation of nanofiber diameter with applied voltage, spinneret-to-collector distance, solution concentration, spinneret diameter, and other factors are investigated. The respective curves are shown in Figures 2.31–2.34.

Near-field electrospinning technology overcomes the disordered and unstable motion of the jet, and the stretching and thinning of the jet only occurs during the stable motion stage. The degree of jet stretching is small, and the diameter of nanofibers are mainly determined by the initial jet diameter. The electric field force acting on the solution droplets increases with the applied voltage, which increases the diameters of the jet and the nanofiber, as shown in Figure 2.31. The variation of nanofiber diameter with the applied voltage is different from that in the traditional electrospinning. In traditional electrospinning, due to the large spinneret-to-collector distance, increasing the voltage will increase the charge density on the jet surface, thereby enhancing the unstable motion of the charged jet, as well as further increasing the stretching rate of the charged jet, and reducing the diameter of the nanofibers [14].

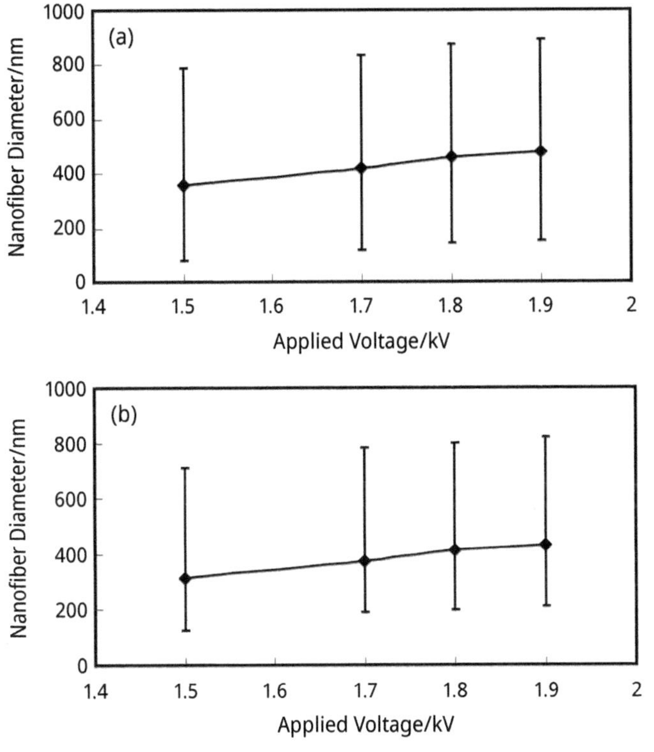

Figure 2.31: Influence of applied voltage on nanofiber diameter with (a) silicon substrate and (b) silicon dioxide substrate (spinneret-to-collection distance is 1 mm, solution concentration is 20%, and nozzle diameter is 40 μm).

Increasing the spinneret-to-collector distance can increase the stretching time and stretching rate of the jet and reduce the diameter of the nanofibers, as shown in Figure 2.32. The viscosity of polymer solution increases with the solution concentration, which hinders

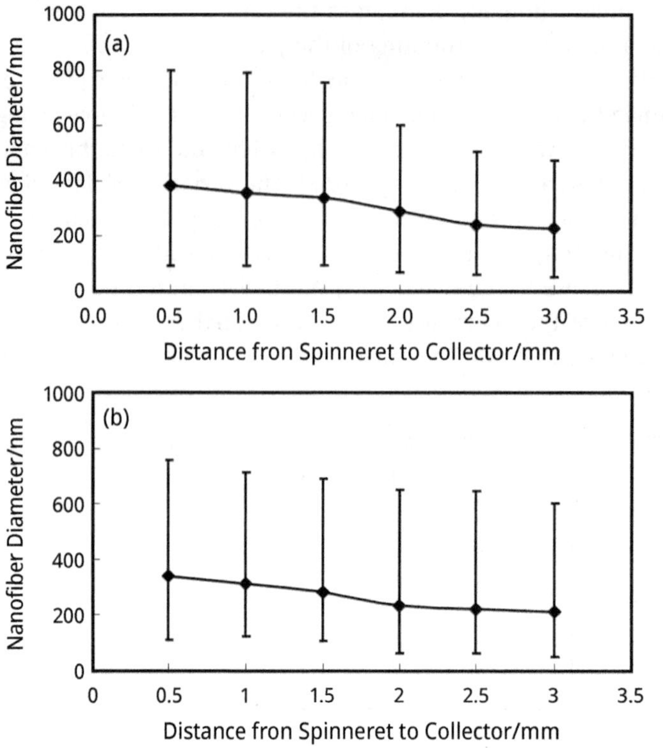

Figure 2.32: Influence of the spinneret-to-collector distance on nanofiber diameters for (a) silicon substrate and (b) silicon dioxide substrate (applied voltage is 1.5 kV, solution concentration is 20%, and spinneret diameter is 40 μm).

the stretching and thinning of the jet, so the fiber diameter increases with the solution concentration, as shown in Figure 2.33. As the volume of the solution droplet at the spinneret and the initial diameter of the jet increase with the increasing spinneret diameter, the diameter of nanofibers increases with the spinneret diameter, as shown in Figure 2.34.

Comparing the diameter of the nanofibers deposited on the two collectors, the fiber diameter on the silicon dioxide substrate is smaller. This is because the charge on the silicon dioxide substrate is discharged at a slower rate, resulting in a higher residual charge density on the deposited nanofibers. This residual charge density has a stronger repulsive effect on the subsequent deposition of nanofibers, leading to further stretching of the jet and a reduction in nanofiber diameter. Compared with traditional electrospinning, the near-field electrospinning process has a shorter spinneret-to-collector distance, resulting in a lower jet stretching rate. However, the residual charges on the collector repel with that on the subsequent jet, which increases the hovering and stretching of the nanofibers in the air, thus reducing the fiber diameter.

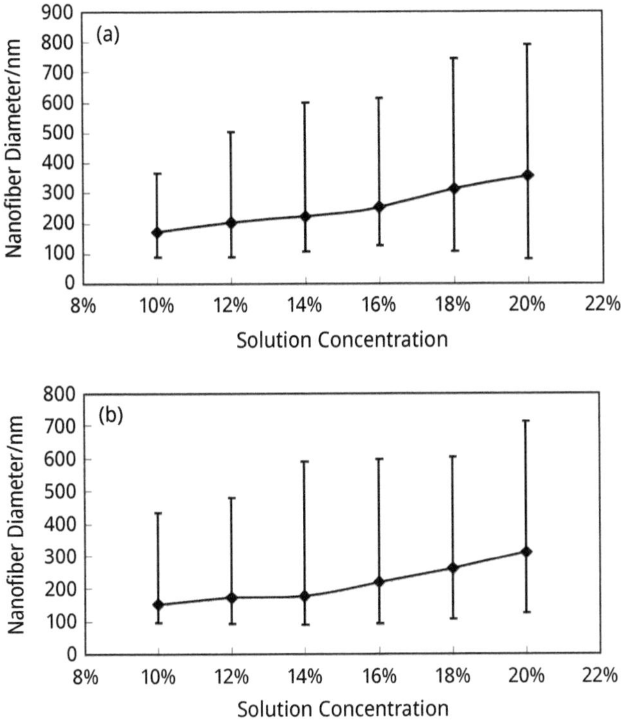

Figure 2.33: Influence of solution concentration on nanofiber diameter for (a) silicon substrate and (b) silicon dioxide substrate (applied voltage is 1.5 kV, spinneret-to-collector distance is 1 mm, and spinneret diameter is 40 μm).

2.1.5.2 Electrohydrodynamic direct-writing liquid film structures

In the EDW process, the jet shows various deposition behaviors on the collector depending on the solution characteristics and the jet solidification situation. If the jet still maintains good flowability when it reaches the collector, it will converge as droplets under the action of surface tension. Under this circumstance, a continuous liquid film pattern will be obtained if the collector is also moving, as shown in Figure 2.35(a). If the jet has already solidified in the air before reaching the collector, its flow characteristics are weakened while its solid characteristics are strengthened. Consequently, entanglement and bending occur during jet deposition, and spiral liquid film patterns or nanofibers can be obtained on the collector, as shown in Figure 2.35(b).

In the experiment, a hollow spinneret with an inner diameter of 0.06 mm and a glass sheet collector is used. The spinneret-to-collector distance is 1.5 mm. The solution used is PVP solution, which is prepared by mixing the PVP powder into blend solvent of deionized water and anhydrous ethanol in a volume ratio of 3:2. The solution concentration ranges from 6% to 12%.

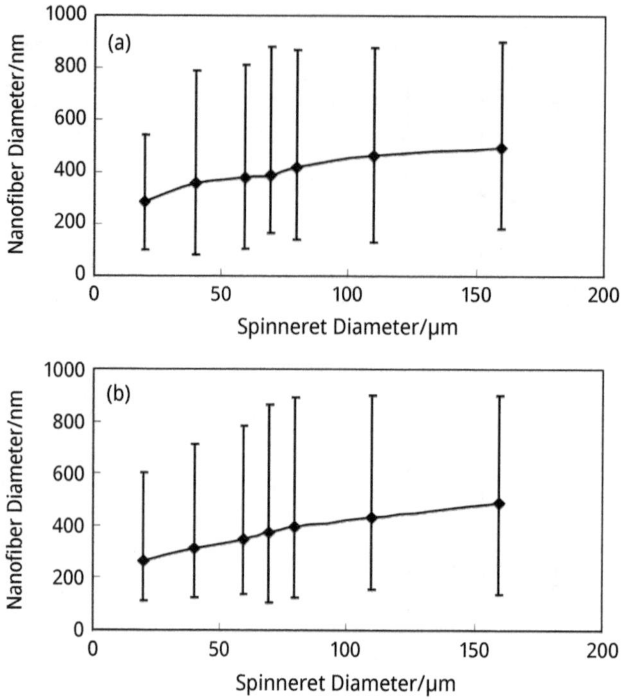

Figure 2.34: Influence of spinneret diameter on nanofiber diameter for (a) a silicon substrate and (b) silicon dioxide substrate (applied voltage is 1.5 kV, solution concentration is 20%, and spinneret-to-collector distance is 1 mm).

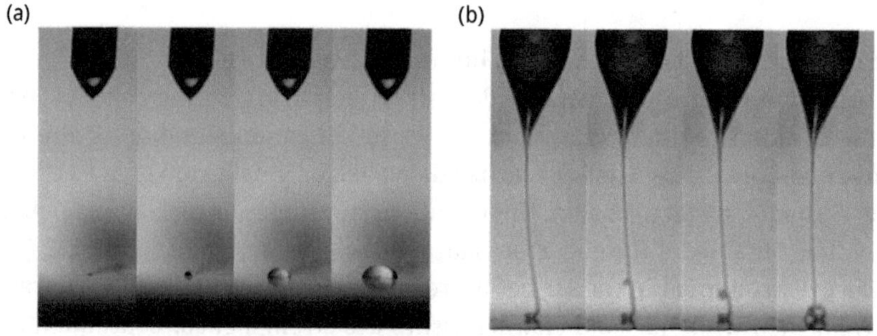

Figure 2.35: Jet deposition process: (a) the jet is converging into a droplet and (b) entanglement and bending of a solidified jet.

Figure 2.36 shows the electrospinning process of an 8% PVP solution at different applied voltages. The PVP solution is in a stable cone-jet mode when the voltage exceeds 1.8 kV. As the applied voltage increased from 1.9 to 2.3 kV, although the axial velocity of the jet accelerates, the average diameter of the prepared line structure maintains stable at about 87–90 μm due to the constant solution supply rate and the constant deposition rate under the stable jet ejection conditions. When the voltage is increased to 2.5 kV, the ejection mode turns from single jet to multiple jet, making it impossible to achieve stable structure fabrication.

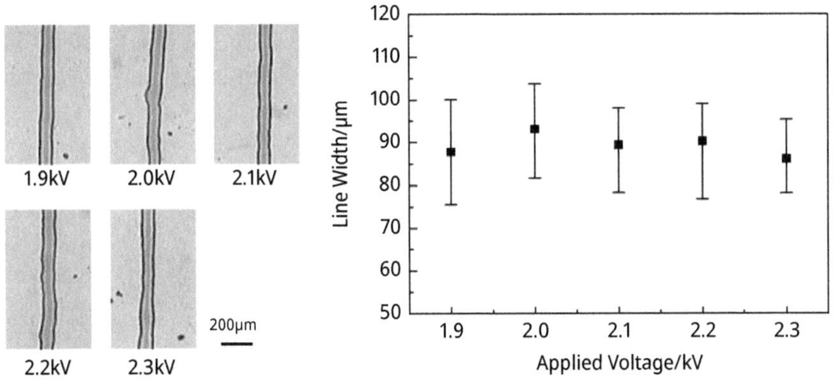

Figure 2.36: Influence of applied voltage on diameter of structures printed with 8% PVP solution (the collector motion speed is 20 mm/s and the liquid supply rate is 70 μL/h).

The influence of solution supply rate on the diameter of the printed structure is shown in Figure 2.37. When the solution supply rate is less than 10 μL/h, the solution exhibits a pulse-jetting mode, leading to a pattern of dot-like bead structure. When the solution supply rate increases to 30 μL/h, the ejection mode changes to a stable cone jet mode, and the prepared pattern is a uniform straight-line structure. As the solution supply rate increases from 30 to 90 μL/h, the average diameter increases from 58 to 95 μm.

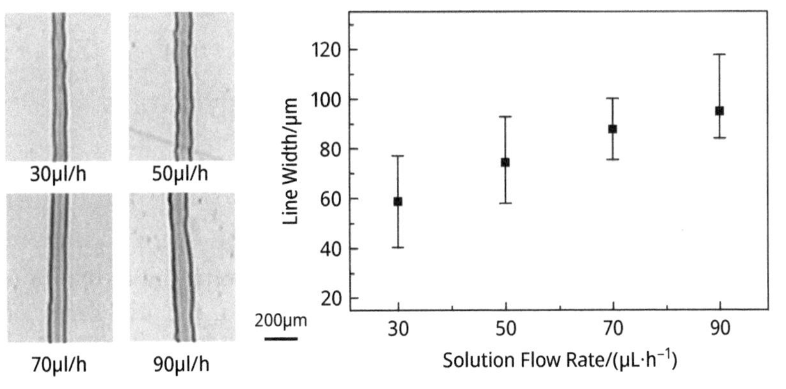

Figure 2.37: Influence of liquid supply rate on diameter of printed structures with 8% PVP solution. The collector motion speed is 20 mm/s and the voltage is 1.9 kV.

Figure 2.38 shows the variation in the diameter of the printed structure at different PVP solution concentrations. PVP solution with high concentration has low fluidity and high viscosity, which hinder the rheology and stretching of the solution during the spinning process, leading to larger jet diameter and smaller ejection speed. In addition, high viscosity and low fluidity also affect the free flow and deformation of the solution on the collector. When the concentration increases from 6% to 10%, the average diameter of the printed structure decreases from 156 to 70 μm due to the influence of the solution characteristics. However, when the solution is increased from 10% to 12%, the change in average diameter is not significant.

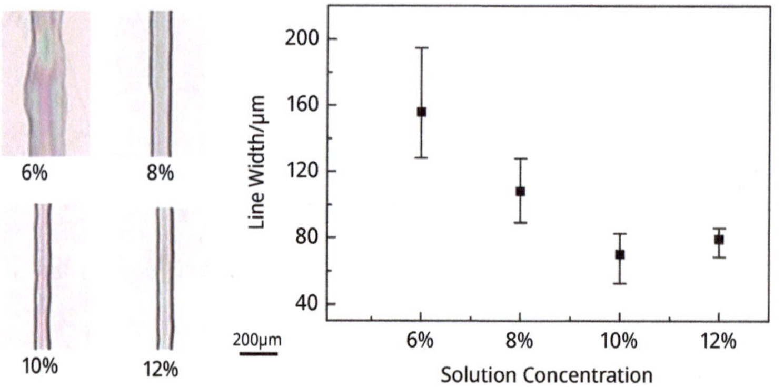

Figure 2.38: Influence of PVP solution concentration on the diameter of printed structure. The applied voltage, solution supply rate, and collector moving speed are 1.9 kV, 30 μL/h, and 20 mm/s, respectively.

Figure 2.39 illustrates the variation in diameter of the prepared structures as the collector speed increases from 5 to 20 mm/s. Due to the faster collector speed, the volume of solution deposited on the unit area of the collector per unit time decreases, so the average diameter decreases from 115 μm at 5 mm/s to 58 μm at 20 mm/s.

2.2 Modeling and simulation analysis of EWD jet dynamics

2.2.1 Simulation model of EDW jet dynamics

2.2.1.1 Force analysis for the EDW jet

The motion and deposition process of the charged jet can be interpreted as the result of the coupling effect of factors such as the electric field force, the charge repulsion force, the viscoelastic force, and the solution surface tension. The force analysis of the charged jet in the EDW process is shown in Figure 2.40.

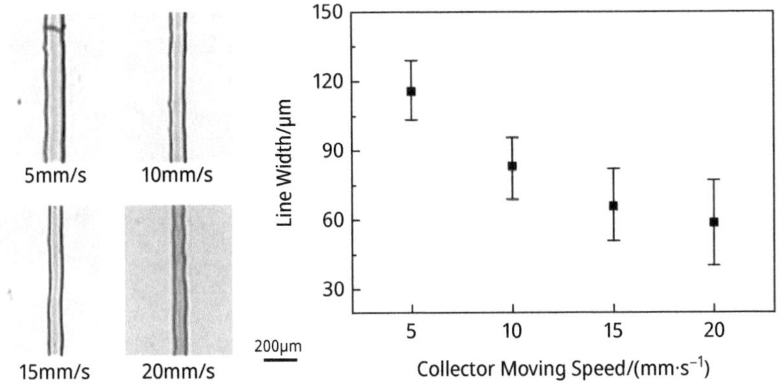

Figure 2.39: Influence of collector speed on the diameter of printed structure with 8% PVP solution (applied voltage is 1.9 kV and liquid supply rate is 30 μL/h.).

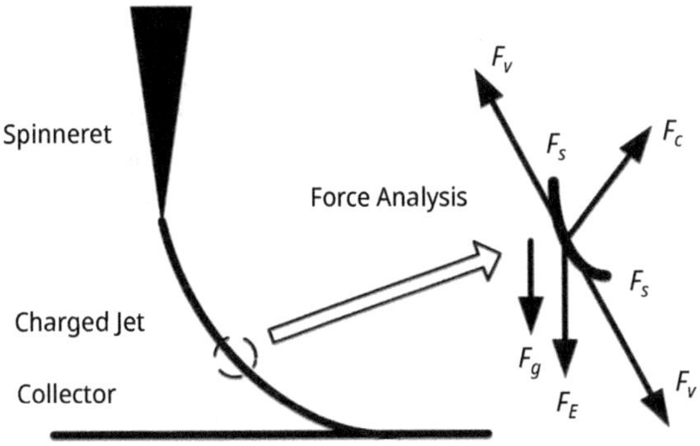

Figure 2.40: Force analysis of charged jet/nanofiber during EDW process (F_E: electric field force; F_C: charge repulsion; F_V: viscoelastic force; F_S: surface tension; F_g: gravity).

1. Electric field force

As the energy source of the spinning jet, the electric field force is the most important force during the electrospinning process. According to the classical electrical formula, the electric field force can be given as follows:

$$\vec{F}_E = q \cdot \vec{E} \tag{2.3}$$

where q is the amount of charge and $\vec{E}$ is the electric field strength. In EDW, the spinneret is perpendicular to the plane of the collector. It is assumed that the spinneret tip is semispherical, and the electric field distribution follows the tip-plane configuration model [15]. According to the space electric field distribution theory of the tip-plane configuration, the distribution of the electric field between the spinneret and the collector in the axial section of the spinneret is shown in Figure 2.41.

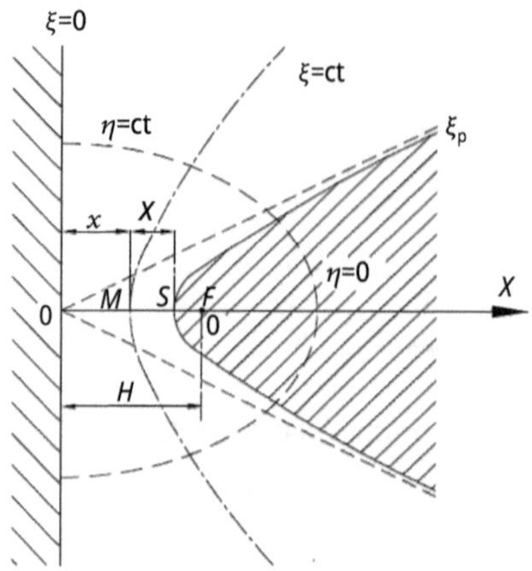

Figure 2.41: Hyperbolic model for the space electric field distribution of the tip-plane configuration [15]. Here, *M* is the space electric field calculation point on the extension line of the axis of the spinneret, *S* is the spinneret tip, *F* is the circle center of the spinneret tip, *H* is the distance from the circle center of the spinneret tip to the collector, *X* is the distance from the spinneret tip to the space electric field calculation point, x is the distance from the calculation point to the collector, and ξ is the polar angle of the confocal elliptical cluster with the circle center as the focus. η is the polar angle of the confocal hyperbolic cluster with the center of the spinneret tip as the focus.

The distributions of electric field lines and equipotential lines in the axial section follow the hyperbolic rule, where the electric field lines form a confocal ellipse cluster with the spinneret tip center as the focus, while the equipotential lines form a cluster of confocal hyperbolas with the center of the spinneret tip as the focus. According to the theory of electric field distribution, the space electric field lines are perpendicular to the equipotential lines, which can be mathematically described as the perpendicularity between the clusters of confocal hyperbolas and confocal ellipse. Its mathematical expression is as follows:

$$\begin{cases} x = a\sin\xi \mathrm{ch}\eta \\ y = a\cos\xi \mathrm{sh}\eta \end{cases} \tag{2.4}$$

In the EDW process, a voltage U is applied to the spinneret tip, and the collector is grounded. Coelho et al. [15] analyzed the potential distribution on the equipotential surface, and the potential can be given as

$$V(\xi) = C \ln \tan(\xi/2 + \pi/4) \tag{2.5}$$

Then the electric field strength can be given as

$$E(\xi,\ \eta) = \frac{C}{a \cos \xi (ch^2\eta - sin^2\xi)^{1/2}} \tag{2.6}$$

The electric field strength at each point on the extension line of the spinneret axis can be further simplified as follows:

$$E(X) = \frac{H \cdot C}{[X(2H - X) + (H - X)a]} \tag{2.7}$$

$$C = \frac{U}{\ln\left[2(H/a)^{1/2}\right]} \tag{2.8}$$

In the above formula, H is the distance from the center of the spinneret tip to the collector, ξ is the polar angle of the confocal elliptical cluster, η is the polar angle of the confocal hyperbolic cluster, U is the applied voltage, h is the distance from the spinneret tip to the collector, a is the diameter of the spinneret tip, X is the distance between the spinneret tip and the calculation point of the space electric field, and C is the electric field constant. The strength of the space electric field increases with the increase of the applied voltage and decreases rapidly with increasing distance from the spinneret. The change in electric field strength on the extension line of the spinneret axis with distance from the spinneret is shown in Figure 2.42.

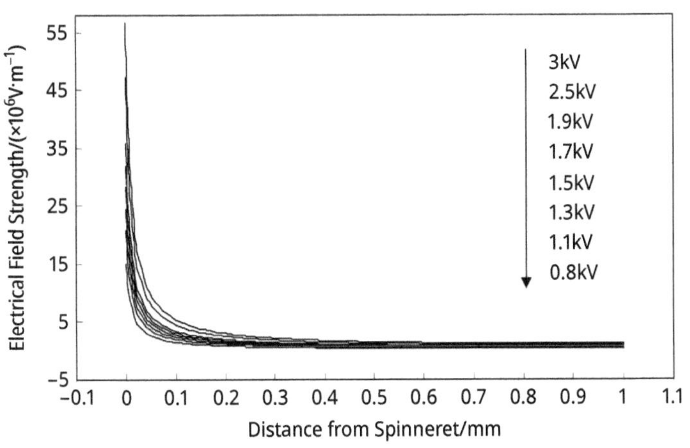

Figure 2.42: Electric field strength on the extension line of spinneret axis.

2. Charge repulsion force

Assuming that the charged body is a particle and the size effect is ignorable, according to Coulomb's law, the charge repulsion force can be expressed as follows:

$$\vec{F}_{C_q} = \frac{1}{4\pi\varepsilon_0} \cdot \frac{q_1 \cdot q_2}{d^2} \vec{R} \tag{2.9}$$

$$d^2 = (x_1 - x_2)^2 + (y_1 - y_2)^2 + (z_1 - z_2)^2 \tag{2.10}$$

where q_1 and q_2 are the charges on two charged bodies interacting with each other; ε_0 is the relative dielectric constant of the medium; d is the distance between the two charged bodies; (x_1, y_1, z_1) and (x_2, y_2, z_2) are the spatial coordinates of the two charged bodies respectively; $\vec{R}$ is a vector label, which can be noted as

$$\vec{R} = \frac{(x_1 - x_2)\vec{i} + (y_1 - y_2)\vec{j} + (z_1 - z_2)\vec{k}}{\sqrt{(x_1 - x_2)^2 + (y_1 - y_2)^2 + (z_1 - z_2)^2}} \tag{2.11}$$

The space charged body will induce inductive charges on the collector and generate additional charge repulsion on other space charged bodies. According to the principle of mirror symmetry in electrostatics [16], the charge repulsion force of the induced charge of the charged body on the collector can be equivalent to the charge repulsion force of the symmetric charged body on the other side of the collector. Suppose that there is a charged body on that other side of the collector, which has the same shape as the charged jet and is mirror-symmetrical to it with respect to the collector, the charges carried by the charged body and the jet are equal in the quantity and opposite in the sign. Given a particle $q_1(x_1, y_1, z_1)$ on the jet and its mirror-symmetric particle on the charged body $q_1'(x_1, y_1, -z_1)$(Fig. 2.43), the charge repulsive force of the particle $q_1'(x_1, y_1, -z_1)$ on q_2 can be expressed as

$$\vec{F}_{Cm} = \frac{1}{4\pi\varepsilon_0} \cdot \frac{-q_1' \cdot q_2}{d'^2} \vec{R} \tag{2.12}$$

$$d'^2 = (x_1 - x_2)^2 + (y_1 - y_2)^2 + (z_1 + z_2)^2 \tag{2.13}$$

During the EDW process, the residual charge on the collector will generate repulsion force on the charged jet. Let the residual charge on the collector be $q_r(x_r, y_r, 0)$, the charge repulsion of residual charge on q_2 can be described as

$$\vec{F}_{C_r} = \frac{1}{4\pi\varepsilon_0} \cdot \frac{q_r \cdot q_2}{d_r^2} \vec{R} \tag{2.14}$$

$$d_r^2 = (x_r - x_2)^2 + (y_r - y_2)^2 + z_2^2 \tag{2.15}$$

Therefore, the charge repulsion imposed to the charged jet is equal to the sum of the above three ones:

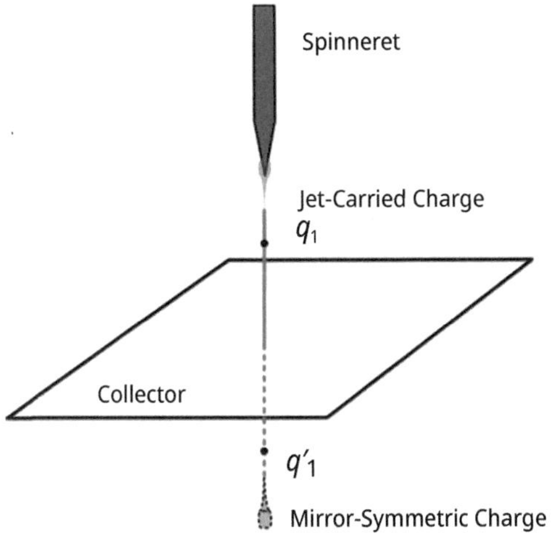

Figure 2.43: Schematic diagram of mirror-symmetric charge repulsion.

$$\vec{F}_C = \vec{F}_{C_q} + \vec{F}_{C_m} + \vec{F}_{C_r} \tag{2.16}$$

3. Viscoelastic force

Electrospinning is a typical rheological process of viscoelastic fluid, where the viscoelastic force affects the ejection, motion, and deposition behavior of the spinning jet. The electrospinning solution shows more fluid characteristics in the initial phase of the ejection, while it exhibits more solid characteristics in the jet or nanofiber deposition stage. Currently, Maxwell's mechanical model is the most commonly used mechanical model to describe the behavior of viscoelastic fluids. In 1991, Joseph [17] conducted a systematic and detailed analysis on the application of Maxwell's mechanical model in viscoelastic fluid. Maxwell's mechanical model uses the basic structure of spring and damper in series to describe the mechanical characteristics of viscoelastic fluid. The basic unit is shown in Figure 2.44. By adjusting the elastic coefficient of the spring and the damping coefficient of the damper in the model unit during calculation, the dynamic characteristics of viscoelastic fluid under different conditions can be reflected.

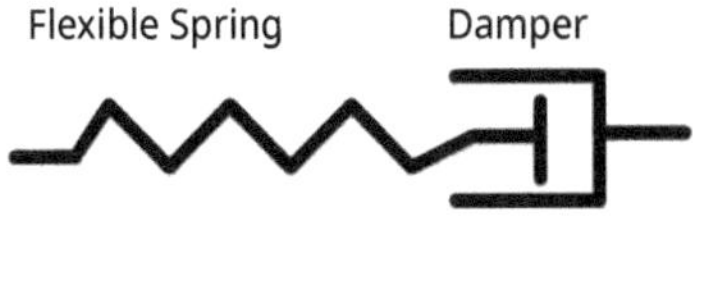

Figure 2.44: Basic unit of Maxwell's mechanical model.

According to the principle of viscoelastic mechanics, the dynamic behavior of the polymer solution jet during the stretching and deformation process under the influence of electric field force can be described by three formulas in Maxwell's mechanical model: viscoelastic mechanical equilibrium, mass balance, and momentum transfer conservation, which are given as follows:

$$\frac{\partial_{\sigma}(s)}{\partial t} = G\frac{1}{\lambda(s)}\cdot\frac{\partial\lambda(s)}{\partial t} - \frac{G}{\mu}\sigma(s) \tag{2.17}$$

$$\frac{D}{Dt}\left[\lambda(s)\pi a^2(s)\right] = 0 \tag{2.18}$$

$$\rho\lambda(s)\pi a^2(s)\frac{DV}{Dt}(s) = \lambda(s)\pi a^2(s)q^2\int_0^s \lambda(s^*)\pi a^2(s^*)\times C\left[\frac{\bar{a}(s,s^*)}{|r(s)-r(s^*)|}\right]\frac{r(s)-r(s^*)}{|r(s)-r(s^*)|^3}d_{s^*}$$
$$+\lambda(s)\pi a^2(s)\rho_{qB}\nabla\Phi + \frac{\partial}{\partial s}\left[\pi a^2(s)\sigma(s)u(s)\right] + \frac{\partial}{\partial s}\left[\pi a(s)u(s)\right] \tag{2.19}$$

where σ is the longitudinal stress inside the fiber, G is the Young's modulus of the fiber, μ is the solution viscosity, ξ is the jet bending coordination coefficient, s is the Lagrange coefficient of fiber solidification, λ is the stretching rate of the fiber, denoted by $\lambda = d\xi/ds$, C is the Coulomb force coefficient of charge, ρ is the mass density of the jet, r represents the coordination matrix, V is the velocity matrix of the jet, defining $D/Dt = \partial/\partial t + V\cdot\nabla$ as the convection offset coefficient of the jet, ρ_{qB} is the charge density of the jet body, Φ is the space potential, and u is the jet axial direction matrix of the jet.

4. Surface tension

The surface tension of liquid mainly takes effect on the solution surface, hindering the changes on the solution surface and reducing its surface energy. It tends to maintain the minimum surface area. In the jet rheological process, the surface tension primarily acts as an obstacle to the stretching effect of the electric field force and the charge repulsion force, driving the jet to contract. The relationship between the solution surface tension and the surface tension coefficient and the change in the surface area can be expressed as

$$\vec{F}_s = \gamma\delta A\vec{R} \tag{2.20}$$

where $\vec{F}_s$ the surface tension of the solution, γ is the surface tension coefficient, and δA is the change in the surface area.

2.2.1.2 Establishment of jet dynamics model

As the jet is a continuous slender body, all the forces it experiences are continuously distributed, which is not conducive to the analysis and calculation of its motion. Currently, the discrete simulation model of the bead-chain structure has become the main research method for the analysis of the electrospinning jet motion [18–20]. By adjusting the coefficients of the Maxwell element between the bead-chain structures, the variation law of the viscoelastic characteristics of the electrospinning jet can be better reflected. Therefore, in this study, based on the viscoelastic mechanics theory, Maxwell's mechanical model is adopted to establish the dynamics model of the EDW jet. The schematic diagram of the model is shown in Figure 2.45.

To facilitate the analysis, the following assumptions are made when establishing the dynamics model of the charged jet:

(1) During the EDW process, no jet bifurcation occurs, only the motion and deposition behavior of a single jet is considered;
(2) It is assumed that the jet is continuous without any fracture, and the radius of the initial jet is r_0;

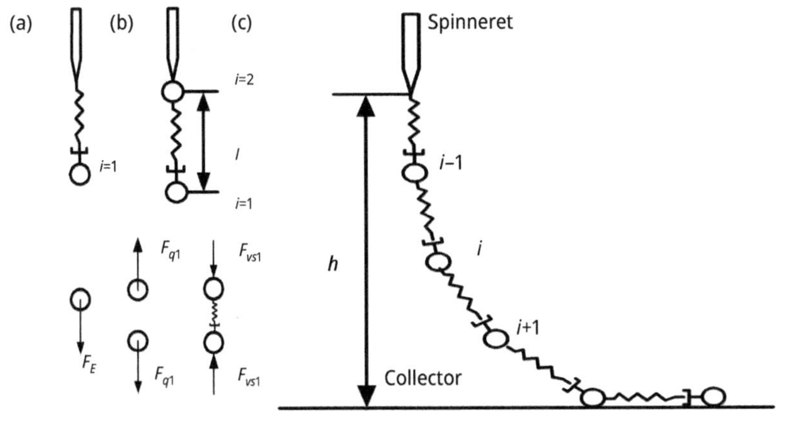

Figure 2.45: Schematic diagram of jet dynamics model: (a) the charged particle moves downward under the action of electric field force; (b) a new particle is generated at the tip of the spinneret as the jet ejects; and (c) the single jet composed of multiple Maxwell elements between the spinneret and the collector.

(3) It is assumed that the jet and nanofibers are composed of discrete particles, with the adjacent particles being connected by Maxwell elements, wherein each Maxwell element is composed of a series of connected springs and dampers;
(4) It is assumed that the charge and mass on the jet are concentrated on individual dispersed particle, the mass of each particle is m and the amount of charge carried by each particle is q_m;
(5) It is assumed that the Maxwell element connecting two adjacent particles is an insulator, which cannot carry out charge migration and transfer, and there is no mass and charge distribution on the Maxwell cell;
(6) The influence of solvent volatilization on jet volume is ignored;
(7) During the EDW process, the voltage between the spinneret and the collector remains constant;
(8) Neglecting the effect of air resistance as it is small during the motion of the jet.
(9) According to the theoretical analysis, the impact of free charge migration on the rheological and movement processes of the jet is small and can be ignored;

(10) Neglecting the polarization effect of polymer molecules under the strong electric field.

The calculation parameters involved in the simulation are described below:

1. Number of particles in jet simulation

During the simulation, the number of particles in the space is denoted as N. At the start of the simulation, it is assumed that there is a particle at the nozzle ($N = 1$). The particle moves toward the collector under the influence of the electric field force and other factors, as shown in Figure 2.45(a); with the distance between the particle and the spinneret increased to be $l_{10} \geq l_0$, a new particle appears at the spinneret ($N = 2$), as shown in Figure 2.45(b); and so on. When the particles arrive at collector, they stop moving downward. Instead, they move along with the collector, as shown in Figure 2.45(c). Note the velocity of the kth particle deposited on the collector as (v_{kx}, v_{ky}, v_{kz}) and the velocity of the collector as (v_{Cx}, v_{Cy}), then

$$v_{kx} = v_{Cx} \ v_{ky} = v_{Cy} \ v_{kz} = 0 \tag{2.21}$$

Each particle is numbered as $1, 2, 3, \ldots, N$ according to its order of the appearance. The mass of each particle does not change during the simulation process and the amount of charge carried by a particle does not change during its movement.

2. Calculation of viscoelastic force

The viscoelastic stress between the $(i + 1)$th particle and the ith particle on the jet [18] can be described as

$$\frac{d\sigma_{i+1,i}}{dt} = G\frac{dl_{i+1,i}}{l_{i+1,i} \cdot dt} - \frac{G}{\mu}\sigma_{i+1,i} \tag{2.22}$$

$$l_{i+1,i} = \sqrt{(x_i - x_{i+1})^2 + (y_i - y_{i+1})^2 + (z_i - z_{i+1})^2} \tag{2.23}$$

And the viscoelastic stress between the $(i - 1)$th and the ith particle on the jet can be noted as

$$\frac{d\sigma_{i-1,i}}{dt} = G\frac{dl_{i-1,i}}{l_{i-1,i} \cdot dt} - \frac{G}{\mu}\sigma_{i-1,i} \tag{2.24}$$

$$l_{i-1,i} = \sqrt{(x_i - x_{i-1})^2 + (y_i - y_{i-1})^2 + (z_i - z_{i-1})^2} \tag{2.25}$$

Then the total viscoelastic force on the ith particle can be given as

$$\vec{F}_{vi} = \pi r_{i+1,i}^2 \sigma_{i+1,i} \vec{R}_{i+1,1} - \pi r_{i-1,i}^2 \sigma_{i-1,i} \vec{R}_{i-1,1} \tag{2.26}$$

where t is the simulation time, G is the elastic coefficient, μ is the viscosity of the polymer solution, σ is the viscoelastic stress, and $\sigma_{i-1,i}$ and $\sigma_{i+1,i}$ are the viscoelastic stresses of the jet between the $(i-1)$th particle and the ith particle and between the ith particle and the $(i+1)$th particle, respectively; r is the jet radius, and $r_{i-1,i}$ and $r_{i+1,i}$ are the radii of the jet between the $(i-1)$th particle and the ith particle and between the $(i+1)$th particle and the ith particle, respectively. The radius of each part of the jet is calculated by the following formulas:

$$r_{i-1,i} = \sqrt{\frac{l_0}{l_{i-1,i}}} \cdot r_0 \tag{2.27}$$

$$r_{i,i+1} = \sqrt{\frac{l_0}{l_{i,i+1}}} \cdot r_0 \tag{2.28}$$

3. Calculation of surface tension

The role of the surface tension is to maintain the shape of the jet, hindering the deformation and stretching of the jet. The surface tension can be represented as

$$\vec{F}_{si} = -\frac{\gamma\pi\left(r'^2\right)_{av} C_i}{\left(x_i^2 + y_i^2\right)^{1/2}} \vec{R} \tag{2.29}$$

where γ is the surface tension coefficient of the polymer solution; $\left(r'^2\right)_{av} = (r_{i-1,i} + r_{i,i+1})^2/4$; and C_i denotes the curvature radius of the circle defined by the particles $(i-1)$, i, $(i+1)$.

4. Electric field force

The electric field force experienced by the ith charged particle can be described as

$$\vec{F}_{Ei} = q_m \vec{E}(x_i, y_i, z_i) \tag{2.30}$$

where q_m is the charge of each particle in the simulation model; $\vec{E}(x_i, y_i, z_i)$ represents the electric field intensity vector at the point (x_i, y_i, z_i) in space.

5. Charge repulsion force

The charge repulsion of the jth charged particle on the ith charged particle can be given as

$$\vec{F}_{qji} = -k\frac{q_j q_i}{l_{j,i}^2} \vec{R} \tag{2.31}$$

$$l_{j,i} = \sqrt{(x_j - x_i)^2 + (y_j - y_i)^2 + (z_j - z_i)^2} \tag{2.32}$$

where q_i and q_j are the charges carried by the ith and jth particles, respectively, and k is the Coulomb constant of the charge.

To better reflect the influence of the collector on the moving charged particles, we have corrected the calculation of the charge repulsive force based on the principle of the space electric field mirror surface, which is given as

$$\vec{F}_{qji} = -k\frac{q_j q_i}{l_{j,i}^2}\vec{R} + k\frac{q_j q_i}{l_{j,i}^{\prime 2}}\vec{R} \tag{2.33}$$

$$l_{j,i}^{\prime} = \sqrt{(x_j - x_i)^2 + (y_j - y_i)^2 + (z_j + z_i)^2} \tag{2.34}$$

where $M < j \leq N$, M is the number of charged particles deposited on the collector, and N is the number of charged particles in the space.

The Maxwell elements between particles are assumed to be insulating, and when particles move in space, the charge they carry remains unchanged. After nanofibers deposit on the collector, the charge carried by the fiber is transferred to the ground through the collector. The speed of charge transfer depends on the charge relaxation time constant $\tau = \xi/(4\pi\delta)$, where ξ and δ are the relative permittivity and conductivity of the collector, respectively. After nanofibers deposit on the collector, the charge of each particle decreases with the increase of deposition time t_d. Therefore, the charge of the kth particle deposited on the collector can be expressed as a decreasing function with respect to the deposition time t_d:

$$q_{kt} = q_m \cdot \exp(-t_d/\tau)/E_c \tag{2.35}$$

where q_{kt} is the amount of charge carried by the particle deposited on the collector at time t_d; E_c is the electric field strength within the collector; t_d is the time when the kth particle deposits on the collector; τ is the charge relaxation time of the collector; and q_m is the initial charge of the particle. Due to the effect of residual charges on the collector, the charge repulsion force exerted by the kth particle, which has been deposited on the collector, on the ith particle of the spatially moving jet can be expressed as

$$\vec{F}_{qki} = -k\frac{q_i q_k}{l_{k,i}^{\prime 2}}\vec{R} = -k\frac{q_i q_k}{l_{k,i}^{\prime 2}} q_m \cdot \exp(-t_d/\tau)/\ E_c\vec{R} \tag{2.36}$$

$$l_{k,i}^{\prime} = \sqrt{(x_k - x_i)^2 + (y_k - y_i)^2 + z_i^2} \tag{2.37}$$

The repulsive force experienced by the ith particle comprises two components: the repulsive forces from other spatially moving particles and the repulsive forces from particles already deposited on the collector. Therefore, the total repulsive force acting on the ith particle can be expressed as follows:

$$\vec{F}_{qi} = \sum_{j=1}^{i-1}\vec{F}_{qji} + \sum_{j=i+1}^{N}\vec{F}_{qji} \tag{2.38}$$

6. Gravity

In accordance with the law of conservation of mass, the mass of each particle remains constant at m throughout the simulation process. Consequently, the gravitational force acting on the particle can be expressed as follows:

$$\vec{F}_g = -mg \cdot \vec{k} \tag{2.39}$$

where g is the gravitational acceleration, $g = 9.8\ m/s^2$ is adopted in the simulation, and $\vec{k}$ is the vector component in the z-axis direction.

7. Dynamic equilibrium equation

Based on the mechanical analysis presented above, the dynamic equilibrium equation governing each particle can be formulated as follows:

$$m\frac{d\vec{v}_i}{dt} = \vec{F}_{vi} + \vec{F}_{Si} + \vec{F}_{Ei} + \vec{F}_{qi} + \vec{F}_g \tag{2.40}$$

where $\vec{v}_i$ is the velocity of the ith particle.

2.2.1.3 Development of jet simulation program

With the advantages of fast operation and high accuracy, the fourth-order Runge-Kutta method [21] has become one of the most commonly used numerical methods for solving differential equations in engineering. Based on the fourth-order Runge-Kutta method, we have developed the simulation program for the motion and deposition behaviors of the jet. The flowchart of the simulation program is shown in Figure 2.46.

Based on the research results of Yarin et al. [22] and Thompson et al. [23–25] on the kinematic characteristics of the charged jet in the electrospinning process as well as the work of Tan [24, 25] and Yuya [26] on the mechanical characteristics of electrospun nanofibers, we make the following assumptions for each parameter in the simulation model: the elastic modulus distribution range of nanofibers is 5–130 MPa, and the jet elastic modulus is $G = 45$ MPa; the jet charge density is $10^{-4} - 10^2$ C/m, the amount of charge carried by each particle is $q_m = 2.83 \times 10^{-13}$ C, and the mass of each particle is $m = 2.83 \times 10^{-12}$ kg.

Silicon substrate and silicon dioxide substrate are used as the collectors. The relative dielectric constant and conductivity of the silicon substrate are set to $\xi = 11.9$ and $\delta = 4.35 \times 10^{-6}$ S/cm. And the relative dielectric constant and conductivity of the silicon dioxide substrate are set to $\xi = 3.9$ and $\delta = 1.5 \times 10^{-11}$ S/cm. The viscosity μ and the surface tension coefficient γ of the solution are determined by the characteristics of the polymer solution. PEO solution with a concentration of 20% is used for subsequent simulation analysis. Therefore, the viscosity of the polymer solution is set to $\mu = 100$ Pa • s and the surface tension coefficient is set to $\gamma = 34$ mN/m. The radius of the solid probe tip is set to $a = 20\ \mu$m, corresponding to the solid probe with the tip diameter of 40 μm used as the nozzle in the experiment. The initial radius of the jet is sent to $r_0 = 5\ \mu$m.

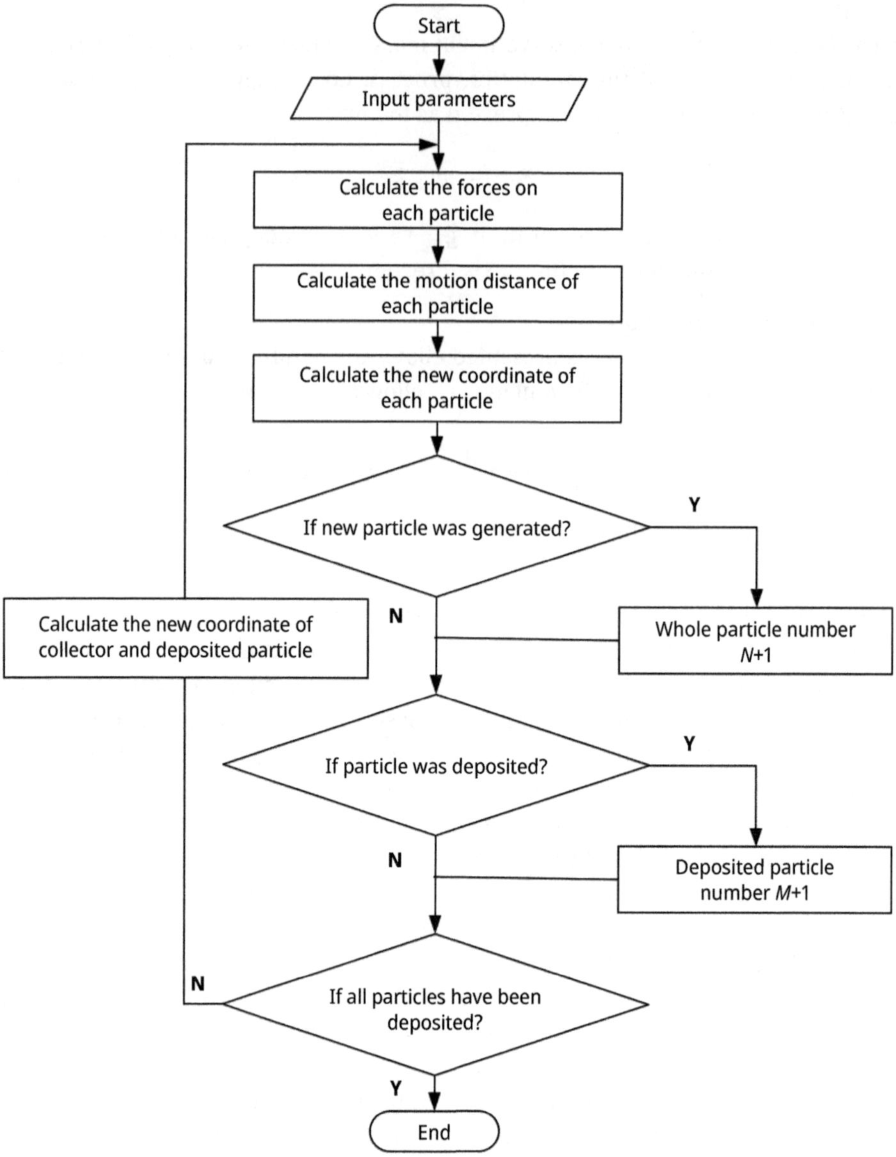

Figure 2.46: Flowchart of jet motion and deposition simulation program.

The nozzle-to-collector distance h , the applied voltage U, the conductivity δ, and the relative permittivity ξ of the collector are used as the input variables of the simulation program.

2.2.2 Simulation of EDW jet motion state

Based on the jet dynamics simulation model, the effects of the spinneret-to-collector distance, the applied voltage, and the collector conductivity on the motion and deposition behavior of the jet are investigated. The motion state of the jet, the deposition morphology of nanofibers on the collector, and the variation rules for the jet stretching ratio, jet velocity, and jet acceleration are obtained by simulation calculation.

2.2.2.1 Simulation of the jet motion state

Using the jet motion simulation program, the spatial coordinates of each particle at every moment are calculated, and the three-dimensional motion state of the jet as shown in Figure 2.47 are plotted.

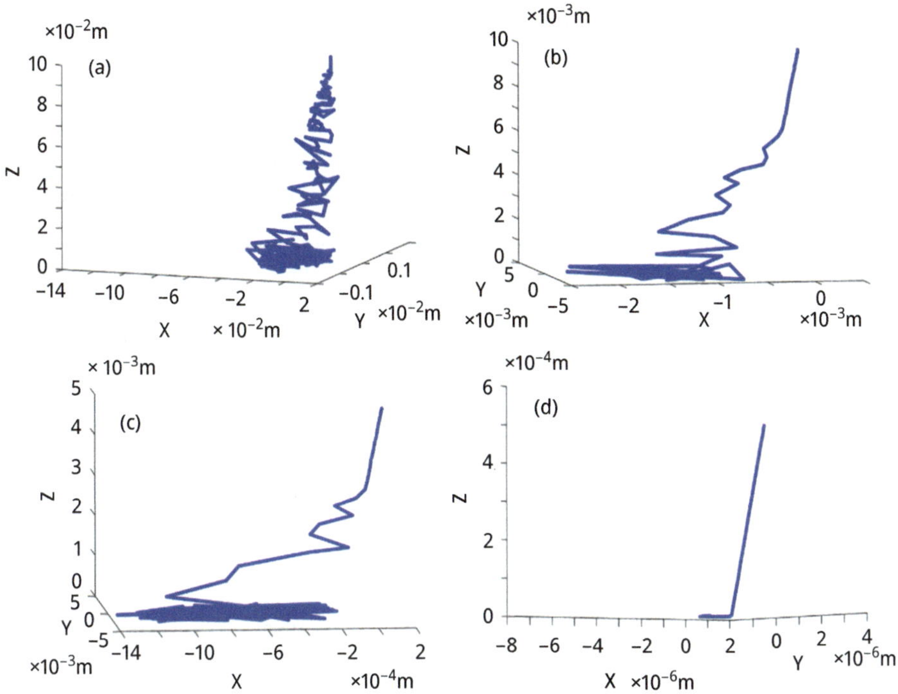

Figure 2.47: Simulation of the jet motion state. The spinneret-to-collector distance and the applied voltage are: (a) 100 mm, 10 kV; (b)10 mm, 5 kV; (c) 5 mm, 2 kV; and (d) 0.5 mm, 1 kV. The collector velocity is 0.05 m/s; the relative dielectric constant and the conductivity of the collector are $\xi = 3.9$ ($\delta = 2.3 \times 10^{-8}$S/cm), respectively.

The three-dimensional motion state of the jet at different spinneret-to-collector distances is shown in Figure 2.47. According to the motion principle of electrospinning jet, due to the influence of factors such as the charge repulsion, the jet exhibits spiral instability after a short period of stable linear motion, as shown in Figure 2.47(a). In Figure 2.47(a)–(c), the spinneret-to-collector distance are 100, 10, and 5 mm, respectively. The simulation results show that, with the decrease of the spinneret-to-collector distance, the proportion of the disordered and unstable motion stage of the jet decreases gradually. Reducing the spinneret-to-collector distance can eliminate the chaotic and unstable motion of the jet. As shown in Figure 2.47(d), when the spinneret-to-collector distance is shortened to 0.5 mm, the jet deposits on the collector before it steps into the chaotic motion phase. The stable straight jet between the spinneret and the collector can provide technical support for EDW of micro-nanostructures. The simulation results are consistent with the experimental results, which well reflects the spatial motion behavior of the jet.

To deeply understand the motion behavior of the jet during the ejecting process, we further analyze the variation law of the length, the maximum velocity, and the stretching ratio of the stable straight jet based on the simulation model. The simulation results of each parameter are shown in Table 2.1.

Table 2.1: Analysis of jet motion simulation results.

Spinneret-to-collector distance (mm)	Applied voltage (kV)	Length of the stable straight jet (mm)	Proportion of the stable straight jet (%)	Maximum velocity	Stretching ratio
0.5	0.6	0.5	100	0.79	102.88
1	0.8	1	100	1.02	116.92
1	1.1	1	100	1.18	138.40
1	1.3	1	100	1.29	150.38
1	1.5	1	100	1.39	158.37
1	1.7	1	100	1.46	162.37
1	1.9	1	100	1.51	181.36
1.5	1.7	1.5	100	1.47	176.56
2	1.5	2	100	1.37	162.18
2	1.7	2	100	1.47	175.17
2.5	1.7	2.5	100	1.46	174.34
3	1.7	2.40	79.92	1.47	179.46
3	2	2.47	82.33	1.59	190.94
3	2.5	2.76	91.99	1.74	213.42
3	3	3	100	1.95	235.91
4	2	2.78	69.48	1.59	193.47
5	2	2.84	56.81	1.60	195.01
6	2	2.94	48.98	1.58	196.04
8	3	3.63	45.37	1.94	242.86
10	3	3.87	38.73	1.95	242.52
20	3	4.24	21.20	1.95	242.70
30	5	4.90	16.32	2.51	314.36
40	5	5.02	12.54	2.51	314.14

Table 2.1 (continued)

Spinneret-to-collector distance (mm)	Applied voltage (kV)	Length of the stable straight jet (mm)	Proportion of the stable straight jet (%)	Maximum velocity	Stretching ratio
50	5	5.21	10.43	2.51	313.93
60	6	5.34	8.90	2.76	343.63
60	8	6.71	11.19	3.18	397.20
70	8	6.31	9.02	3.18	397.12
80	8	5.65	7.07	3.17	397.05
90	8	5.83	6.48	3.18	396.59
100	7	4.90	4.90	2.97	370.32
100	8	5.35	5.35	3.18	396.78
100	9	5.59	5.59	3.37	421.06
100	10	5.90	5.90	3.56	443.51
100	11	6.30	6.30	3.73	466.05

The parameters investigated in Table 2.1 are defined as follows.

1. Length of the stable straight jet

In the simulation process, the coordinates of the nozzle are noted as $(0,0,h)$, where h is the spinneret-to-collector distance. Given a certain point on the jet with the spatial coordinates of (x_i, y_i, z_i), the distance from the point to the extension line of the spinneret is $l_s = \sqrt{x_i^2 + y_i^2}$. When $l_s > 20\ \mu\text{m}$, the particle is considered to have left the stable straight jet phase. Let the coordinates of the last particle s that leaves the stable straight jet phase be (x_s, y_s, z_s), the height difference between the particle s and the spinneret tip is defined as the stable straight jet length, which can be represented as $W_s = h - z_s$.

2. Proportion of the stable straight jet

The proportion of the stable straight jet is defined as the ratio of the stable straight jet length to the spinneret-to-collector distance, which can be calculated by $R_s = (h - z_s)/h \times 100\%$.

3. Maximum moving speed

The maximum moving speed refers to the component of the jet velocity along the z-axis v_z when the particle deposits on the collector in the stable ejection stage.

4. Jet stretching ratio

When particles deposit on the collector, the distance between two adjacent particles can be given as

$$l_{dk} = \sqrt{(x_k - x_{k-1})^2 + (y_k - y_{k-1})^2} \tag{2.41}$$

The ratio between the average distance of adjacent particle on the collector and the initial distance between the particles can be referred to as the jet stretching ratio mentioned in Table 2.1, which can be described as

$$\lambda_r = \frac{\bar{l}_d}{l_0} = \frac{\frac{\sum_{k=2}^{M} l_{dk}}{M-1}}{l_0} = \frac{\sum_{k=2}^{M} l_{dk}}{(M-1)\cdot l_0} \tag{2.42}$$

where $(x_k,\ y_k,\ 0)$ is the coordinate of the particle deposited on the collector, M is the total number of particles deposited on the collector, and l_0 is the initial distance between adjacent particles in the simulation of the jet motion.

Other simulation conditions include the collector is stationary and the silicon substrate is used with a relative dielectric constant of $\xi = 11.9$ and conductivity of $\delta = 4.35 \times 10^{-6}$ S/cm.

To facilitate subsequent discussion and analysis, two additional variables are defined:

(1) Distance from the spinneret

The vertical distance between a certain particle on the jet $p(x_p,\ y_p,\ z_p)$ and the spinneret is noted as the particle-to-spinneret distance l_{ps}, which can be given as

$$l_{ps} = h - z_p \tag{2.43}$$

where h is the spinneret-to-collector distance. This distance can also be interpreted as the vertical distance the jet travels in space.

(2) Ejection time

The time it takes for particle p to move in space after being ejected from the nozzle is referred to as the ejection time, denoted by t_j.

2.2.2.2 Simulation of stable straight jet length

The stable straight jet between the spinneret and the collector is fundamental to EDW technology and crucial for controlling the deposition behaviors of micro-nanostructures. Table 2.1 displays the length and proportion of the stable straight jet under different process parameters. The impact of various process parameters on the length of the stable straight jet is further discussed, as shown in Figures 2.48–2.51.

Increasing the strength of the space electric field accelerates the movement of particles toward the collector, helping to overcome the interference factors such as electrostatic repulsion and prolonging the length of the stable straight jet. Therefore, the length and proportion of the stable straight jet increase with the applied voltage, as shown in Figure 2.48. The proportion of the stable straight jet reaches 100%, and there is no chaotic or unstable jet movement when the voltage is increased to 3.0 kV.

The repulsive forces between the like charges on the charged jet, as well as the repulsive force produced by the residual charges on the collector, are the main reasons for the disordered and unstable motion of the jet. When the spinneret-to-collector distance is small, the repulsive force from the residual charges significantly affects the charged jet. Increasing the spinneret-to-collector distance reduces the influence of these residual charges, thereby increasing the length of the stable straight jet. When the spinneret-to-collector distance increases from 3 to 6 mm, the length of the stable straight jet increases, but the proportion decreases. This relationship is shown in Figure 2.49.

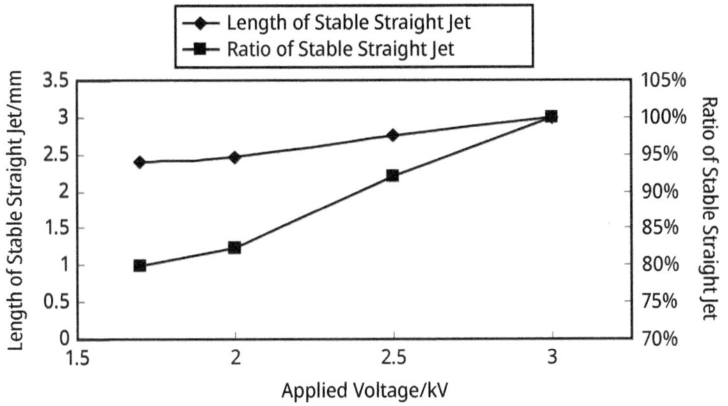

Figure 2.48: Impact of applied voltage on length and proportion of stable straight jet at a spinneret-to-collector distance of 3 mm, silicon substrate.

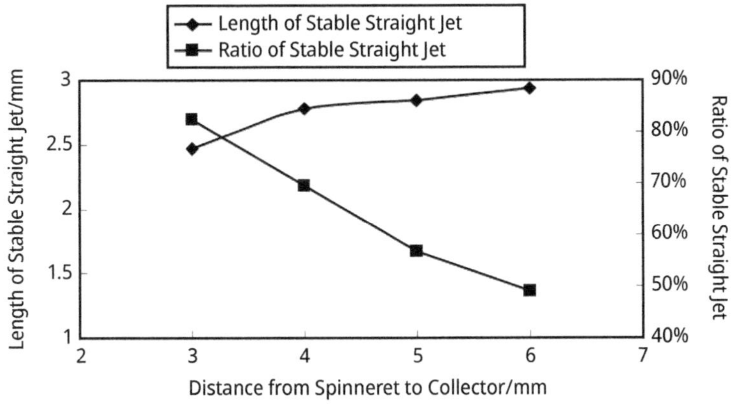

Figure 2.49: Influence of spinneret-to-collector distance on the length and the proportion of stable straight jet at an applied voltage of 2 kV, silicon substrate.

To study the variations in stable straight jet, the influence of process parameters on the length and proportion of stable straight jet under non-near-field conditions (where the spinneret-to-collector distance is greater than 60 mm) is further analyzed. The results are shown in Figures 2.50 and 2.51.

Under the non-near-field condition, the effect of the repulsive force from the residual charges on the jet is reduced, and the repulsive force between the like charges on the jet becomes the main factor affecting the chaotic and unstable jet motion. The velocity of the jet can be accelerated by increasing the applied voltage, which is beneficial to reducing the influence of the charge repulsion force and other factors. At a spinneret-to-collector distance of 100 mm, the length and proportion of the stable straight jet increase with the applied voltage, as shown in Figure 2.50. Increasing the distance between the spinneret and the collector reduces the space electric field strength, decreases the repulsive force between the like charges on the jet, and increases the disordered unstable motion of the charged jet, resulting in a decrease in the stable straight jet length. When the spinneret-to-collector distance is increased from 60 to 100 mm, both the length and the proportion of the stable straight jet decrease with the increase of the spinneret-to-collector distance, as shown in Figure 2.51.

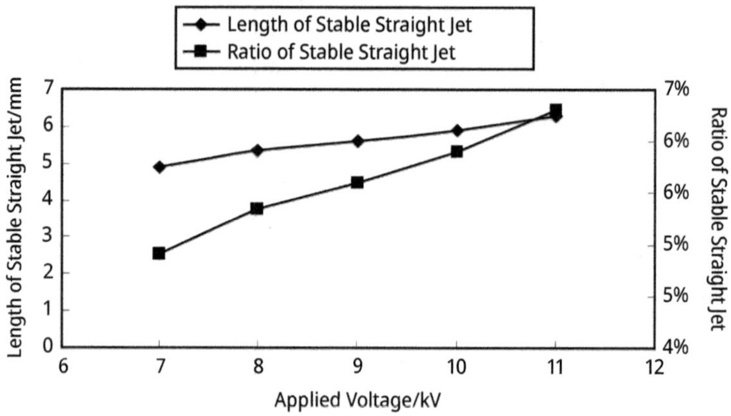

Figure 2.50: Influence of applied voltage on length and proportion of stable straight jet under non-near-field conditions at a spinneret-to-collector distance of 100 mm, silicon substrate.

2.2.2.3 Simulation of jet moving velocity

1. Motion velocity of the jet front

After ejecting from the spinneret, the jet accelerates toward the collector under the influence of the electric field force. The motion velocity of the jet front is simulated and analyzed. Figures 2.52 and 2.53 show the variation patterns of the jet front's motion velocity with distance from the spinneret and the ejection time at different spinneret-to-collector distances. Figures 2.54 and 2.55 show the variation patterns of the jet front velocity with the distance from the spinneret and the ejection time when different voltages are applied.

Under the acceleration effect of the electric field force, the velocity of the jet front increases with both the distance from the spinneret and the ejection time. For a given spinneret-to-collector distance, increasing the voltage enhances the electric field

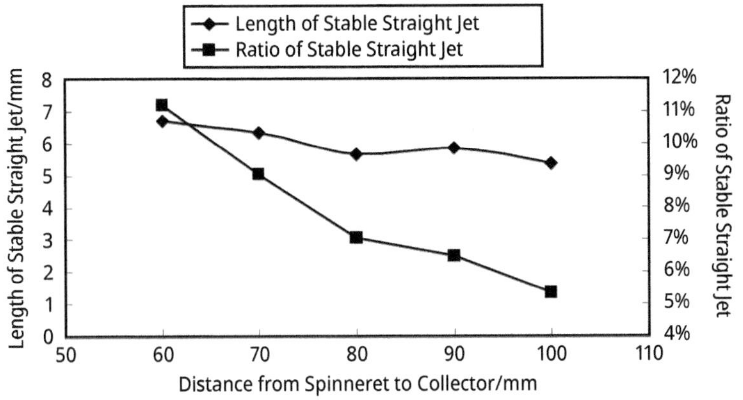

Figure 2.51: Influence of spinneret-to-collector distance on length and proportion of stable straight jet at an applied voltage of 8 kV under non-near-field conditions, silicon substrates.

strength, thereby increasing the velocity of the jet front. Conversely, with the same applied voltage, an increase in the spinneret-to-collector distance weakens the electric field, reducing the moving speed of the jet front. The time required for the jet front to move from the spinneret to the collector decreases with higher applied voltage and increases with greater spinneret-to-collector distance.

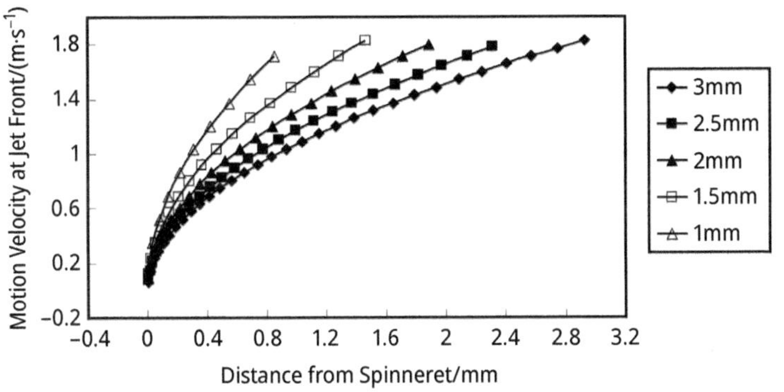

Figure 2.52: Variation in velocity of jet front with distance from the spinneret at different spinneret-to-collector distances (applied voltage is 1.7 kV, silicon substrate).

The jet front accelerates toward the collector under the action of the electric field force. The velocity reaches its maximum value when it deposits on the collector. In our research, the vertical downward component of the jet front velocity at the time it deposits on the collector is defined as the maximum moving speed of the jet front. Figure 2.56 illustrates the influence of the applied voltage and the spinneret-to-collector distance on the maximum moving speed. The maximum moving speed increases with

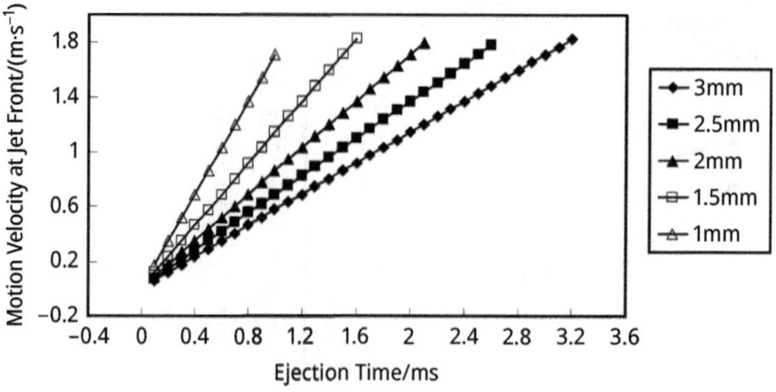

Figure 2.53: Variation in jet front velocity with ejection time at different spinneret-to-collector distances (applied voltage of 1.7 kV, silicon substrate).

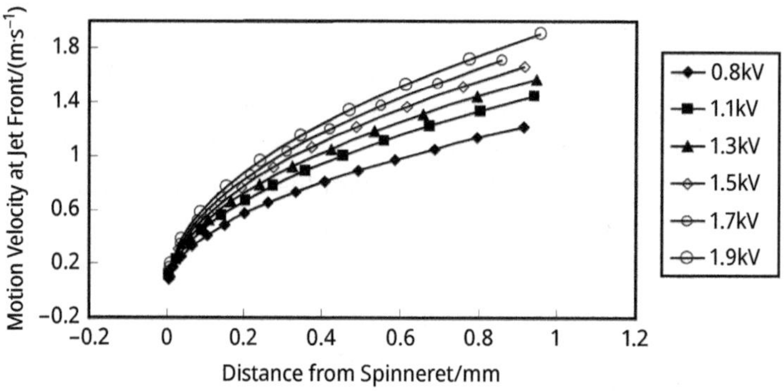

Figure 2.54: Variation of the motion speed of the jet front with distance from spinneret at different applied voltages (spinneret-to-collector distance of 1 mm, silicon substrate).

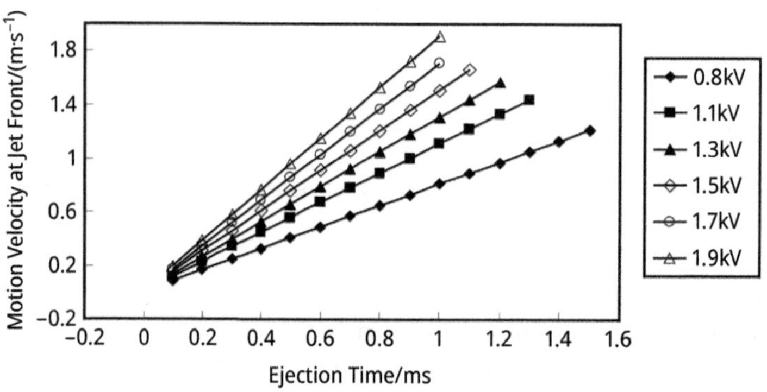

Figure 2.55: Variation of motion speed of jet front with ejection time at different applied voltages (spinneret-to-collector distance of 1 mm, silicon substrate).

the applied voltage, while the spinneret-to-collector distance shows no obvious influence on the maximum moving speed. Rising the applied voltage increases the electric field strength, thus increasing the acceleration and velocity of the charged jet. When the voltage is kept constant, increasing the spinneret-to-collector distance can increase the time available for jet acceleration. However, this also decrease the electric field strength, thus reducing the acceleration of the charged jet.

Based on the simulation model, the variation law of jet front's motion acceleration is further analyzed. As illustrated in Figure 2.57, the acceleration of the jet front decreases rapidly with increasing distance from the spinneret.

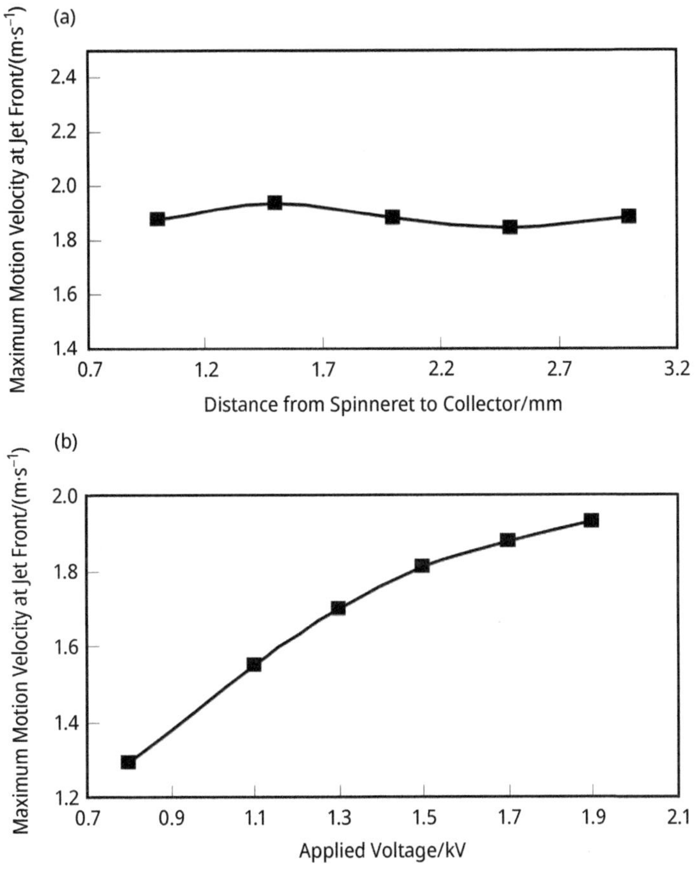

Figure 2.56: The influence of process parameters on maximum velocity of jet front: (a) maximum motion velocity of the jet front versus spinneret-to-collector distance at the applied voltage of 1.7 kV and (b) maximum motion velocity of the jet front versus applied voltage at the spinneret-to-collector distance of 1 mm (silicon substrate).

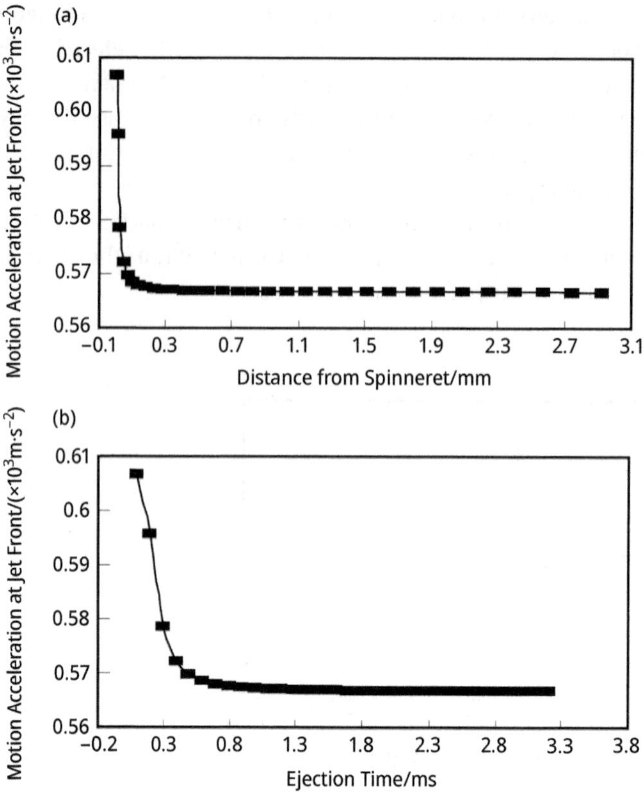

Figure 2.57: The influence of process parameters on motion acceleration of jet front: (a) acceleration of the jet front versus the distance from the spinneret and (b) acceleration of the jet front versus ejection time (spinneret-to-collector distance is 3 mm, the applied voltage is 1.7 kV, silicon substrate).

The electric field force is the primary driving force for the motion and rheology of the charged jet in the EDW process. The jet accelerates toward the collector under the influence of this force. According to the calculations, the electric field strength is highest at the spinneret and decreases rapidly with increasing distance from it. Closer to the collector, both the electric field strength and its gradient become smaller. The electric field above the conductive collector can be approximated as uniform. Figure 2.57 (a) illustrates that after ejection from the spinneret, the jet front accelerates toward the collector. When the jet front moves to 0.05 mm from the spinneret, its acceleration decreases from $0.606{\times}10^3$ m/s^2 to $0.572{\times}10^3$ m/s^2 and continues to decrease gradually as it approaches the collector. When the jet front nears the collector's surface (at 0.5 mm from the spinneret), its acceleration is $0.567{\times}10^3$ m/s^2. The variation in the jet front's acceleration with distance from the spinneret closely mirrors the changes in electric field strength shown in Figure 2.42, indicating that electric field strength is the main factor influencing the jet front's velocity.

The simulation results align with the experimental observations captured by the high-speed camera, showing that the velocity of the jet front increases with the distance from the spinneret, the jetting time, and the applied voltage.

2. Jet motion velocity

Based on the simulation model, we further analyzed the distribution of jet motion velocity during the stable ejection phase. Figures 2.58 and 2.59 show the variation of jet motion velocity with distance from the spinneret at different applied voltages and spinneret-to-collector distances under near-field conditions (spinneret-to-collector distance less than or equal to 3 mm). The simulation results indicate that the jet moving speed increases with the distance from the spinneret. At a certain distance from the spinneret, the jet velocity increases with higher applied voltage but decreases as the spinneret-to-collector distance increases.

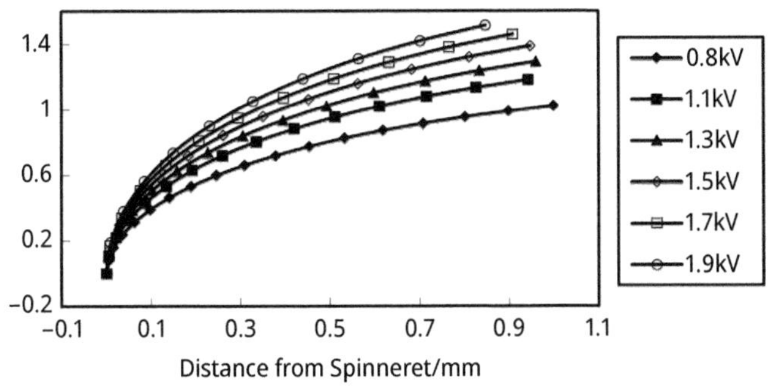

Figure 2.58: Variation of jet motion velocity with distance from spinneret at different applied voltages (spinneret-to-collector distance is 1 mm, silicon substrate).

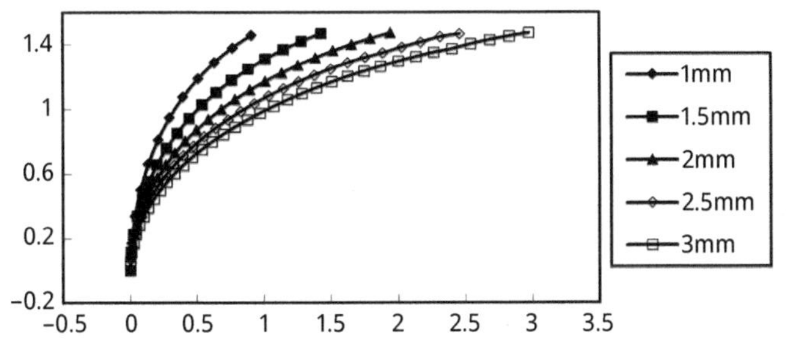

Figure 2.59: Variation of jet velocity with distance from spinneret at different spinneret-to-collector distances (applied voltage is 1.7 kV, silicon substrates).

Figures 2.60 and 2.61 show the variation of the acceleration of the jet motion with the distance from the spinneret during the stable ejection stage under near-field conditions. The acceleration of the jet motion decreases with the increase of the distance from the spinneret. For the same distance from the spinneret, the acceleration increases with the increase of the applied voltage and decreases with the increase of the spinneret-to-collector distance.

Increasing the voltage can increase the electric field strength, thereby increasing the electric field force acting on the jet and increasing the moving speed and acceleration of the jet for given spinneret-to-collector distance. Increasing the distance between the spinneret and the collector reduces the electric field strength, so the moving speed and acceleration of the jet decrease with the increase of spinneret-to-collector distance under the same voltage. The above simulation results show that the moving speed and acceleration of the jet in the stable jet ejection stage are smaller than those of the jet front. This is because the viscoelastic force of the deposited jet on the subsequent jet lowers the jet motion acceleration after the jet front deposits on the collector.

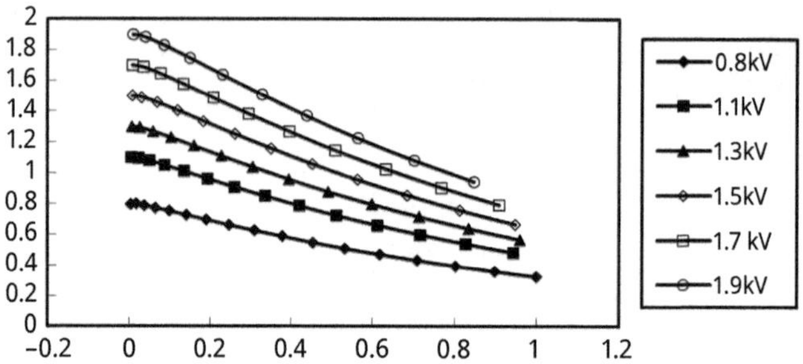

Figure 2.60: Variation of jet motion acceleration with distance from the spinneret at different voltages (spinneret-to-collector distance is 1 mm, silicon substrate).

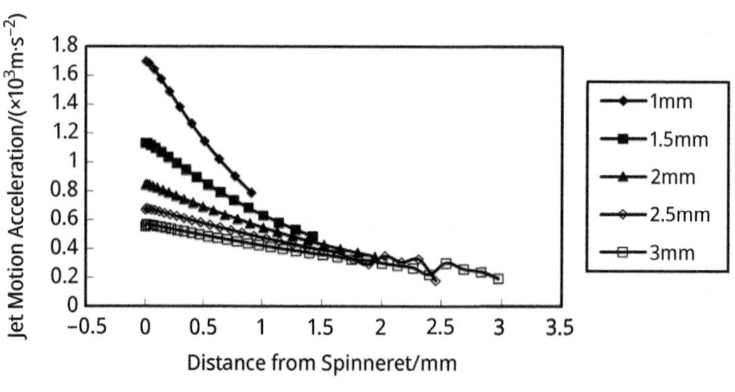

Figure 2.61: Acceleration of jet motion as a function of distance from the spinneret for different spinneret-to-collector distances (applied voltage is 1.7 kV, silicon substrate).

As a comparative study, the variations of jet moving speed under non-near-field conditions are simulated and analyzed. As shown in Figures 2.62 and 2.63, the jet moving speed increases with the distance from the spinneret. For given spinneret-to-collector distance, the jet moving speed increases with the increase of the applied voltage. With the same applied voltage, the jet moving speed decreases with the increase of spinneret-to-collector distance. The jet moving speed under the non-near-field condition is obviously greater than that under the near-field condition.

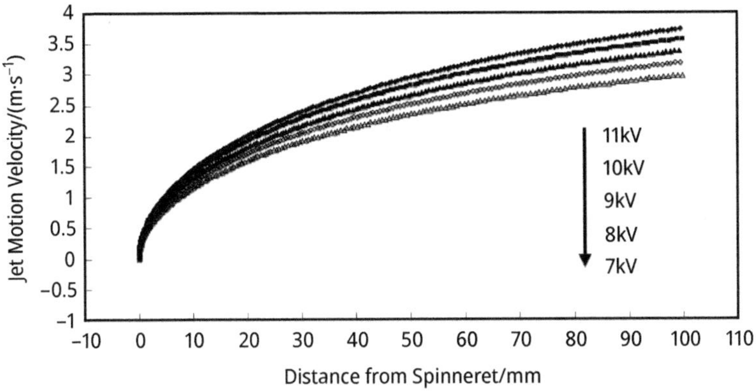

Figure 2.62: Variation of jet motion speed with distance from the spinneret at different voltages under non-near-field condition (spinneret-to-collector distance is 100 mm, silicon substrate).

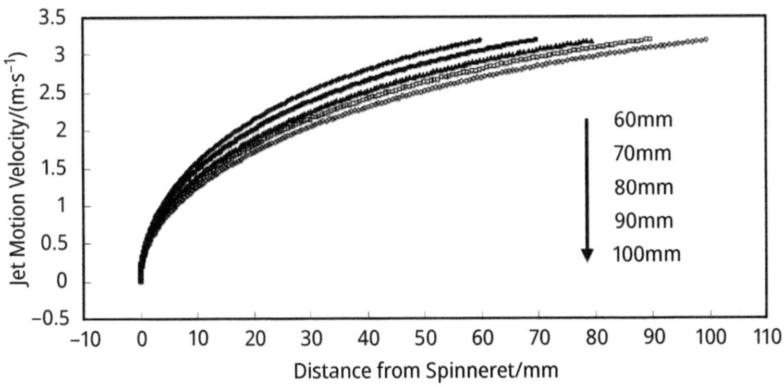

Figure 2.63: Variation of jet motion speed with distance from spinneret at different spinneret-to-collector distances under non-near-field conditions (applied voltage is 8 kV, silicon substrate).

The variation laws of jet motion velocity and acceleration with ejection time in the stable ejection stage under the near-field condition are shown in Figures 2.64–2.67. Under the action of the electric field force, the jet motion velocity increases with the ejection time, and the acceleration of jet motion decreases as the electric field force acting on the jet decreases rapidly as the ejection goes on. For the same ejection time, the motion velocity and acceleration of the jet increase with the applied voltage and decrease with the increase of the spinneret-to-collector distance.

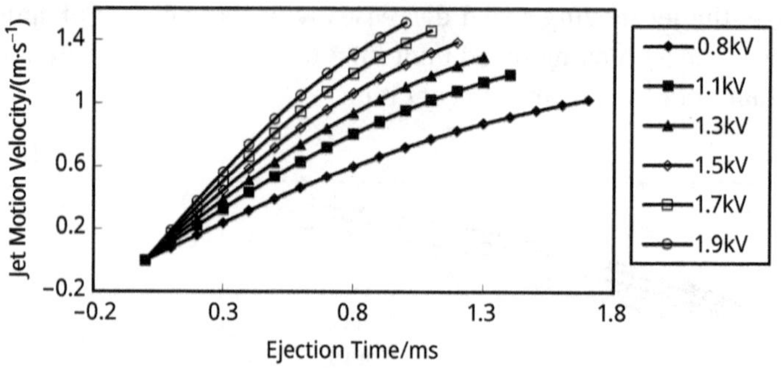

Figure 2.64: Variation of jet motion velocity with ejection time at different applied voltages (spinneret-to-collector distance is 1 mm, silicon substrate).

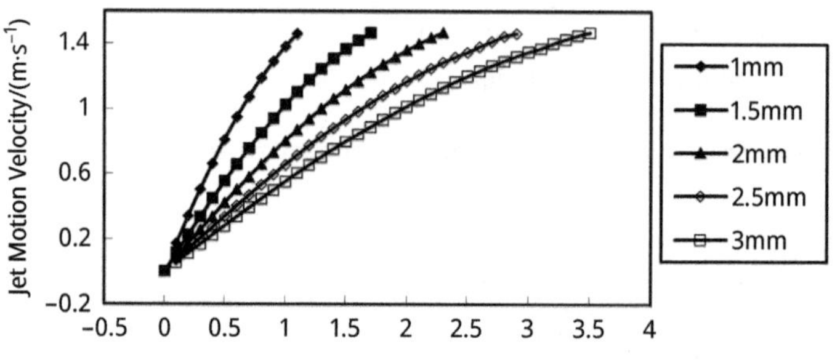

Figure 2.65: Variation of jet motion velocity with ejection time at different spinneret-to-collector distances (applied voltage is 1.7 kV, silicon substrate).

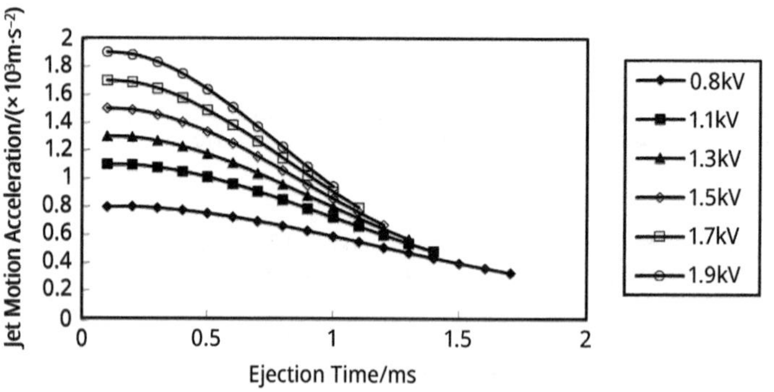

Figure 2.66: Variation of jet motion acceleration with ejection time at different applied voltages (spinneret-to-collector distance is 1 mm, silicon substrate).

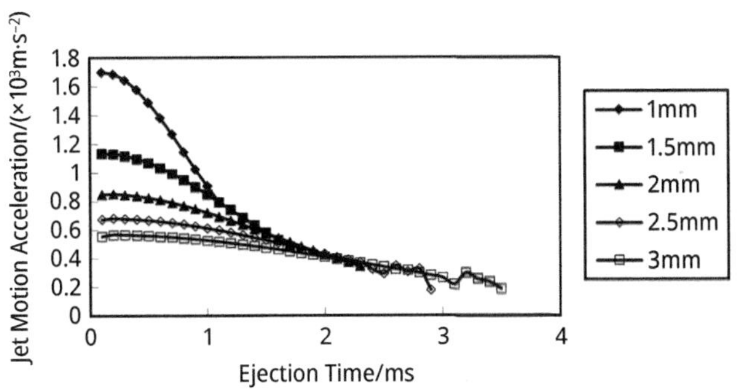

Figure 2.67: Variation of jet motion acceleration with ejection time at different spinneret-to-collector distances (applied voltage is 1.7 kV, silicon substrate).

For comparison, we examine the variation in jet motion velocity and acceleration with ejection time under non-near-field conditions, as illustrated in Figures 2.68 and 2.69. Similar to that under near-field conditions, the jet motion velocity increases continuously with increasing ejection time. At a constant spinneret-to-collector distance, the jet motion velocity at each ejection moment rises as the applied voltage increases. Conversely, for a fixed applied voltage, the motion velocity decreases as the spinneret-to-collector distance increases.

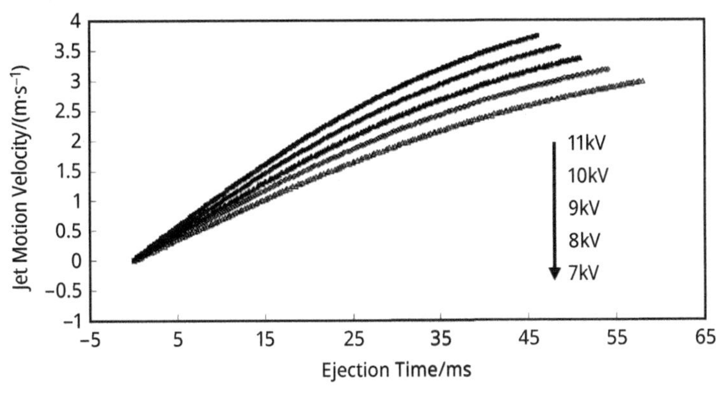

Figure 2.68: Variation of jet motion velocity with ejection time at different applied voltages under non-near-field conditions (spinneret-to-collector distance of 100 mm, silicon substrate).

The simulation results show that increasing the spinneret-to-collector distance extends the jet's motion distance, thus increasing its motion time in space. Increasing the applied voltage accelerates the jet motion, which reduces its travel time.

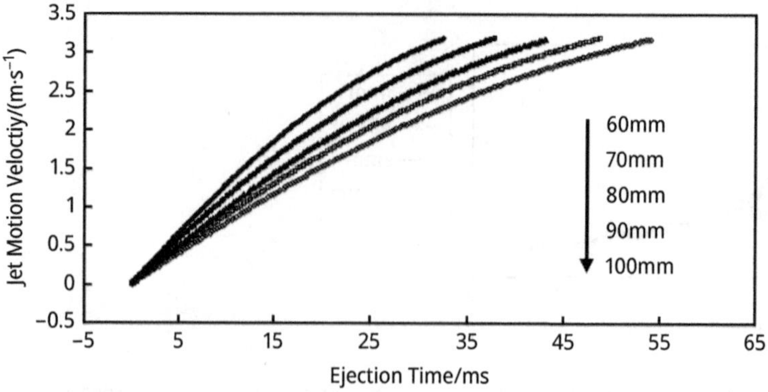

Figure 2.69: Variation of jet motion velocity with ejection time at different spinneret-to-collector distances under non-near-field conditions (applied voltage of 8 kV, silicon substrate).

3. Maximum jet motion velocity

We simulate and analyze the influence of two process parameters, the applied voltage and the spinneret-to-collector distance, on the maximum jet motion velocity, as shown in Figures 2.70 and 2.71. The simulation results in Figure 2.70(a) indicate that under the near-field condition, the spinneret-to-collector distance has little influence on the maximum motion velocity. The reason lies in that although increasing the spinneret-to-collector distance will increase the moving distance of the jet, it simultaneously reduces the space electric field strength and the jet acceleration. For an identical spinneret-to-collector distance, the maximum jet motion velocity increases with the applied voltage, as shown in Figure 2.70 (b). This is mainly because increasing the voltage can increase the electric field strength and the jet acceleration, thereby increasing the maximum jet motion velocity.

Figure 2.71 illustrates the impact of applied voltage and spinneret-to-collector distance on the maximum jet motion velocity under non-near-field conditions. Under these conditions, the trend in maximum jet motion velocity is similar to that observed under near-field conditions. Specifically, the maximum motion velocity increases with higher applied voltages, whereas it remains relatively unaffected by changes in the spinneret-to-collector distance.

The simulation results reveal that the jet motion velocity under near-field conditions (with a maximum of 1.5 m/s) is lower than that under non-near-field conditions (with a maximum of 4 m/s). This jet motion velocity is crucial in determining the collector's matching speed necessary for EDW of straight nanofibers. Consequently, reducing the jet motion velocity creates more favorable technical conditions for the controllable deposition of electrospun nanofibers.

2.2.2.4 Simulation of steady-state jet morphology

Figure 2.72 shows the comparison between the simulation and the experimental results of the steady-state jet morphology under different voltages using a hollow nozzle

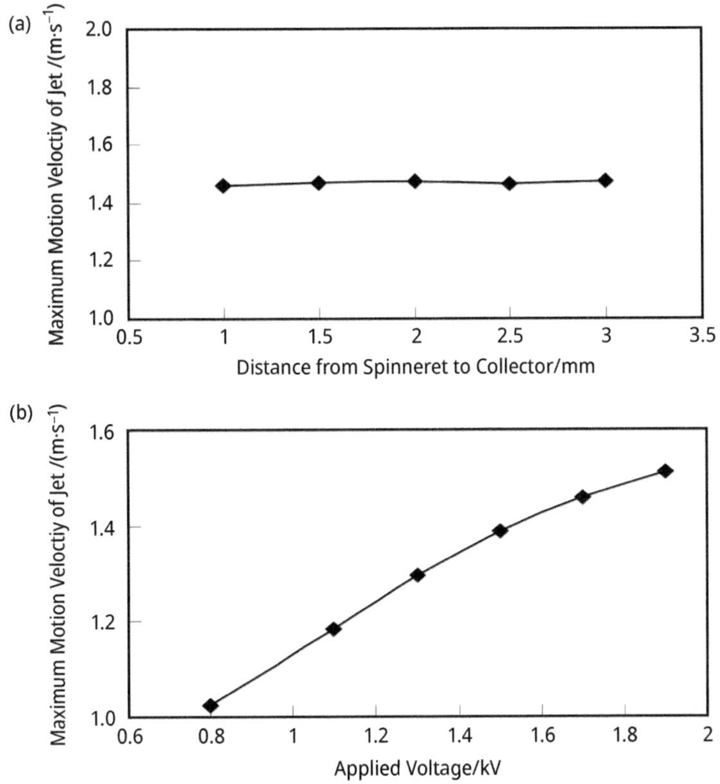

Figure 2.70: Impact of process parameters on maximum motion velocity of the jet under near-field condition: (a) impact of spinneret-to-collector distance, the applied voltage is 1.7 kV; and (b) influence of applied voltage, the spinneret-to-collector distance is 1 mm (silicon substrate).

with an inner diameter of 0.6 mm. Here R is the radius of the current jet position, R_0 is the radius of the nozzle, and Z is the distance between the spatial position and the nozzle.

The simulation results and experimental results differ significantly in the Taylor cone region but are more consistent in the jet region. In addition, the smaller the volume of Taylor cone, the closer the simulation results are to the experimental results. The reason for this phenomenon may be that in the simulation process, the charge of each particle is assumed to be constant, whereas in reality, this assumption only holds true for the jet part and does not apply to the Taylor cone part. The EDW currents mainly consist of convection current and Ohmic bulk conduction current. In the Taylor cone part, both these two current components vary, but in the simulation process, the charge of the particles mainly simulates the convection current. The jet is simplified in this model, resulting in a significant difference from the experimental results in the Taylor cone region. However, the simulation results still match the

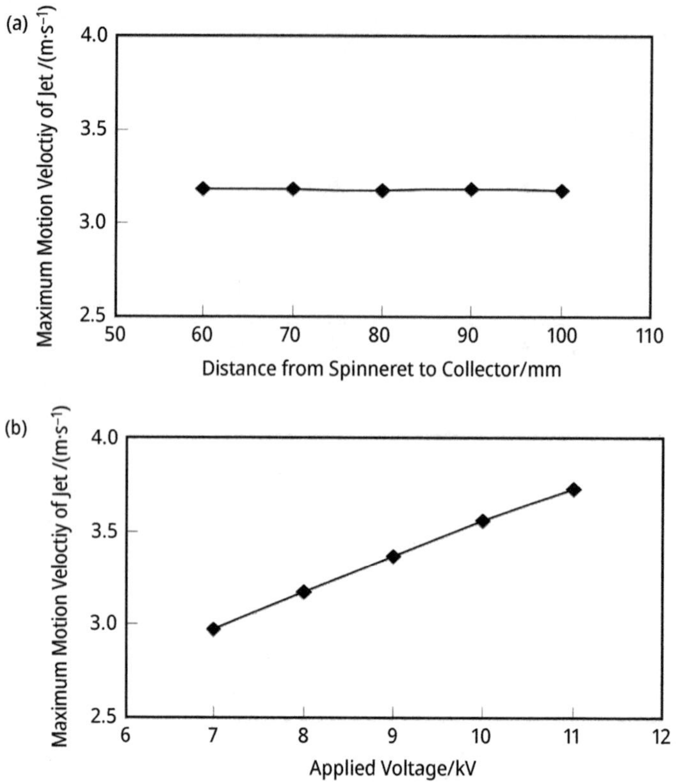

Figure 2.71: Influence of process parameters on the maximum velocity of the jet under non-near-field conditions: (a) influence of spinneret-to-collector distance at applied voltage of 8 kV; and (b) influence of applied voltage at spinneret-to-collector distance of 100 mm (silicon substrate).

experimental results in the jet part, indicating that it still shows some predictive capability for the steady-state jet morphology.

Based on this simulation model, we examine the influence of applied voltage, solution supply rate, and spinneret-to-collector distance on the jet morphology. The results are presented in Figures 2.73–2.75.

As the applied voltage increases, the space electric field strengthens, enhancing the electric force on the jet and accelerating its velocity. Consequently, at a constant solution supply rate, the volume of the Taylor cone decreases, and the jet diameter becomes thinner. Figure 2.73 illustrates the simulation outcomes for different voltages, showing that the variation in jet diameter with applied voltage aligns with the experimental findings depicted in Figure 2.15.

Figure 2.74 shows the variation of the jet morphology with the solution supply rate. Under constant conditions of other parameter values, the volume of the Taylor cone increases with increasing solution supply rate which is consistent with the experimental

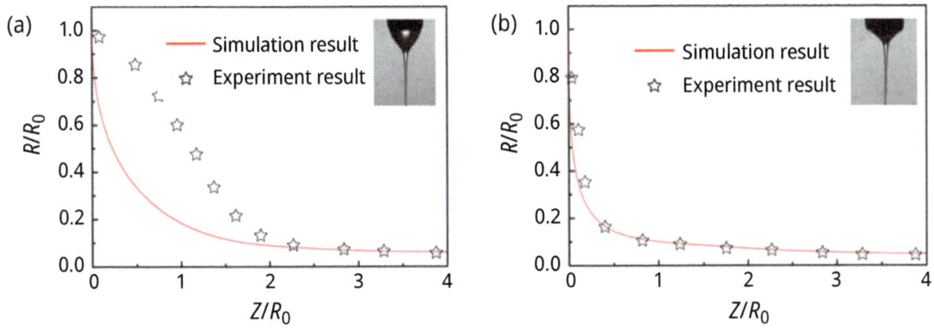

Figure 2.72: Simulation and experimental results of jet morphology under different voltages: (a) 3.1 and (b) 3.6 kV (solution supply rate is 400 μL/h and nozzle-to-collector distance is 6 mm).

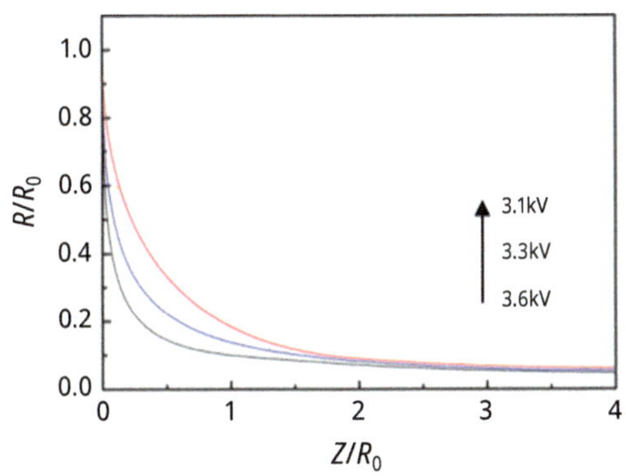

Figure 2.73: Simulation results of jet morphology under different applied voltages (liquid supply rate is 400 μL/h and spinneret-to-collector distance is 6 mm).

results shown in Figure 2.16. Additionally, the diameter of the jet also increases with increasing liquid supply velocity.

The spinneret-to-collector distance will affect the space electric field strength. Figure 2.75 shows that increasing the spinneret-to-collector distance weakens the electric field strength, reduces the electric field force imposed on the jet, and decreases the stretching rate. Therefore, the jet diameter increases with the spinneret-to-collector distance at the same position.

As shown in the above simulation results, due to the short motion distance, the stretching ratio of the EDW jet is small, and the size of the jet diameter is more dependent on its initial diameter (the jet diameter at the tip of the Taylor cone, as shown in Figure 2.76). In traditional electrospinning, the stretching and refining of the jet mainly occurs during the unstable whipping stage; while in EDW, as the spinneret is closer to the collector, the jet will deposit directly on the collector after a short linear motion. To further reduce the final diameter of the EDW jet, measures should be

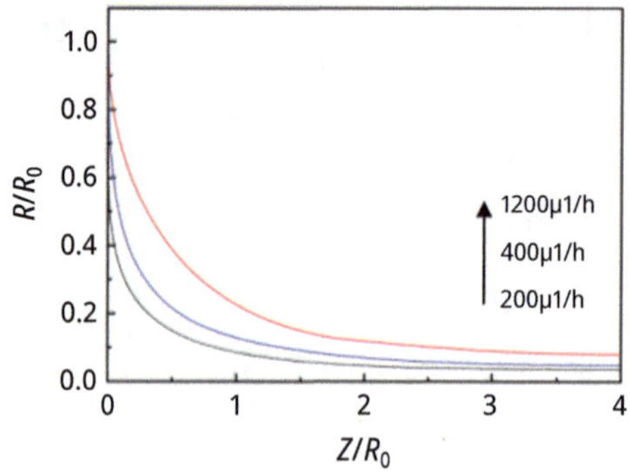

Figure 2.74: Simulation results of jet morphology at different liquid supply rates (nozzle-to-collector distance is 6 mm and applied voltage is 3.4 kV).

taken to reduce the initial diameter, including reducing the spinneret diameter and increasing the applied voltage. In addition, increasing the distance between the spinneret and the collector can also promote the jet refinement.

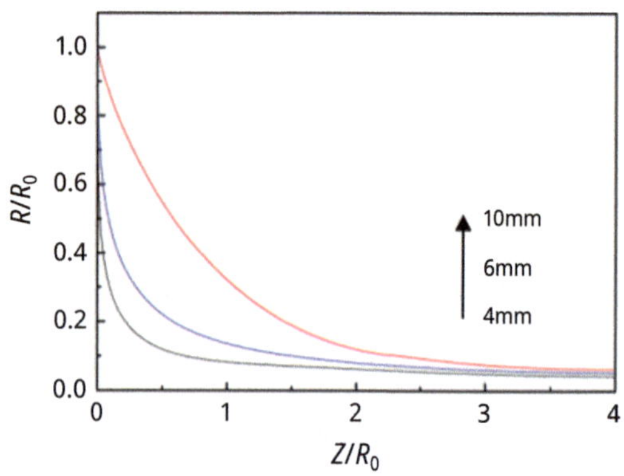

Figure 2.75: Simulation results of jet morphology at different spinneret-to-collector distances (liquid supply rate is 400 μL/h and applied voltage is 3.4 kV).

2.2.3 Simulation of nanofiber deposition behavior

2.2.3.1 Simulation of nanofiber deposition morphology

Based on the simulation model, the deposition behaviors of nanofibers on both the silicon dioxide and silicon substrates are analyzed, and the simulation results are shown in Figures 2.77 and 2.78. The silicon dioxide is of low conductivity, so the residual charge density is high, which has a greater repulsion effect on the subsequent

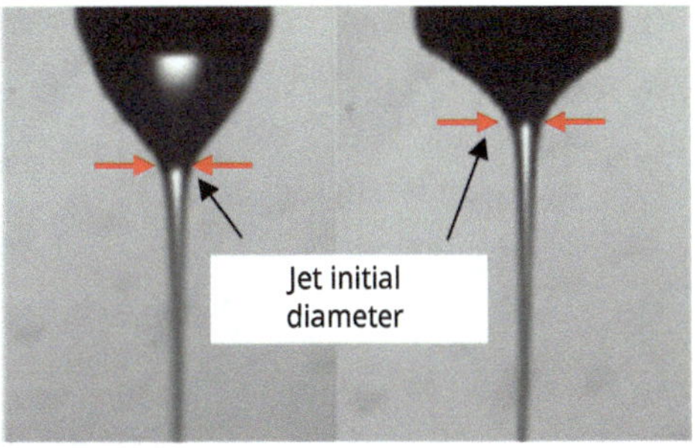

Figure 2.76: Diagram of initial jet diameter.

charged jet. The charged jet is guided to swing in the plane perpendicular to the motion direction of the collector, and wavy nanofibers, as shown in Figure 2.77(a), are obtained. The SEM image of the wavy nanofibers prepared on the silicon dioxide substrate under the same experimental conditions is shown in Figure 2.77(b). The simulation results are consistent with the experimental results.

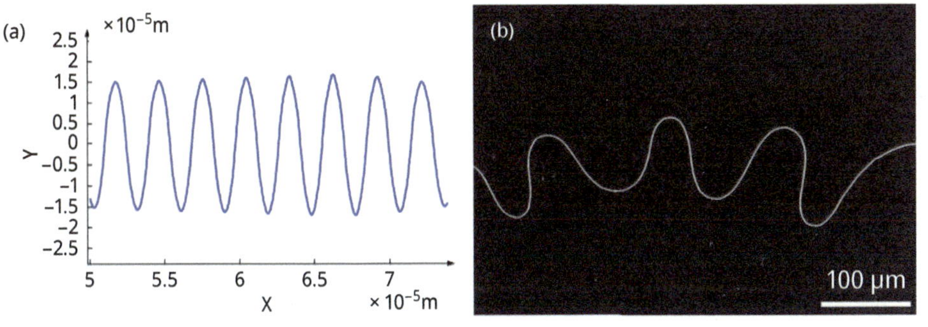

Figure 2.77: Nanofibers deposited on silicon dioxide substrate: (a) simulation results and (b) experimental results (collector motion velocity is 0.25 m/s, spinneret-to-collector distance is 1 mm, applied voltage is 1.7 kV, and solution concentration is 20%).

Figure 2.78 shows the influence of the motion velocity of the silicon substrate on the deposition morphology of nanofibers. The high conductivity of the silicon substrate results in less charge repulsion from the already deposited nanofibers during the jet deposition process, leading to spiral deposition of the sequential nanofibers. When the collector moving speed is low (0.10 m/s), the helical whipping behavior of the charged jet results in a mutually overlapping loop structure as shown in Figure 2.78(a). As the moving speed increases, the influence of the charged jet's spiral whipping gradually decreases, resulting in a reduce in the degree of overlapping loop nanofibers, as shown in Figure 2.78(b) (the collector velocity of 0.15 m/s); as the collector speed is further increased, a discrete single-loop nanofiber structure is obtained as shown in Figure 2.78(c). When the collector speed is

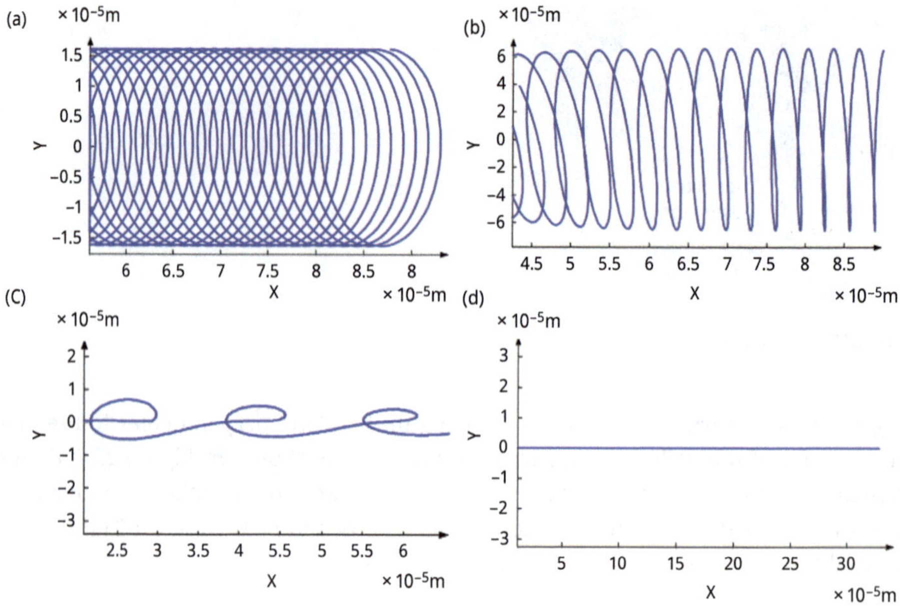

Figure 2.78: Simulation results of nanofibers deposited on silicon substrate. The collector motion velocities are: (a) 0.10; (b) 0.15; (c) 0.25; and (d) 0.40 m/s (spinneret-to-collector distance is 1 mm, applied voltage is 1.7 kV, and 20% solution concentration).

increased to a certain value (such as 0.40 m/s), the influence of the jet spiral whipping can be overcome, and the linear nanofiber structure is obtained as shown in Figure 2.78 (d). The simulation results of electrospun nanofibers deposited on silicon substrate are highly consistent with the experimental results.

It can be seen from the experimental results that when the collector is stationary or moving at a low speed, the whipping of the charged jet leads to bending and spiral deposition of the fibers, forming a multi-loop nanofiber structure. The same results are also obtained through simulation calculations, as shown in Figure 2.79. The conductivities of the collector in Figure 2.79(a)–(c) are 4.35×10^{-6} S/cm, 1.00×10^{-9} S/cm, and 1.50×10^{-11} S/cm, respectively. The residual charge density on the collector and the charge repulsion force exerting on the charged jet increase with the decrease of the collector conductivity, so the whipping amplitude of the charged jet and the deposition range of nanofibers increase accordingly. The diameters of the distribution area of the multi-loop nanofiber structures in Figure 2.79 are 28, 32, and 80 μm, respectively.

2.2.3.2 Analysis of jet stretching ratio

Figures 2.80 and 2.81 show the simulation results on the influence of the spinneret-to-collector distance and, as well as the applied voltage, on the jet stretching ratio under

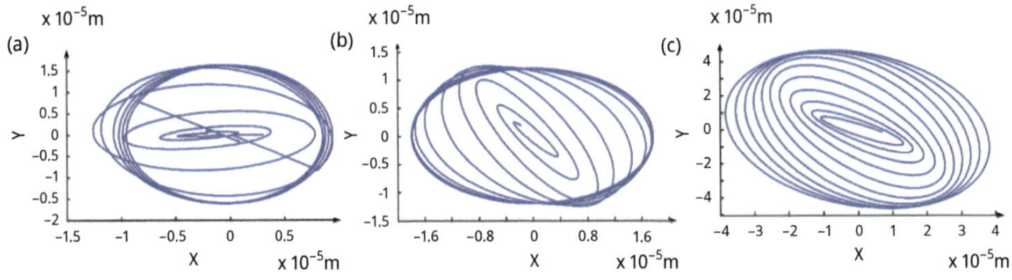

Figure 2.79: Simulation results of multi-loop nanofiber structures deposited on collectors with different conductivities. The relative dielectric constant and conductivity of each collector are: (a) $\xi = 11.9$, $\delta = 4.35\times10^{-6}$ S/cm; (b) $\xi = 5.6$、$\delta = 1.00\times10^{-9}$ S/cm; (c) $\xi = 3.90$; and (d) $\delta = 1.50\times10^{-11}$ S/cm (solution concentration is 20%, collector speed is 0.003 m/s, spinneret-to-collector distance is 1 mm, and applied voltage is 1.7 kV).

near-field conditions and non-near-field conditions, respectively. The jet stretching ratio mainly reflects the degree to which the jet is stretched and the variation in the nanofiber diameter. In the near-field electrospinning process, the jet stretching ratio (100–200) is significantly smaller than that of traditional electrospinning (350–500). This suggests that the degree of jet stretching is smaller in near-field electrospinning, and therefore the reduction in nanofiber diameter mainly depends on the initial diameter of the jet. Under the near-field conditions, the distance of jet motion is small. Increasing the spinneret-to-collector distance between the spinneret and the collector can increase the distance and time of jet movement, thereby improving the degree of jet stretching. Increasing the applied voltage can increase the electric field force acting on the jet, increasing the jet stretching ratio, as shown in Figure 2.80. Under non-near-field conditions, the increase in applied voltage significantly increases the jet stretching ratio, while the change in spinneret-to-collector distance shows little impact on it, as shown in Figure 2.81.

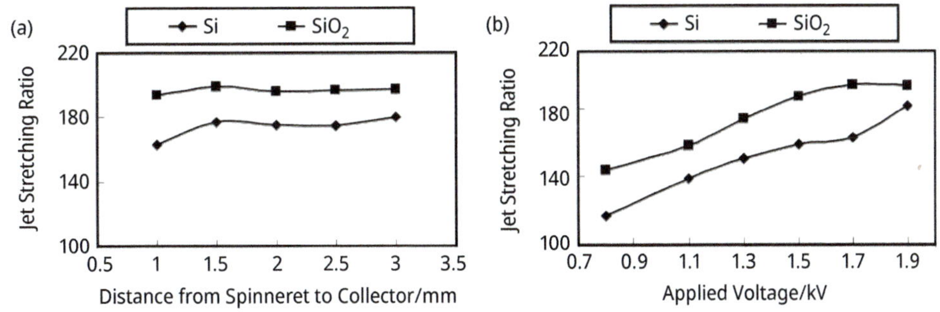

Figure 2.80: Jet stretching ratio under near-field conditions: (a) impact of spinneret-to-collector distance on jet stretching ratio at an applied voltage of 1.7 kV and (b) impact of applied voltage on the stretching ratio at a spinneret-to-collector distance of 1 mm.

The charged jet accelerates toward the collector under the influence of electric field force and undergoes continuous stretching and refining. To gain a deeper understanding of the stretching dynamics during the deposition process, the stretching ratio of the spatially moving jet is defined. The distance between two adjacent particles on the moving jet can be expressed as

$$l_{i,i+1} = \sqrt{(x_i - x_{i+1})^2 + (y_i - y_{i+1})^2 + (z_i - z_{i+1})^2} \tag{2.44}$$

Then the stretching ratio for each part can be defined as

$$S_{ri} = \frac{l_{i,i+1}}{l_0} \tag{2.45}$$

where $M \le i < N$. The simulation results of the jet stretching ratio varying with the distance from the spinneret are shown in Figure 2.82. The jet stretching ratio increases with both applied voltage and distance from the spinneret. However, the acceleration decreases as the distance from the spinneret increases. This behavior occurs because, under near-field conditions, the electric field force is the primary driver of jet stretching. As the jet moves further away from the spinneret, the electric field force acting on the jet diminishes, resulting in a reduced degree of jet stretching.

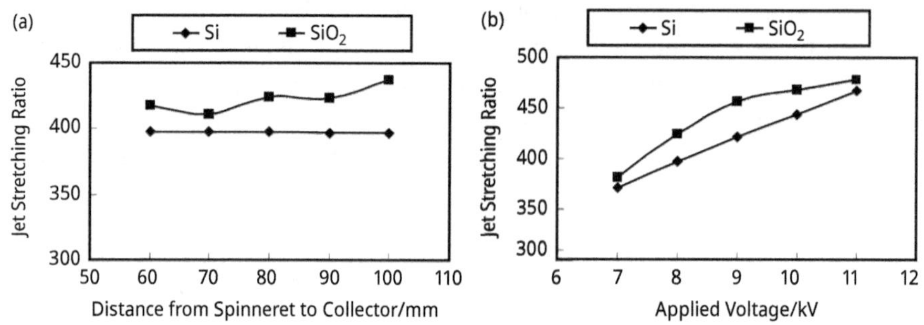

Figure 2.81: Jet stretching ratio under non-near-field conditions: (a) impact of spinneret-to-collector distance on jet stretching ratio at an applied voltage of 8 kV and (b) impact of applied voltage on the stretching ratio at a spinneret-to-collector distance of 100 mm.

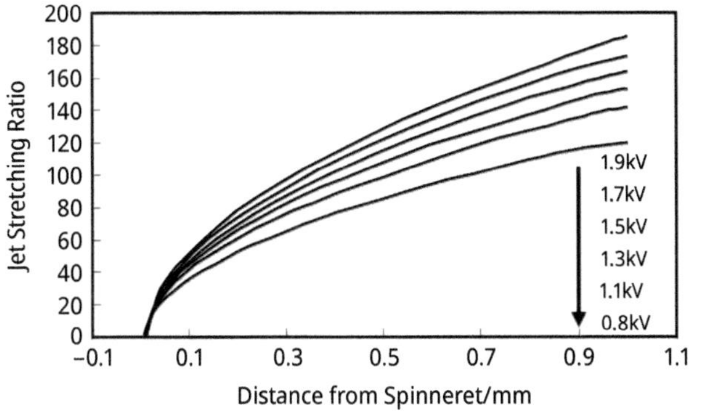

Figure 2.82: Variation of jet stretching ratio with distance from the spinneret (spinneret-to-collector distance is 1 mm, silicon substrate).

References

[1] Megelski S, Stephens J S, Chase D B, Rabolt J F. Micro- and nanostructured surface morphology on electrospun polymer fibers. Macromolecules 2002, 35(22), 8456–8466.

[2] Moseler M, Landman U. Formation, stability, and breakup of nanojets. Science 2000, 289(5482), 1165–1169.

[3] Yoerdem O S, Papila M, Menceloglu Y Z. Effects of electrospinning parameters on polyacrylonitrile nanofiber diameter: An investigation by response surface methodology. Mater & Design 2008, 29(1), 34–44.

[4] Cheng Y, Zou B, Wang C, Liu Y, Fan X. Formation mechanism of Fe_2O_3 hollow fibers by direct annealing of the electrospun composite fibers and their magnetic, electrochemical properties. Crystengcomm 2011, 13(8), 2863–2870.

[5] Manee-In J, Nithitanakul M, Supaphol P. Effects of solvent properties, solvent system, electrostatic field strength, and inorganic salt addition on electrospun polystyrene fibres. Iran Polym J 2006, 15(4), 341–354.

[6] Jin R G, Ma X Q. Polymer Rheology. Shanghai, China, East China University of Science and Technology Press, 2012.

[7] Garg K, Bowlin G L. Electrospinning jets and nanofibrous structures. Biomicrofluidics 2011, 5(1), 013403.

[8] Reneker D H, Yarin A L, Zussman E, Xu H. Electrospinning of nanofibers from polymer solutions and melts. Adv Appl Mech 2007, 41(07), 43–195.

[9] Zussman E, Yarin A L, Bazilevsky A V, Avrahami R, Feldman M. Electrospun polyaniline/poly (methyl methacrylate)-derived turbostratic carbon micro-/nanotubes. Adv Mater 2006, 18(3), 348–353.

[10] Melcher J R, Taylor G I. Electrohydrodynamics: A Review of the role of interfacial shear stresses. Annu Rev Fluid Mech 1969, 1(1), 111–146.

[11] Saville D A. Electrohydrodynamics:The Taylor-Melcher leaky dielectric model. Annu Rev Fluid Mech 1997, 29(1), 27–64.

[12] Bisht G S, Canton G, Mirsepassi A et al.. Controlled continuous patterning of polymeric nanofibers on three-dimensional substrates using low-voltage near-field electrospinning. Nano Lett 2011, 11(4), 1831–1837.

[13] Lu D, Jiang P. Solid State Physics. Peking: Higher Education Press, 2011.

[14] Beachley V, Wen X. Effect of electrospinning parameters on the nanofiber diameter and length. Mater Sci Eng., C 2009, 29(3), 663–668.
[15] Coelho R, Debeau J. Properties of the tip-plane configuration. J Phys D Appl Phys 1971, 4(9), 1266–1280.
[16] Wu Z H. Basic Electrostatics. Peking: Peking University Press, 2010.
[17] Joseph D D. Fluid Dynamics of Viscoelastic Liquids. New York: Springer, 1990.
[18] Reneker D H, Yarin A L. Electrospinning jets and polymer nanofibers. Polymer 2008, 49(10), 2387–2425.
[19] Kowalewski T A, Barral S, Kowalczyk T. Modeling Electrospinning of Nanofibers. Neu-Isenharg: Springer, 2008.
[20] Zeng Y C, Sun Y F, Wang X H. Numerical approach to modeling fiber motion during melt blowing. J Appl Polym Sci 2011, 119(4), 2112–2123.
[21] Bai S F. Introduction to Numerical Computation. Peking: Higher Education Press, 2004.
[22] Reneker D H, Yarin A L, Fong H, Koombhongse S. Bending instability of electrically charged liquid jets of polymer solutions in electrospinning. J Appl Phys 2000, 87(9), 4531.
[23] Thompson C, Chase G G, Yarin A L. Effects of parameters on nanofiber diameter determined from electrospinning model. Polymer 2007, 48(23), 6913–6922.
[24] Tan E P S, Goh C N, Sow C H, Lim C T Tensile test of a single nanofiber using an atomic force microscope tip. Appl Phys Lett 2005, 86(7), 073115.
[25] Tan E P S, Ng S Y, Lim C T. Tensile testing of a single ultrafine polymeric fiber. Biomaterials 2005, 26(13), 1453–1456.
[26] Yuya P A, Wen Y, Turner J A, Dzenis Y A, Li Z. Determination of Young's modulus of individual electrospun nanofibers by microcantilever vibration method. Appl Phys Lett 2007, 90(11), 111909.

Chapter 3
Electrical characteristics of EDW jet

3.1 Study on the electrical characteristics of EDW jet

3.1.1 Electrohydrodynamic direct-writing current detection system

The current passing through the collector can reflect the current variation in the electrohydrodynamic direct-writing (EDW) process. Therefore, a high-precision multimeter is connected in series between the collector and the negative electrode of the power supply to build up a current detection system, as shown in Figure 3.1(a).

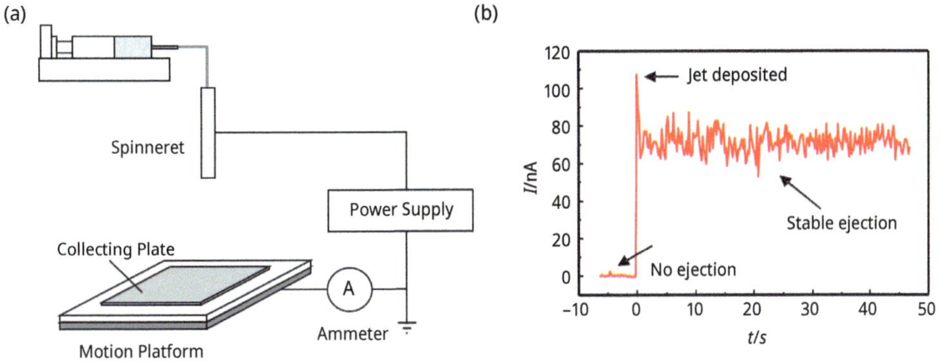

Figure 3.1: EDW current detection system: (a) schematic diagram of the current detection system; and (b) schematic diagram of current waveform (glycerin solution, the inner diameter of the spinneret is 0.6 mm, the liquid supply rate is 1,200 μL/h, and the spinneret-to-collector distance is 6 mm).

Figure 3.1(b) shows the current variation recorded since the initial ejection of the solution to its steady state. The fundamental noise signal of the current is less than 1 nA before the voltage is applied. During the initial stage of the solution ejection, the surface charge density of the droplet structure at the jet front is high, and a peak current will appear when it deposits on the collector. At this time, the jet is in an unstable motion process. When the jet front deposits on the collector, the ejection will self-adjust and tend to be stable. In the stable ejection stage, the surface charge density of the jet is less than that of the droplet at the frontend in the initial ejection stage, so the current in the stable ejection stage is lower than the peak current in the initial ejection stage.

https://doi.org/10.1515/9783110797879-003

3.1.2 Current characteristics under different ejection modes

The jet ejection process is a process of solution and charge migration. When the solution is jetting, the charge moves with the solution to the collector, forming a current loop. The current is a direct reflection of the charge movement. By detecting the micro-nano current during EDW process, the current characteristics under different jet ejection states are studied.

3.1.2.1 Current under different ejection modes

Typical ejection modes of the charged jet include dripping, pulsed jet, stable cone jet, multi-jet, etc. The corresponding current signals are shown in Figures 3.2–3.5.

3.1.2.1.1 Dripping mode

Figure 3.2 shows the current in the dripping mode. Because of the low voltage, the electric field force is insufficient to overcome the surface tension to form a conical jet. The solution drops under the action of gravity. During the dripping process, the solution separates from the spinneret, so there is almost no detectable current signal on the collector, as shown in Figure 3.2(a). If the voltage is increased, the charge carried by the solution increases, and the electric field force the solution experiences will increase. Under the dripping mode, when the voltage reaches a certain threshold, fluctuation in current can be observed, as shown in Figure 3.2(b).

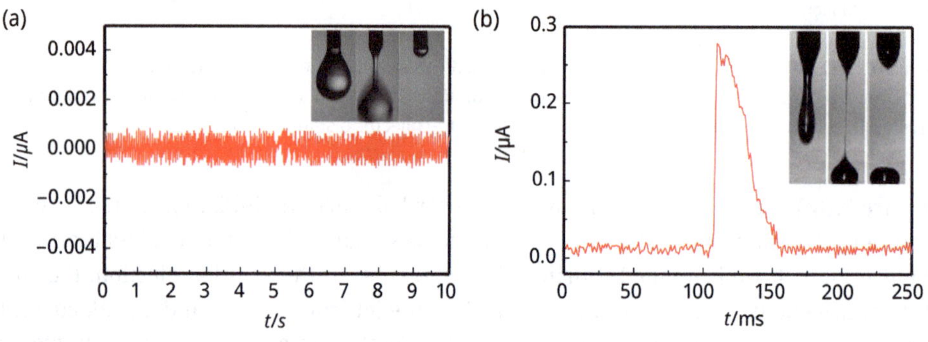

Figure 3.2: Current characteristics of dripping mode. The applied voltage, the spinneret-to-collector distance and the liquid feeding rate are: (a) 2 kV, 6 mm, 400 μL/h; and (b) 3.1 kV, 6 mm, 400 μL/h.

3.1.2.1.2 Pulsed jet mode

Figure 3.3(a) shows the detected current during high-frequency pulsed ejection of glycerol solution. It can be seen that the period and amplitude of the current are relatively stable, with the printing frequency of about 250 Hz. This high-frequency pulsed ejection under DC voltage is caused by the imbalance among the solution ejection speed, the charge migration speed and the solution supply rate [1].

3.1.2.1.3 Stable cone jet mode

EDW technology mainly makes use of the stable cone jet mode, which is of good controllability and stability, to prepare micro-nanostructures. Figure 3.3(b) shows the current signal under the stable cone jet ejection state. In this mode, the current is relatively stable and changes slightly with time.

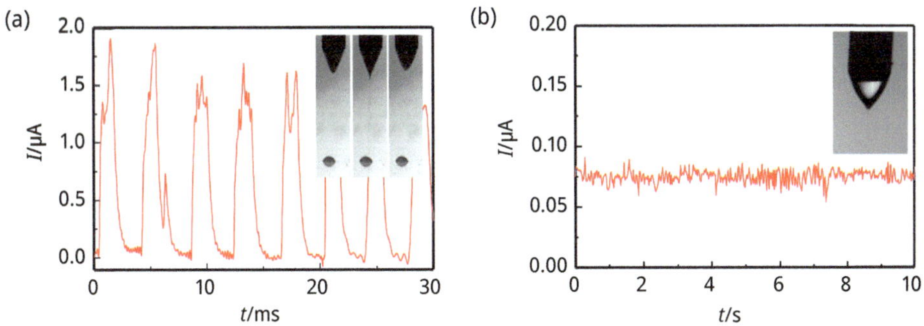

Figure 3.3: Current during jet ejection in high-frequency pulse ejection mode and stable cone jet mode: (a) high-frequency pulse ejection mode, the applied voltage is 3.3 kV, spinneret-to-collector distance is 6 mm, and liquid supply is 400 μL/h; and (b) stable cone jet mode, applied voltage is 4.0 kV, spinneret-to-collector distance is 6 mm, and liquid supply rate is 1,000 μL/h.

3.1.2.1.4 Multi-jet mode

In the cases of liquid with low surface tension or high applied voltage, the multi-jet mode may occur, as illustrated in Figure 3.4(a). In this mode, multiple jets form at the spinneret, while the ejection position and state are unstable. The current amplitude fluctuates significantly, as shown in the corresponding current waveform.

3.1.2.1.5 Breakdown discharge mode

If the applied voltage or electric field is too large, air breakdown would be triggered, resulting in the discharge and conduction between the spinneret and the collector. There are multiple forms of breakdown discharge, such as the breakdown between the spinneret and the collector, the breakdown between the solution front end and the collector, etc. When air is broken down, it becomes a conductor, and the current amplitude formed in this mode is much larger than that of the stable jet mode, as shown in Figure 3.4(b). Air breakdown and discharge will cause some damage to both the EDW process and equipment, which should be avoided in practical applications.

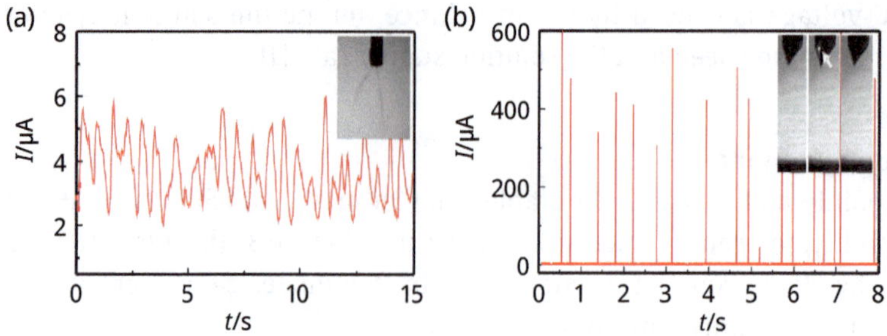

Figure 3.4: Current in multi-jet ejection and air breakdown discharge modes: (a) multi-jet mode, applied voltage is 5.5 kV, spinneret-to-collector distance is 6 mm, liquid supply speed is 1,000 μL/h; (b) air breakdown mode, applied voltage is 7 kV, spinneret-to-collector distance of 6 mm, liquid supply rate is 600 μL/h.

3.1.2.1.6 Suspended bead-on-string structures

The current can not only reflect the ejection mode, but also reflects the deposition behavior of the jet on the collector. Figure 3.5(a) shows the current when the glycerol/polyvinylpyrrolidone (PVP) solution generates a suspended bead-on-string structure. At the very beginning, there is no bead structure, and the current is relatively stable. After the formation of bead-on-string structures, the current will fluctuate with the movement of the structure. As the beads merge into the main droplet, the surface charge they carry will also transfer. Therefore, the current through the collector increases as the beads merge, as indicated by the arrow in the figure.

3.1.2.1.7 Jet spiral deposition

Figure 3.5(b) shows the current characteristics of the jet during the spiral deposition. As the jet undergoes spiral oscillation, the amplitude of the current changes accordingly. As indicated by the arrows in the figure, a peak current value appears each time the jet completes one spiral deposition, allowing the jet deposition situation to be obtained through the current curve.

3.1.2.2 Analysis of current detection results for different ejection modes

Among the above various ejection modes and deposition behaviors, the jet ejection mode can be roughly distinguished through data analysis of the current signal collected by the current detection system using several statistical parameters.

3.1.2.2.1 Coefficient of variation (CV)

The coefficient of variation (CV) is the ratio of the standard deviation to the arithmetic mean and is primarily used to compare the variability of different data sets. CV is a

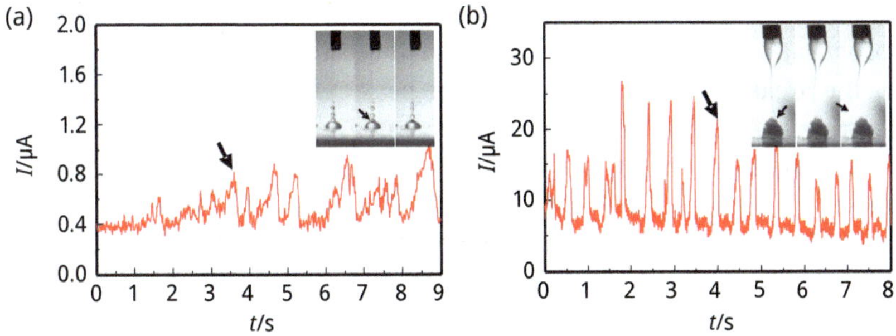

Figure 3.5: The current during formation of suspended bead-chain structure and spiral deposition: (a) suspended bead-on-string structures appears on the jet for glycerol/PVP solution, the applied voltage is 4.0 kV, spinneret-to-collector distance is 6 mm, and liquid supply rate is 80 μL/h. The arrow indicates the peak current that occurs after one bead merges into the main droplet; and (b) jet spiral deposition for 16% PEO solution, applied voltage is 2.0 kV, spinneret-to-collector distance is 6 mm, and liquid supply rate is 20 μL/h. The arrow indicates the peak current that occurs during one round of jet spiral deposition.

dimensionless number that can eliminate the influence of measurement scale and dimension, which can be calculated as follows:

Arithmetic mean of current data samples:

$$I_{\text{avg}} = \frac{1}{n}\sum_{i=1}^{n} I_i \tag{3.1}$$

Standard deviation of the samples:

$$S = \sqrt{\frac{\sum_{i=1}^{n}\left(I_i - I_{\text{avg}}\right)^2}{n-1}} \tag{3.2}$$

Coefficient of variation:

$$CV = \frac{S}{I_{avg}} = \sqrt{\frac{\sum_{i=1}^{n}\left(I_i - I_{avg}\right)^2}{n-1}} \tag{3.3}$$

3.1.2.2.2 Median (MED)

The median (MED) is a statistical measure that represents the middle value in a set of data when it is organized in ascending or descending order. If there is an even number of sample values, the median is typically defined as the average of the two middle values.

In the dripping mode illustrated in Figure 3.2(a), the mean current is $I_{\text{avg}} \approx 0$. Considering the inherent measurement error in the system (such as the system error $I_{\text{err}} \leq 1$ nA in this experiment), it can be assumed that there is no current passing through the system, when $I_{\text{avg}} \leq I_{\text{err}}$, indicating the dripping mode. The dripping mode shown in Figure 3.2(b) is similar to the pulsed jet mode and the breakdown discharge mode in terms of waveform. It has large current fluctuations. Defining the maximum coefficient of variation of the jet flow CV_{max}, for these cases, $CV > CV_{max}$, $I_{avg} > I_{err}$. However, for the breakdown discharge mode, the peak current is much greater than the current during the stable electrospinning process. Therefore, the maximum current $I_{critical}$ can be defined, and if $I > I_{critical}$, it can be considered as the breakdown discharge mode.

When the jet is in the stable ejection mode, $I \approx I_{avg}$, so the coefficient of variation of the current must be within a certain range, that is, $CV \le CV_{max}$. Meanwhile, for the current in the stable ejection mode, there is $I_{err} > I_{avg}$, and the median $MED \approx I_{avg}$.

In this experiment, the current data in Figures 3.2–3.5 are analyzed, and the findings are presented in Table 3.1. Based on the above comprehensive analysis of the current, the jet ejection mode during a specific time period can be simply judged according to the current situation within that time period. Within the scope of this experimental data, the current characteristics of glycerol solution under different ejection modes are shown in Table 3.2.

Table 3.1: The characteristic data of current signal for different modes.

Figure	Mode	I_{avg} (nA)	CV	MED (nA)	I_{max} (nA)	I_{min} (nA)
Figure 3.2(a)	Dripping	0.058	6.76	0.102	0.923	−0.718
Figure 3.2(b)	Dripping	36.2	1.77	13.7	276.9	0.682
Figure 3.3(a)	Pulsed jet	594.2	1.03	282.9	1,906.3	0.746
Figure 3.3(b)	Stable jet	74.926	0.05	75.351	90.754	53.639
Figure 3.4(a)	Multi-jet	3,642.9	0.26	2,032.7	5,938.3	2,032.7
Figure 3.4(b)	Breakdown	5,532.1	8.30	0	>6,00,000	−0.223
Figure 3.5(a)	Bead-on-tring structure	533.64	0.27	493.34	1037.5	303.7
Figure 3.5(b)	Spiral deposition	8,906.7	0.46	7,200	26,666.7	3733.3

Table 3.2: Current characteristics of glycerol solution for different ejection modes.

Ejection mode	Current characteristics		
Dripping/pulsed jet	$I_{max} \le 0.6$ mA	$I_{avg} \le 1$ nA (dripping)	
		$CV > 0.1$, $I_{min} \le 1$ nA, $I_{max} > 1$ nA (dripping/pulsed jet)	
Stable jet		$I_{avg} > 1$ nA	$CV \le 0.1$, $MED \approx I_{avg}$
Multi-jet/unstable ejection			$I_{min} > 1$ nA, $CV > 0.1$
Breakdown	$I_{max} > I_{critical}$		

3.1.3 Electrical characteristics of EDW jet

3.1.3.1 Current variation process during EDW

Using a solid probe for liquid dipping and supply, the entire process of current variation during EDW process is recorded in real-time. The current waveform is shown in Figures 3.6–3.8.

When the high-voltage DC power supply is switched on to launch the ejection, the current signal experiences its first fluctuation due to the impact of the strong electric field at the moment the voltage is applied, marked as 0 s in Figure 3.6. After the voltage is applied for a period of time (ejection delay time), the charged jet ejects from the spinneret and deposits on the collector. As the charged jet approaches the collector, it induces the migration of charge on the collector, resulting in oscillating current signals. When the frontend of the jet deposits on the collector, the large amount of charges it carries are transferred through the collector, causing a detectable peak current. As the jet ejection continues, the current amplitude gradually decreases until the ejection ends, and the current tends to zero.

The time from the loading of voltage to the detection of peak current is defined as the current delay time, which is approximately equal to the sum of the ejection delay time and the time for the jet front to move to the collector. The ejection delay time (1–5 s) is much larger than the motion time of the jet front (1–10 ms), therefore, the current delay time is approximately equal to the ejection delay time. The current delay time and the ejection delay time have the same variation law, which increase with the increase of solution concentration. As shown in Figures 3.6–3.8, and the current delay time is 2.0 s, 1.8 s, and 2.7 s for the solution concentration of 10%, 16%, and 20%, respectively.

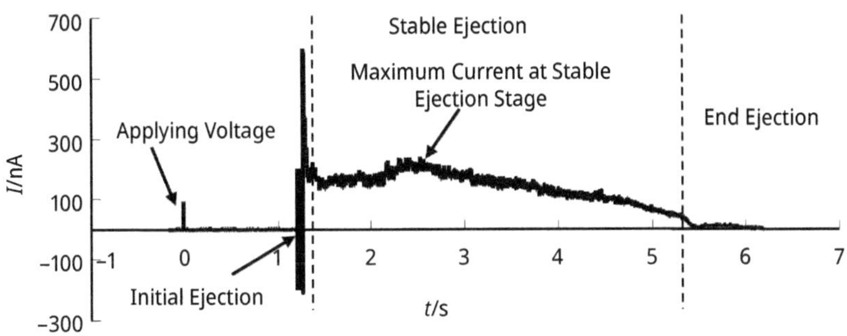

Figure 3.6: Current variation with the solution concentration of 10% (applied voltage is 1.7 kV, spinneret-to-collector distance is 1 mm, spinneret diameter is 40 μm, silicon substrate).

According to the current characteristics, the EDW process can be divided into three stages: initial ejection stage, stable ejection stage, and ending ejection stage.

The first stage is the initial ejection stage, in which the current oscillates.

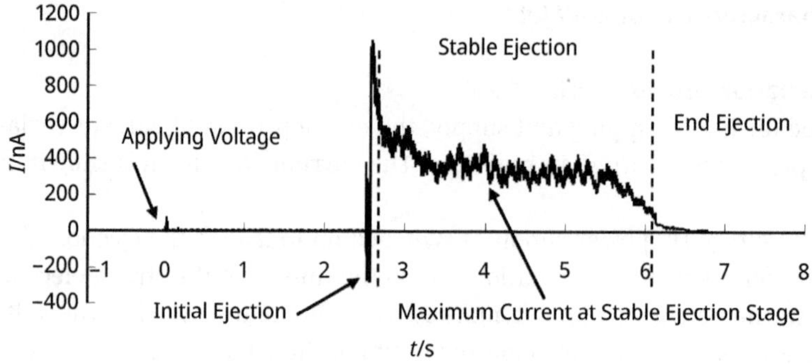

Figure 3.7: Current variation with the solution concentration of 16% (applied voltage is 1.7 kV, spinneret-to -collector distance is 1 mm, spinneret diameter is 40 μm, silicon substrate).

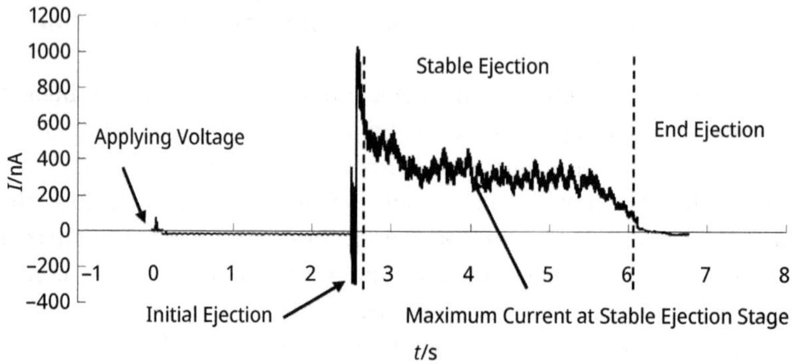

Figure 3.8: Current variation with the solution concentration of 20% (applied voltage is 1.7 kV, spinneret-to-collector distance is 1 mm, spinneret diameter is 40 μm, silicon substrate).

When the jet ejects, it carries away the accumulated charge on the surface of the droplets. At the front end of the jet, a larger diameter bead-like structure forms at the front end of the jet, with a high charge density. As the charged jet approaches the collector, the induction effect of the charges induces the migration of the charges on the collector. The whipping of the jet causes repeated migration of the charges on the collector, and the current signal of violent oscillation is detected. Once the jet deposits on the collector, the charge carried by the jet will be migrated. As the front end of the jet has a high charge density, the peak current will appear at the initial ejection stage, with a characteristic peak current of about 400–1,000 nA.

At this stage, the jet is in an unstable rheological and motion process, especially, the front end of the jet has not been fully stretched. The bead structure at the jet front is of large diameter and small surface area. As the solvent volatilizes slowly, the solvent content is relatively high when the jet deposits on the collector, resulting in the formation of polymer blocks as shown in Figure 3.9(a), making it impossible to prepare uniform nanofibers.

The second stage: stable ejection stage, with a slow decrease in the current.

As the front end of the jet deposits on the collector, the jet enters the stable ejection stage. In this stage, the jet is fully stretched, and the diameter and the surface charge density of the jet are smaller than those of the bead structure at the jet front. When the dipping supply mode is used, as the jetting process and solvent evaporation proceed, the volume of droplets at the needle tip will gradually decrease, and the jetting process will tend to stop. The nanofibers depositing on the surface of the collector per unit time and the charge they carry will gradually decrease with the progress of ejection, and the current will also show a slow decrease.

In the stable ejection stage, the jet is uniformly and fully stretched, the diameter is small, the specific surface area is large, and the solvent volatilizes fast, leading to the formation of uniform solid nanofiber as shown in Figure 3.9(b). The stable ejection stage, which lasts for 3 to 6 s (Figures 3.6–3.8), is the main process to direct writing of uniform nanofibers.

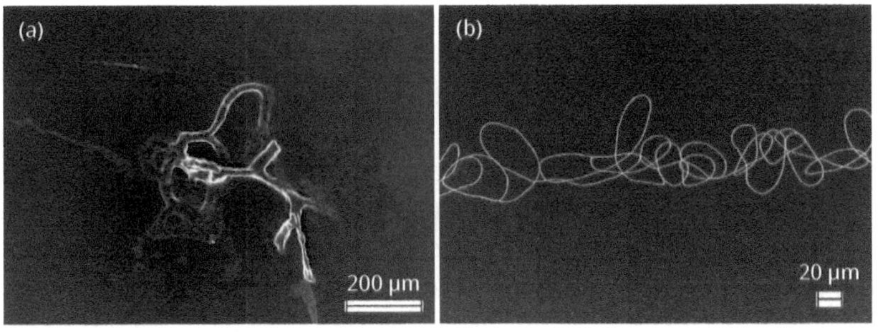

Figure 3.9: Polymer structures obtained at different ejection stages: (a) polymer block formed by deposition of the jet front end on the collector; and (b) uniform solid nanofibers prepared in the stable ejection stage (solution concentration is 18%, applied voltage is 1.7 kV, spinneret-to-collector distance is 1 mm, spinneret diameter is 40 μm, silicon substrate).

The third stage: the end of the ejection stage, with rapid decay in current.

As the jetting continues and the solvent evaporates, the jet ejection gradually stops and the current rapidly decreases. As shown in Figures 3.6–3.8, the current decreases sharply from 50 to 10 nA at the end of the ejection phase, and the duration of this phase is the shortest among the three phases of current variation. The jet between the spinneret and the collector has solidified, so no more nanofibers could be prepared.

3.1.3.2 Theoretical analysis of electrical characteristics of jet

Electrospinning current reflects the coupling effect between the viscoelastic fluid and the electric field, which mainly includes three parts [2]: free electron flow current I_F, Ohmic bulk conduction current I_O, and convection current I_C. Therefore, it can be described as follows:

$$I_E = I_F + I_O + I_C \tag{3.4}$$

I_E refers to the total current flowing through the jet; free electron flow current I_F mainly refers to the current formed by the migration and movement of free charges on the jet surface and the surrounding space. Ohm bulk conduction current I_O mainly refers to the current formed by the directional flow of charges due to the application of an electric field. The convection current I_C mainly refers to the current formed by the movement of charges the solution carries when it flows.

Under the strong electric field, the charges in the jet play a role in the formation and evolution process of the jet. The free moving charges in the jet are small in quantity and easy to lose due to the small diameter and large specific surface area of the jet. The amount decreases rapidly after the jet ejects from the tip of the Taylor cone. As the jet steps into the stable ejection stage, the number of free charges is already very small, so its influence on the jet rheology is very limited. Yarin et al. [2] suggested that the free charges in the Taylor cone region mainly came from the corona discharge of the surrounding air, and the research results showed that the migration behavior of free charges in the jet can be ignored in the electrospinning process.

According to the leaky dielectric model, the polymer solution shows conductivity under the action of the strong electric field, and the internal charge carriers can move under the external electric field. The charge carried by the carriers move directionally to form an Ohmic bulk conduction current, which can be described as

$$I_O = nq_c A v_{\text{drift}} \tag{3.5}$$

where

$$v_{\text{drift}} = \frac{\Lambda E}{F_c} \tag{3.6}$$

So there is,

$$I_O = nq_c A \frac{\Lambda E}{F_c} = \frac{nq_c \Lambda}{F_c} AE \tag{3.7}$$

According to the principle of solution conduction, formula (3.7) for the Ohmic bulk conduction current of the jet can be modified by using the electrical conductivity κ of polymer solution as follows:

$$I_O = \kappa AE \tag{3.8}$$

where n is the number of charge carriers per unit volume of the polymer solution, q_c is the amount of charges carried by each charge carrier, A is the cross-sectional area of the jet, v_{drift} is the motion speed of the charge carriers, Λ is the molar conductivity, E is the external electric field strength, F_c is the Faraday constant, and κ is the conductivity of the polymer solution.

Let the radius of the jet be r, the cross-sectional area of the jet is $A = \pi r^2$, Eq. (3.8) can be expressed as

$$I_O = \kappa \pi r^2 E \tag{3.9}$$

According to the leaky dielectric model, the electrospinning polymer solution holds a weak charge transfer ability of insulator [3, 4]. Assuming that the charge carried by the jet mainly concentrates on the jet surface, the convection current can be interpreted as the amount of charge passing through the cross section of the jet per unit time. Therefore, the convection current can be calculated by

$$I_C = 2\pi r v_{\text{jet}} \rho_{qS} \tag{3.10}$$

where v_{jet} is the jet motion speed, and ρ_{qS} is the charge density on the jet surface.

In the electrospinning experiment, not all the charges concentrate on the jet surface, and the migration of the bulk charges also has an impact on the change of the current. Therefore, the bulk charge density ρ_{qB} can be used to replace the surface charge density ρ_{qS}, and the formula for the convection current can be modified as

$$I_C = 2\pi r v_{\text{jet}} \rho_{qB} \tag{3.11}$$

Yarin et al. [5] studied the dual characteristics of conductor and insulator of electrospinning polymer solution. They believed that the conduction characteristics of the two kinds of charges in the spinning jet were independent of each other. The spinning current I_E can be expressed as a direct linear superposition of the Ohmic bulk conduction current I_O and the convection current I_C:

$$I_E = I_O + I_C \tag{3.12}$$

That is,

$$I_E = \kappa \pi r^2 E + 2\pi r v_{\text{jet}} \rho_{qB} \tag{3.13}$$

Theron et al. [6] studied the influence of applied voltage U, solution flow rate Q, solution concentration c, and spinneret-to-collector distance h on the jet fluid's charge density ρ_{qB}, and described the relationship between the parameters with the power law:

$$\begin{cases} \rho_{qB}(U) = k_{\sigma U} U^{P_{\sigma U}} \\ \rho_{qB}(Q) = k_{\sigma Q} Q^{P_{\sigma Q}} \\ \rho_{qB}(c) = k_{\sigma c} c^{P_{\sigma c}} \\ \rho_{qB}(h) = k_{\sigma h} h^{P_{\sigma h}} \end{cases} \tag{3.14}$$

where $k_{\sigma U}$, $k_{\sigma Q}$, $k_{\sigma c}$, $k_{\sigma h}$, $P_{\sigma U}$, $P_{\sigma Q}$, $P_{\sigma c}$, and $P_{\sigma h}$ are the coefficients of the power law. The results show that the jet charge density ρ_{qB} increases with the increase of applied voltage U and solution concentration c, but decreases with the increase of solution flow rate Q and spinneret-to-collector distance h.

The electric field strength and the solution conductivity are the main factors affecting the change of the current. Increasing the applied voltage increases the electric field strength, which promotes the charge accumulation and increases the charge density of the jet. The increase of solution concentration enhances the ability of the jet to carry charge and helps increase the charge density of the jet. The electric field strength decreases with the increase of spinneret-to-collector distance, which reduces the amount of charge accumulated on the solution.

3.1.3.3 Factors affecting the EDW current

To better understand the variation in the EDW current, we further analyze the influence of parameters such as applied voltage, solution concentration, and spinneret-to-collector distance on the peak current during the initial ejection stage and the average current during the stable ejection stage.

3.1.3.3.1 Peak current in initial ejection stage

In the initial ejection stage, the charge carried by the jet front migrates through the collector after deposition, forming a peak current as shown in Figures 3.6–3.8. The diameter and surface charge density of the bead-like structure at the jet front increase with the solution concentration and applied voltage, and more charges will migrate through the collector after depositing on the collector, resulting in a higher peak current. Therefore, the peak current in the initial ejection stage grows with the increase of solution concentration and the applied voltage, as shown in Figure 3.10.

3.1.3.3.2 Average current during the stable ejection phase

The peak current at the beginning of the jet ejection mainly reflects the migration of the charge carried by the jet front. After the jet deposits onto the collector, it enters into the stable ejection stage, which is the main stage for the preparation of uniform nanofibers. Therefore, the current in this stage can better reflect the charge migration law in the ejection process of uniform nanofibers. To facilitate the analysis and discussion, we define the current average I_{avg} in this phase, which can be given as

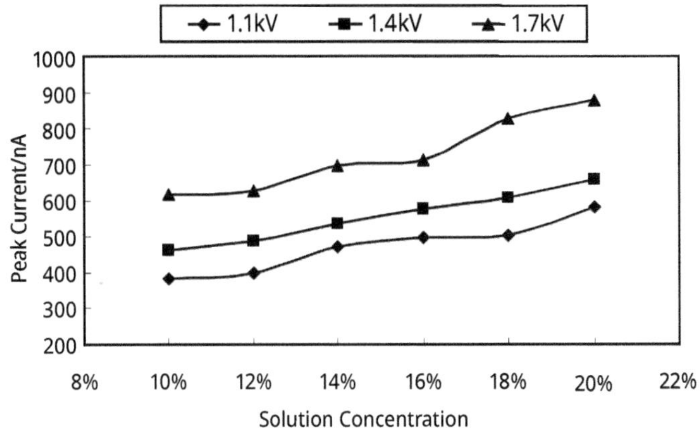

Figure 3.10: Influence of solution concentration and applied voltage on peak current (spinneret diameter is 40 μm, spinneret-to-collector distance is 1 mm, silicon substrate).

$$I_{avg} = \frac{1}{T - T_0} \int_{T_0}^{T} I_E' \, tdt \tag{3.15}$$

where T_0 is the starting time of the stable ejection stage, that is, the current peak moment; T is the ending time of the stable ejection phase; I_E' is the real-time current. We analyze the influence of applied voltage, spinneret-to-collector distance, and solution concentration on the average current, and the corresponding curves are shown in Figures 3.11–3.13 respectively. The characteristic average current in the stable ejection stage is 200–400 nA.

According to the analysis above, both the electric field strength and the jet charge density increase with the applied voltage, thus increasing the average current during the stable ejection stage, as shown in Figure 3.11. The influence of the spinneret-to-collector distance on the EDW process is mainly reflected by its influence on the electric field strength. The electric field strength decreases with the increase of the distance from the spinneret to the collector, and the current average decreases accordingly, as shown in Figure 3.12. The rise of solution concentration enhances the charge carry ability of the solution, thereby increasing the charge density of jet, and causing the average current in the stable ejection stage to increase with the increase of solution concentration, as shown in Figure 3.13.

3.1.3.3.3 Orthogonal analysis for average current

The effects of various process parameters in the EDW process are systematically studied using orthogonal experimental method. The jet ejection process is a typical multi-factor coupling process that is affected by various factors such as applied voltage, solution concentration, solution viscosity, solution conductivity, and the spinneret-to-collector

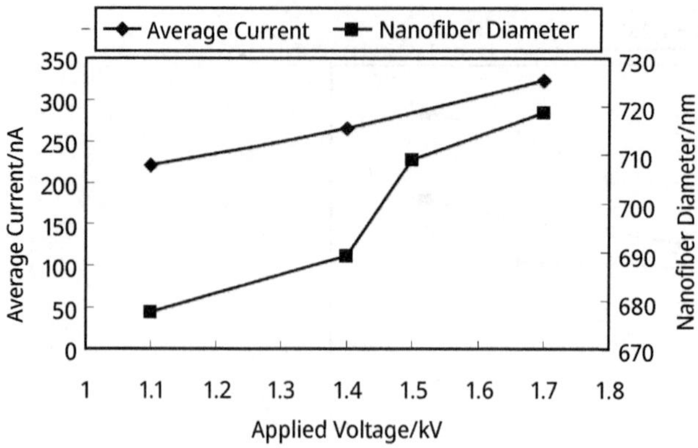

Figure 3.11: Influence of applied voltage on average current and nanofiber diameter (spinneret diameter is 40 μm, solution concentration is 20%, and spinneret-to-collector distance is 1 mm, silicon substrate).

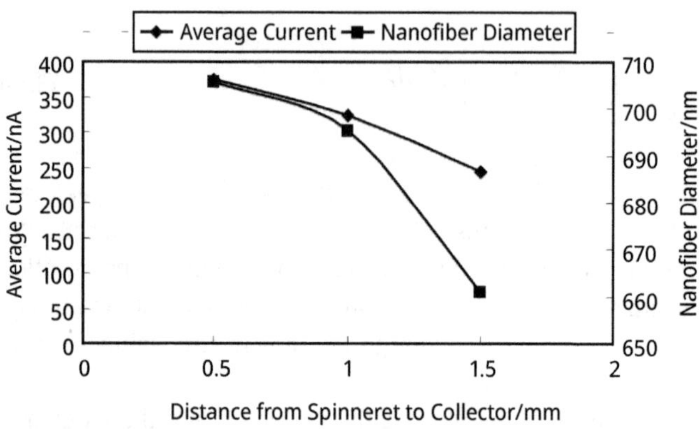

Figure 3.12: Influence of spinneret-to-collector distance on average current and nanofiber diameter (spinneret diameter is 40 μm, 20% solution concentration, and applied voltage is 1.7 kV, silicon substrate).

distance. Understanding the influence of each factor is conductive to deepen the understanding of the electrospinning mechanism and to improve the control of the jet ejection process. The orthogonal experimental method has the advantages of simple operation and convenient experimental design, which is suitable for the analysis of the experimental process of multi-factor coupling.

Based on the principle of orthogonal experimental method, we designed the $L9(3^3)$ orthogonal experimental table as shown in Table 3.3 to investigate the influence of the three controllable process parameters, namely the solution concentration, the spinneret-to-collector distance and the applied voltage on the average current in the stable ejection stage. In the experimental analysis, the three impact factors are divided into three experimental levels, which are denoted as solution concentration A

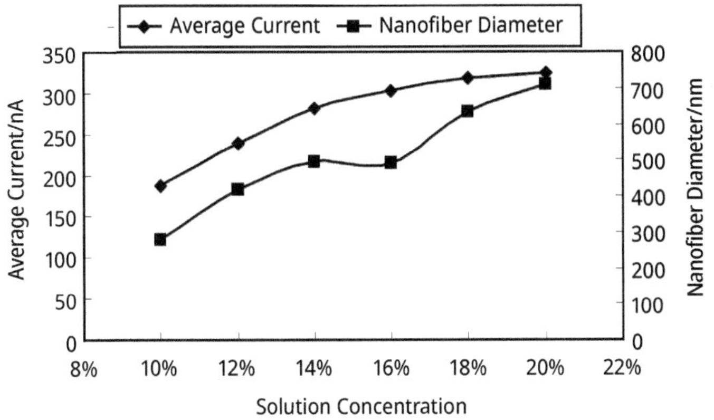

Figure 3.13: Influence of solution concentration on average current and nanofiber diameter (spinneret diameter is 40 μm, applied voltage is 1.7 kV, and spinneret-to-collector distance is 1 mm, silicon substrate).

Table 3.3: L9(3^3) orthogonal experimental design and experimental results for average current in stable ejection phase.

No.	Process factors			Average current (nA)
	Solution concentration	Spinneret-to-collector distance (mm)	Applied voltage (kV)	
1	A1 12%	B1 0.5	C1 1.1	185.77
2	A1 12%	B2 1.0	C2 1.4	185.01
3	A1 12%	B3 1.5	C3 1.7	201.06
4	A2 16%	B1 0.5	C2 1.4	295.23
5	A2 16%	B2 1.0	C3 1.7	302.82
6	A2 16%	B3 1.5	C1 1.1	165.75
7	A3 20%	B1 0.5	C3 1.7	375.14
8	A3 20%	B2 1.0	C1 1.1	221.57
9	A3 20%	B3 1.5	C2 1.4	226.11

(A1: 12%; A2: 16%; A3: 20%); spinneret-to-collector distance B (B1: 0.5 mm; B2: 1.0 mm; B3: 1.5 mm); and applied voltage C (C1: 1.1 kV; C2: 1.4 kV; C3:1.7 kV).

According to the orthogonal experimental table, the experiments are conducted in each group sequentially, and the average current values are obtained as shown in the last column of Table 3.3. Based on the principle of orthogonal experiment, the average current values measured from each group of experiments are analyzed, and the analysis results are shown in Table 3.4.

For the convenience of discussion, the average current value measured in the experiment at the *j*th level of the *i*th impact factor is denoted as I_{ij}, where $i = A, B, C$, $j = 1, 2, 3$. In Table 3.4, M_j represents the sum of the average current values for each influence factor measured at the *j*th level, that is, $M_1 = \sum I_{i1}$, $M_2 = \sum I_{i2}$, $M_3 = \sum I_{i3}$. m_j

Table 3.4: Analysis of L9 (3^3) orthogonal experiment results on average current in stable ejection stage.

Experiment level	Average current classification statistics (nA)		
	A: solution concentration	**B: Spinneret-to-collector distance**	**C: Applied voltage**
M1	571.83	856.14	573.09
M2	763.80	709.40	706.35
M3	822.81	592.91	879.01
m1	190.61	285.38	191.03
m2	254.60	236.47	235.45
m3	274.27	197.64	293.00
R	83.66	87.74	101.97

represents the average current value of each factor at the jth level, that is, $m_j = M_j/n$, where n is the number of test level for each factor; Defining the range of the experimental result $R = \max(m_j) - \min(m_j)$, which reflects the influence level of each factor on the experimental results. According to the analysis of L9(3^3) orthogonal experimental results shown in Table 3.4, the influence level of each factor on the average current is $C > B > A$, that is, the applied voltage has the greatest impact on the average current, followed by the spinneret-to-collector distance, and the solution concentration has the lowest level of influence.

The applied voltage is the most important process parameter in EDW process, which not only determines the distribution of space electric field strength, but also affects the charge density of the jet. It also plays an important role in the variation of the current. Regulating the applied voltage is the most effective means to control the jet ejection state and the diameter of the nanofiber distribution.

3.1.3.4 The variation law of EDW current

Electrohydrodynamic direct-writing mainly uses cone jet mode to prepare microstructures. In order to understand the variation law of the current in the cone jet mode, the influence of parameters such as applied voltage, solution characteristics, spinneret-to-collector distance, and solution supply rate on the average current during the stable ejection stage are investigated.

A capillary spinneret with an inner diameter of 0.6 mm, and a variety of solutions with different physical properties are used in the experiments, as shown in Table 3.5.

3.1.3.4.1 Relationship between average current and applied voltage

Figure 3.14 shows the current situation of pure glycerol solution jetting at different applied voltages. With a solution supply rate of 400 μL/h, the average current remains stable at about 44 nA regardless of the voltage. As the feeding rate increases to

Table 3.5: Solutions used in the experiments and their physical properties.

Solution	Viscosity (mPa/s)	Conductivity (μS/cm)	Surface tension (mN/m)
Glycerol	1,412	<0.01	63
Glycerol/NaCl	1,412	0.54 ~ 9.14	63
3% PEO	35	136.40	65
9% PEO	3,492	149.20	56
12% PEO	8,367	163.50	50

800 μL/h, the average current exhibits a linear relationship with the voltage. The figure clearly shows that the slope of the fitting line for the average current versus applied voltage increases with the solution supply rate, indicating that the influence of the applied voltage on the average current is also increasing.

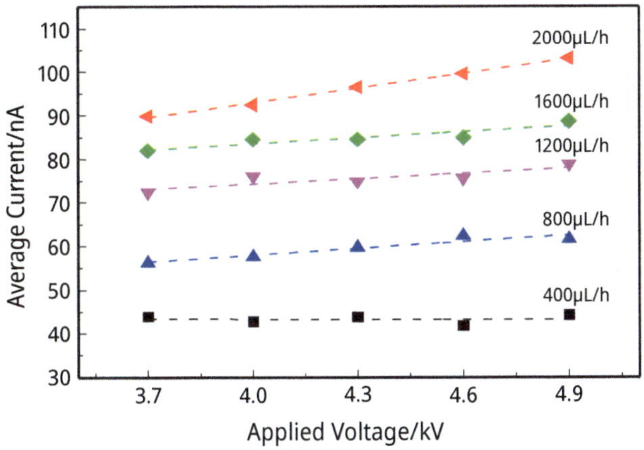

Figure 3.14: Current of glycerin solution jetting at different applied voltages (spinneret-to-collector distance is 4 mm).

A small amount of NaCl is added to pure glycerol to increase the solution conductivity, which rises to 2.60 μS/cm. Figure 3.15 shows the current of glycerol/NaCl solution ejecting at different applied voltages. For a solution supply rate of 400 μL/h and 800 μL/h, the average current increases linearly but slightly with the increase of the applied voltage. For example, at 400 μL/h, the average current is 135 nA at 3.7 kV and 145 nA at 4.9 kV, while at 800 μL/h, the average current was 156 nA at 3.7 kV and 173 nA at 4.9 kV. With the increase of the solution supply rate, the average current increases sharply when the applied voltage exceeds a certain threshold. For example, at a feeding rate of 1,200 μL/h, the average current is about 200 nA at applied voltage of less than 4.6 kV; it rises to 617 nA

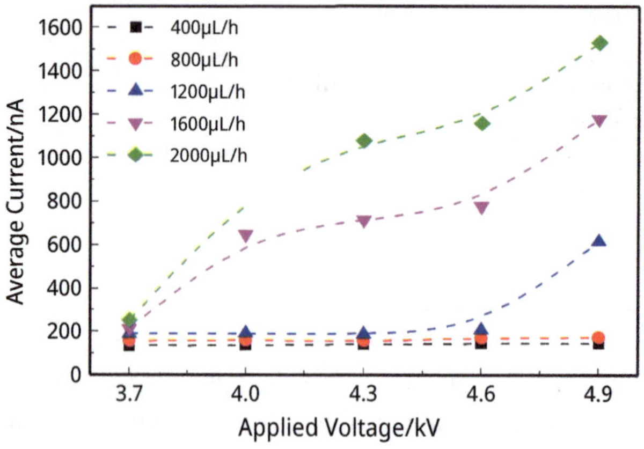

Figure 3.15: Current of glycerol/NaCl solution ejecting at different applied voltages (spinneret-to-collector is 4 mm).

when the voltage is increased to 4.9 kV. When the feeding rate exceeds 1,600 μL/h, the average current increases significantly with the increase of applied voltage.

Figure 3.16 shows the current of PEO solution jetting at different voltages, which indicates that the average current increases with the applied voltage under various conditions, and higher solution concentration leads to greater slope of the corresponding average current-voltage curve. In addition, under the same other conditions, the higher the solution supply rate, the greater the corresponding average current.

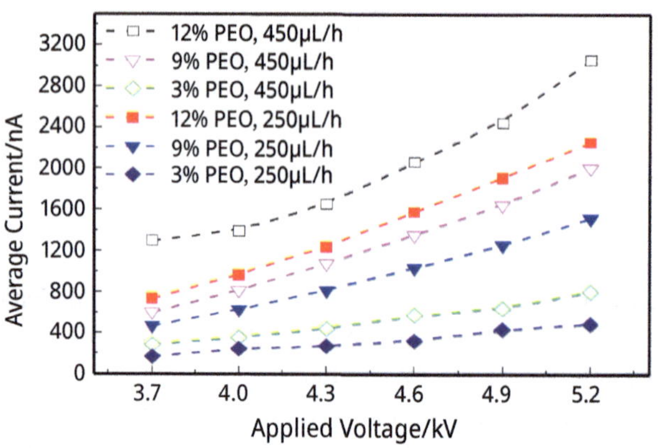

Figure 3.16: Current of PEO solution during ejection at different applied voltages (spinneret-to-collector distance is 15 mm).

3.1.3.4.2 Relationship between average current and solution supply rate

For pure glycerol solution, the average current increases with the increase of the solution supply rate, as shown in Figure 3.17. Through data fitting, a relationship between the average current I and the solution supply rate Q can be set up as $I \sim aQ^b$, where the values of (a, b) are (2.646, 0.464), (2.293, 0.488), (2.233, 0.494), (1.890, 0.519), (1.823, 0.529), respectively, for applied voltages of 3.7kV, 4.0kV, 4.3 kV, 4.6 kV and 4.9 kV.

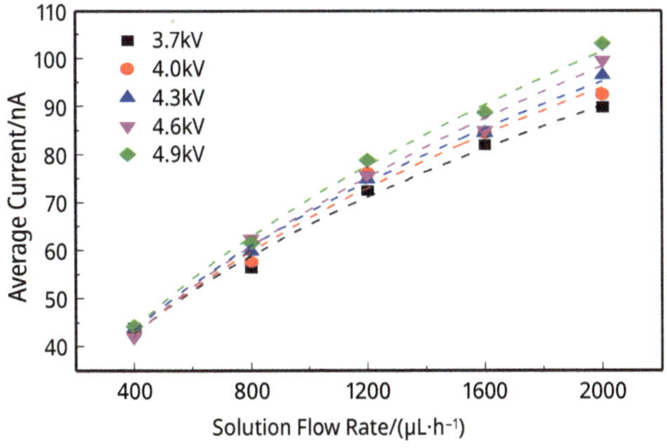

Figure 3.17: The current of pure glycerol solution during ejection at different solution flow rates (spinneret-to-collector distance is 4 mm).

Figure 3.18 shows the current of glycerol/NaCl solution with conductivity of 2.60 μS/cm ejecting at different solution supply rates. Unlike the pure glycerol solution shown in Figure 3.17, with an increased solution conductivity, the average current and the solution supply rate no longer exhibit an obvious exponential relationship similar to $I \sim aQ^b$. However, the overall trend is still that the average current grows with the solution supply rate.

Figure 3.19 shows the current of 12% polyethylene oxide(PEO) solution ejecting at different solution supply rates. The experimental results show that the average current increases with the increase of the solution supply rate and is linearly related to it.

3.1.3.4.3 Relationship between average current and spinneret-to-collector distance

For pure glycerol solution, the variation in spinneret-to-collector distance shows little effect on the average current. At a solution supply rate of 400 μL/h, the average current measured for spinneret-to-collector distances of 4 mm and 6 mm with the same applied voltage is of little difference, as exhibited in Figure 3.20.

Figure 3.21 shows the current of glycerol/NaCl solution with a conductivity of 2.60 μS/cm ejecting at different spinneret-to-collector distances. In the experiment for

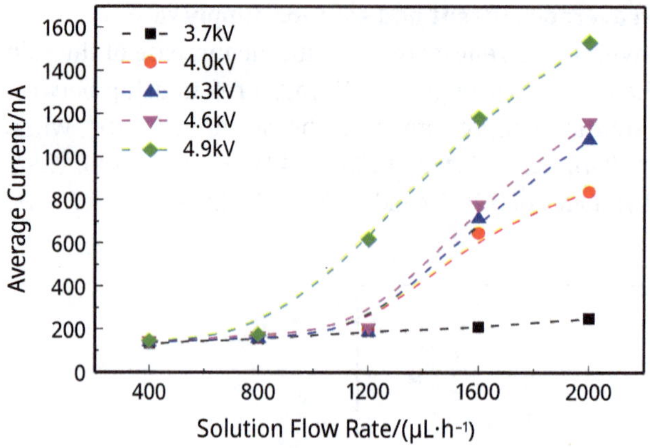

Figure 3.18: Current of glycerol/NaCl solution ejecting at different solution feeding rates (spinneret-to-collector distance is 4 mm).

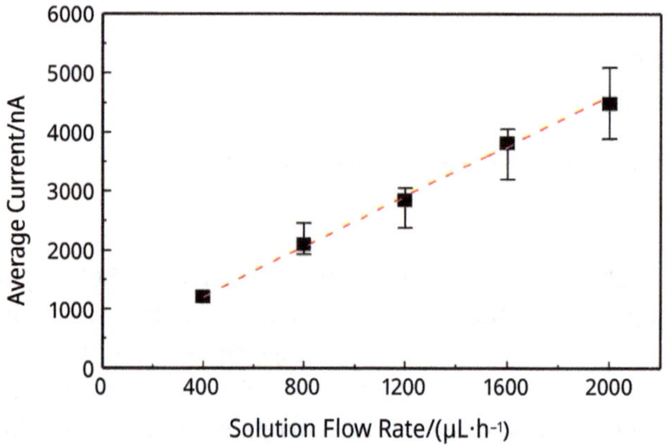

Figure 3.19: Current of 12% PEO solution during ejection at different solution flow rates (spinneret-to-collector distance is 15 mm, and applied voltage is 3.7 kV).

pure glycerol solution, the variation in the spinneret-to-collector distance does not significantly affect the average current (Figure 3.20). However, the addition of NaCl increases the solution conductivity, and the average current decreases with the increase of the spinneret-to-collector distance.

Figure 3.22 shows the current of PEO solution ejecting at different spinneret-to-collector distances. Experiments are conducted at spinneret-to-collector distances of 10 mm and 15 mm, respectively, with a solution supply rate of 450 μL/h under different experimental parameters. The results show that the average current decreases with the increase of the spinneret-to-collector distance.

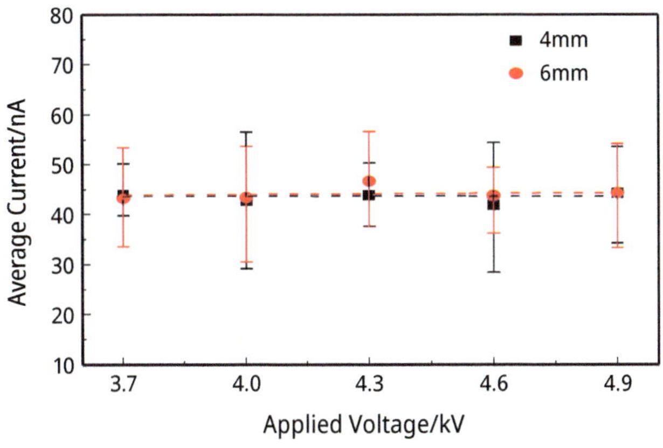

Figure 3.20: Current of glycerin when ejecting at different spinneret-to-collector distances. (solution supply rate is 400 μL/h).

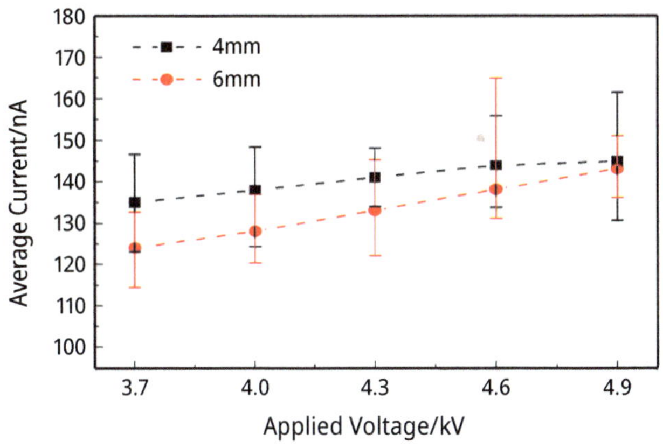

Figure 3.21: Current of glycerin/NaCl solution when ejecting at different spinneret-to-collector distances (solution supply rate is 400 μL/h).

3.1.3.4.4 Relationship between average current and solution conductivity

Glycerol solutions with conductivities of 0.54 μS/cm, 1.41 μS/cm, 3.88 μS/cm, and 9.14 μS/cm are obtained by adding different masses of NaCl into glycerol. Under the same experimental conditions, the current during the ejection of the four solutions with different conductivities is shown in Figure 3.23. It is obvious that the average current increases with the increase of the solution conductivity.

3.1.3.4.5 Relationship between the current and PEO solution concentration

Figure 3.24 shows the current of PEO solution with different concentrations. Experimental results show that the higher the concentration of PEO solution, the greater the EDW

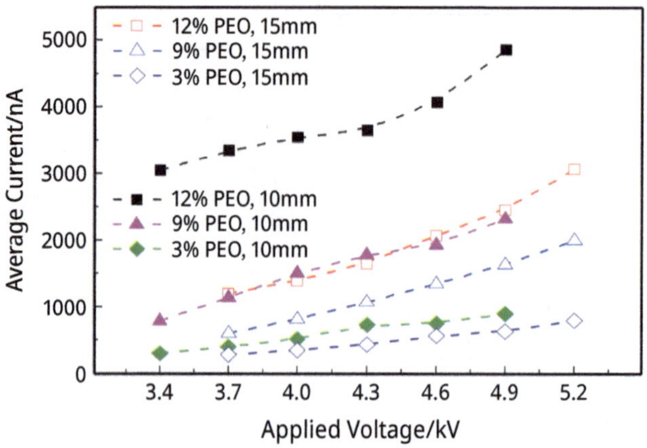

Figure 3.22: The average current when the PEO solution is ejecting at different spinneret-to-collector distances (solution feeding rate is 450 μL/h).

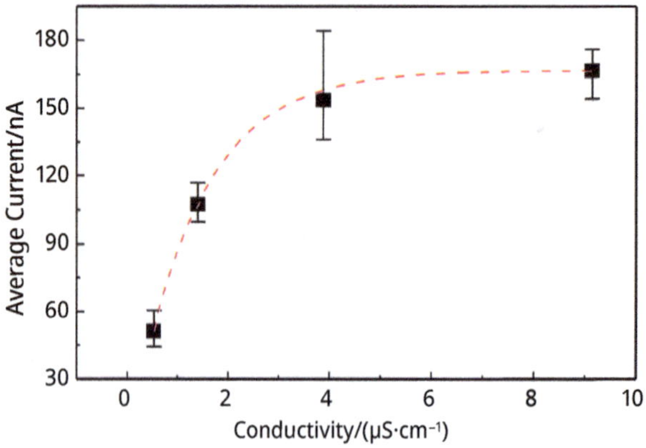

Figure 3.23: Current of glycerol/NaCl solution with different conductivities (the applied voltage is 4.0 kV, the liquid supply rate is 400 μL/h, and the spinneret-to-collector distance is 6 mm).

current; and the increasing concentration enhances the effect of the applied voltage on the current. In fact, this result can also be clearly reflected in Figures 3.16 and 3.22.

3.1.3.5 Correlation analysis between EDW current and process

In the theoretical model of electrospraying or electrospinning, as the spinneret is relatively far from the collector (tens of centimeters), the electrospraying jet splits into discrete droplets in the air, while the electrospinning jet solidifies into fibers. Therefore, the Ohmic bulk conduction current $I_O = \pi r^2 \kappa E \approx 0$, that is, the total current is $I \approx I_C = 2\sigma Q/r$. However, for EDW, the distance between the spinneret and the collector is shorter, and a fluid "channel" will form between the spinneret and the collector after jet ejection,

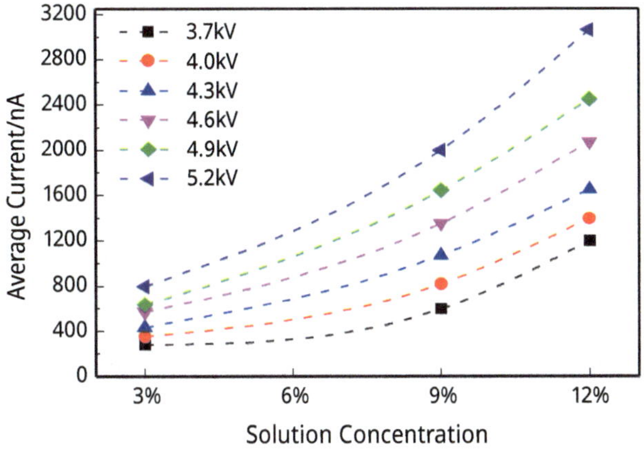

Figure 3.24: Current during ejection of PEO solution with different concentrations (the spinneret-to-collector distance is 15 mm, and the liquid flow rate is 450 μL/h).

forming a complete current loop. Therefore, in the EDW process, $I_O = \pi r^2 \kappa E \neq 0$, that is, the Ohmic bulk conduction current cannot be completely ignored.

We first estimate the order of magnitude of I_O during the experiment. Assuming that the jet is a slender cylinder, in the experiment, the distance from the spinneret to the collector is between 4 and 15 mm. Taking the jet length L as 10 mm, the jet radius $r = 10\,\mu\text{m}$, and the applied voltage $U = 4\,\text{kV}$, with electric field strength $E \approx U/L$, the relationship between I_O and solution conductivity can be expressed as

$$I_O \approx \frac{\pi r^2 U k}{L} \approx 0 \tag{3.16}$$

Correlation of relevant orders of magnitude is shown in Figure 3.25.

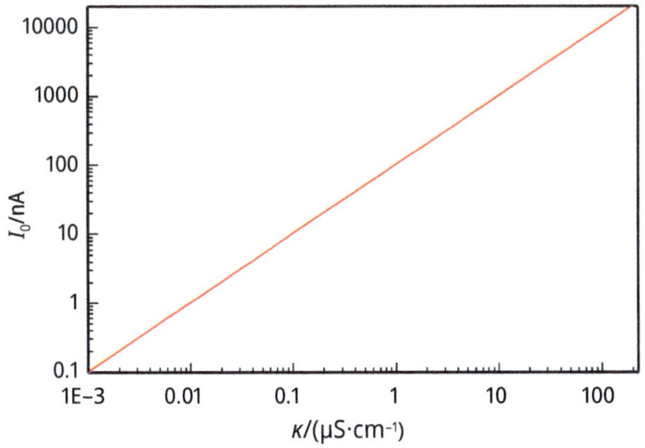

Figure 3.25: Correlation of Ohmic bulk conduction current with solution conductivity.

In this session, the current characteristics of glycerol and PEO solution will be further analyzed, and the correlation of EDW current with process parameters will be studied.

3.1.3.5.1 Current characteristics during the ejection of glycerin

The Ohmic bulk conduction current passing through the jet is mainly determined by the jet diameter, the solution conductivity, and the applied voltage. As shown in Figure 3.14, for pure glycerol solution, the jet radius is less than 10 μm at a solution supply rate of 400 μL/h, the conductivity of the solution is less than 0.01 μS/cm, so there is $I_0 < 0.1$ nA. However, the total current measured is above 40 nA, indicating that $I_0 \ll I_C$, and

$$I \approx I_C = 2\sigma Q/r \tag{3.17}$$

As the solution supply rate increases, the jet diameter becomes larger, the Ohmic bulk conduction current I_0 increases, and there is $I_0 \sim r^2$. Therefore, with the increase of the liquid supply speed, the total current I gradually shows a positive correlation with the applied voltage (Figure 3.17), while I shows little correlation with the spinneret-to-collector distance (Figure 3.20).

3.1.3.5.2 Current characteristics during the ejection of glycerol/NaCl solution

When the conductivity of glycerol solution is increased to 2.60 μS/cm by adding NaCl, I_0 is about 10^0~10^3 nA depending on the solution supply speed and the applied voltage, as shown in Figure 3.15. At this time, $I_0 \ll I_C$ is no longer true and I_0 cannot be ignored. Therefore, the applied voltage, the spinneret-to-collector distance have a significant impact on the total current I. Increasing the solution conductivity can significantly increase the current during the ejection process. However, it is found in the experiment that electrospraying occurs when the conductivity of glycerol/NaCl solution reaches 9.14 μS/cm, so the change in current is relatively small compared to that for solution conductivity of 3.88 μS/cm (Figure 3.23).

3.1.3.5.3 Current characteristics during PEO solution ejection

The impact of process parameters on the current is more significant when PEO solution is used for electrospinning. According to the results shown in Figure 3.16, Figure 3.19 and Figure 3.22, the total current I increases with the applied voltage and liquid supply speed. In the ejection process of PEO solution, I_0 is approximately (10^2–10^3) nA, which is much greater than the convection current, and has been dominant in the total current. So, there is:

$$I \approx I_o = \pi r^2 \kappa E = QE_k/v_{\text{jet}} \tag{3.18}$$

Suppose that the variation in the jet velocity v_{jet} is small with respect to the solution supply rate and applied voltage, Eq. (3.18) can be presented as

$$I \sim EQ \tag{3.19}$$

That is, the current shows linear correlation with both the solution supply rate and the applied voltage (or electric field strength).

To sum up, there is a correlation between the EDW current and the process parameters, so the current can be controlled by adjusting parameters such as the applied voltage, and the solution supply rate. In the EDW process, the Ohmic bulk conduction current cannot always be ignored, which sometimes even dominates the total current. The experimental results verify that the current have shown the following relationship with the process parameters:

$$\begin{cases} I \sim Q^{1/2} & (I_0 \ll I_C) \\ I \sim EQ & (I_0 \gg I_C) \end{cases} \tag{3.20}$$

3.1.3.6 EDW current and jet diameter

Based on the experimental study of the influence of process parameters on electrospinning current, the relationship between the current and the jet diameter is further analyzed. During the EDW process, the jet diameter gradually decreases as it travels from the spinneret to the collector. To facilitate comparison, the jet diameter at 3 mm from the spinneret is measured (as shown in Figure 3.26). When the jet moves in the air, the diameter shows no much change, and the stretching and refining effect is not obvious. Moreover, the closer to the collector, the smaller the change in its diameter. Therefore, it can be considered that the jet diameter at 3 mm from the spinneret is approximately equal to the diameter of the jet at the collector.

When disturbed, the jet will fluctuate, and the rheological behavior of the solution will be changed, thus affecting the current. Figure 3.26 shows the relationship between the average current and the jet diameter. With the voltage and the solution supply rate remains unchanged, the average current will increase with the increase of jet diameter.

In fact, according to $I = \pi r^2 \kappa E + 2\sigma Q/r$, the relationship between the current and the jet diameter can be given as follows:

$$I \sim d^2 \kappa E/4 + d\sigma \tag{3.21}$$

Under the premise that both the external electric field strength and the solution delivery rate remain constant, consideration is given to the impact of external disturbances on the jet diameter:

$$\begin{cases} I \sim d & (I_0 \ll I_C) \\ I \sim d^2 & (I_0 \gg I_C) \end{cases} \tag{3.22}$$

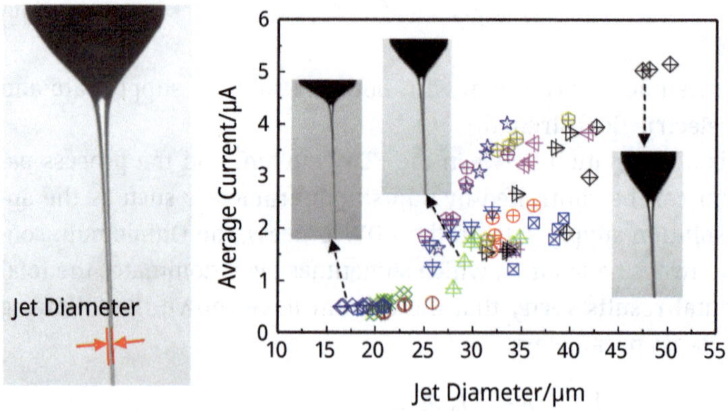

Figure 3.26: The average current and the jet diameter (solution concentration is 12% and spinneret-to-collector distance is 15 mm).

The relation between the average current and the jet diameter under different parameters are also studied. Experimental results in Figure 3.27 show that the jet diameter decreases and the average current increases with increasing voltage. This conclusion does not conflict with the conclusion of Eq. (3.22), as the precondition of Eq. (3.22) is that the electric field strength and the solution supply rate are constant, and the jet diameter changes due to interference, so the average current changes accordingly. Figure 3.27 shows the variation of jet diameter and average current caused by the change of applied voltage. With the increase of applied voltage, although the diameter of the jet decreases, the electric field strength increases, and the charge density on the jet surface increases, which together make the average current increase.

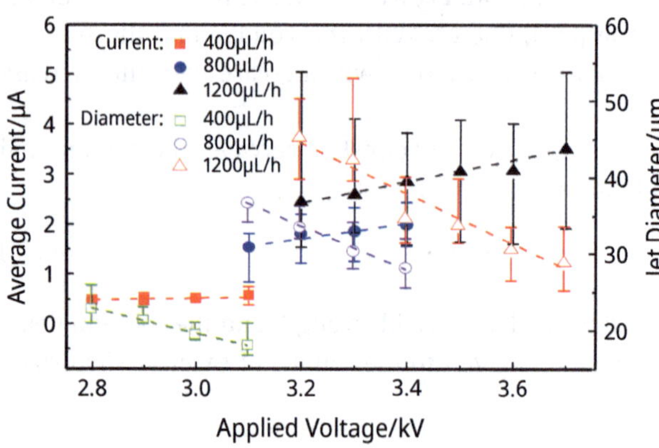

Figure 3.27: Variation of average current and jet diameter under different experimental conditions (solution concentration is 12%, and spinneret-to-collector distance is 15 mm).

The above experiments and theoretical analysis both confirm that the change in EDW current directly reflects the changes in the jetting behavior, and the jet diameter can be effectively controlled by adjusting the current, which provides a theoretical and experimental basis for the control of steady-state jet in EDW based on current feedback control.

3.1.3.7 EDW current and diameter of deposited structure

The EDW current reflects the charge transfer behavior in the jet. It is also an effective variable to monitor the jet ejection state and the nanofiber diameter variation [7–10]. To better understand the influence of charge transfer on the preparation of electrospun nanofibers, the relation between the average current and the nanofiber diameter is analyzed. The influence of process parameters on the average current and nanofiber diameter is shown in Figures 3.11–3.13. The results show that the diameter of nanofibers increases with the applied voltage and solution concentration, and decreases with the increase of the spinneret-to-collector distance. The variation in the average current follows a similar pattern to that in the nanofiber diameter.

The increase in average current is an important reflection of the increase in jet charge density. A higher jet charge density can not only increase the charge repulsion force between the adjacent parts of the jet, but also increase the instability factors during the jet ejection process, which affects the consistency of the ejection process, resulting in a larger nanofiber diameter distribution range.

To better understand the ejection behavior of EDW jet, the current variations during the stable ejection stage are investigated. By observing the fluctuation of the current in the ejection process, we find that the current stability during the ejection of glycerol is higher than that of PVP solution. Taking the current variation shown in Figure 3.28(a) as an example, the current fluctuation of glycerin solution is relatively stable during the whole ejection process, with a maximum variation amplitude of about ±10% from the average value. However, the current of PVP solution fluctuated obviously during the spinning process, with maximum variation amplitude of about ±40% from the average value.

To facilitate the discussion, the current change rate ΔI and the change rate of deposited structure diameter ΔW are defined as follows:

$$\Delta I = \frac{I_{\max} - I_{\min}}{I_{\text{avg}}} \tag{3.23}$$

$$\Delta W = \frac{W_{\max} - W_{\min}}{W_{\text{avg}}} \tag{3.24}$$

where the subscripts max, min, and avg represent the maximum, minimum, and average values, respectively. The correlation between ΔI and ΔW is analyzed by direct writing of 8% PVP solution, as shown in Figure 3.29. The diameter variation rate of the deposited structure ΔW tends to increase with the current change rate ΔI, that is,

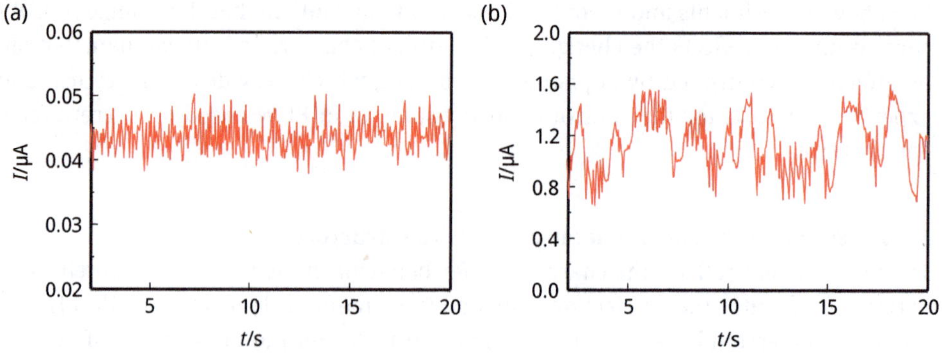

Figure 3.28: Current change in EDW process (a) for glycerin solution at applied voltage of 4.0 kV, solution feeding rate of 400 μL/h, and spinneret-to-collector distance of 6 mm; (b) for 8% PVP solution at applied voltage of 1.9 kV, solution supply rate of 70 μL/h, and spinneret-to-collector distance of 1.5 mm.

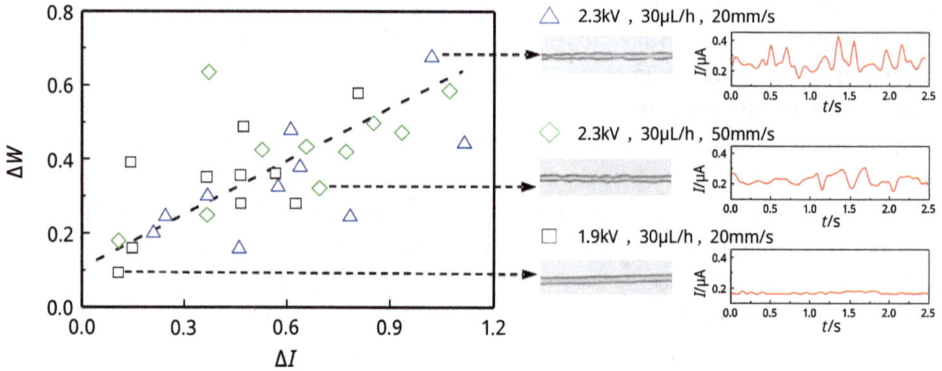

Figure 3.29: Current change rate and diameter change rate of the deposited structure in EDW process.

the change rule of the current stability is similar to that of the diameter uniformity of the direct-written structure. The more stable the current, the better the uniformity of the diameter.

The current variation during EDW process can effectively reflect the jet motion behavior. The jet is unstable due to the influence of many factors during the ejection process, which affects the uniformity of the directly written structure. The results depicted in Figure 3.29 indicate that the variation in current can reflect the uniformity of the deposited structure, which provides an effective online monitoring means for the precise control of jet stability and uniformity of the deposited structure.

3.2 Development of EDW micro-current monitoring system and test of electrical characteristics

3.2.1 Development of EDW micro-current monitoring system

3.2.1.1 Function analysis for the micro-current monitoring system

The EDW current monitoring system uses virtual instrument technology to achieve an organic combination of computer control modules and instrument hardware through software. Virtual instrument is a technology that is based on the powerful computing of computers. It relies on software design platforms, utilizes hardware for measurement and communication, and allows for operations on a virtual panel. In this work, the LabVIEW created by National Instruments (USA) is used as the system's software platform. LabVIEW stands out as the most powerful and widely adopted graphical programming language and development environment, known for its user-friendly operation. Leveraging a graphical programming approach, it simplifies the creation of programs through an intuitive block diagram interface, driven by data flow logic. This method replaces traditional code with visual blocks, making it accessible to both novice and experienced programmers. It has a user-friendly interface and high compatibility, allowing for the portability and maintenance of various programs. National Instruments also provides corresponding hardware facilities, empowering users to design the tailored systems that fulfill specific control functions and unique requirements. LabVIEW comes equipped with an extensive collection of signal processing function packages and driver programs, facilitating tasks such as data acquisition, analysis, storage, and communication via protocols like RS-232 and PXI (PCI extensions for instrumentation — here, "PCI" refers to "protocol control information" - related extensions for instrumentation systems). Its dataflow modeling capability enables seamless implementation of multithreaded parallel computing, enhancing processor efficiency. Widely applied in data acquisition and equipment control scenarios, LabVIEW is well-suited to meet the demands of EDW current monitoring system.

To achieve the remote control of high-voltage power supply, and the detection and recording of current during the EDW process, a current monitoring system is constructed, through which the data transmission between the host computer and the high-voltage power supply is achieved, the voltage setting of the high-voltage power supply, and the reading of the feedback voltage and current values is accomplished. In addition, as the current value is generally in nanoampere and microampere levels, a micro-current detection system is introduced into the monitoring system, where the current signal is displayed on the main interface in real time through the data acquisition card and recorded for subsequent in-depth analysis of the electrical characteristics of the jet.

The EDW current monitoring system comprises a control program and a micro-current detection system. Through the host computer, instructions are sent to the high-voltage power supply of the slave computer to control the start-up and shutdown of the high-voltage power supply, set the voltage value, and meanwhile, the output voltage and current values returned by the high-voltage power supply can be read. The signals collected by the data acquisition card are read and converted into actual current values. The voltage and current values returned by the high-voltage power supply and the current signals collected by the data acquisition card are displayed in real-time, facilitating observation and regulation by managers. And the data is saved in documents for convenient data processing. The schematic diagram of the EDW current monitoring system is shown in Figure 3.30.

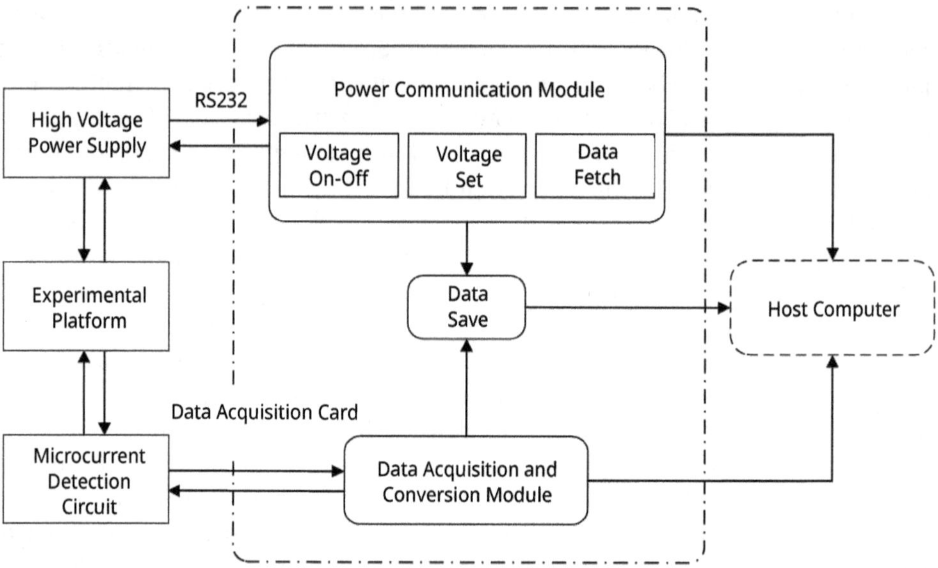

Figure 3.30: Schematic diagram of EDW current monitoring system.

3.2.1.2 Development of high-voltage power supply communication module

A communication connection is established between the host computer and the high-voltage power supply through a USB-RS 232 data cable. Based on the communication protocol MODBUS-RTU of the high voltage DC power supply DW-P603-1ACF1, using the VISA interface provided by the LabVIEW software to configure the serial port, a control program for the high voltage DC power supply is developed as the subroutine of the main program to send instructions to the high voltage DC power supply to achieve remote control from the host computer. The voltage control subroutine includes three modules: on-off control of the high voltage DC power supply, setting of voltage and reading of the output voltage and current values. The three parts of the subroutines completed according to the communication protocol of the high-voltage power supply is shown in Figures 3.31–3.33. After the program is started, a command should be sent to the high-voltage power supply to switch on the high voltage. Throughout the program running process, the voltage can be adjusted online in real-time according to the demand, the voltage and the current value output by the high-voltage power supply can be transmitted to the host computer.

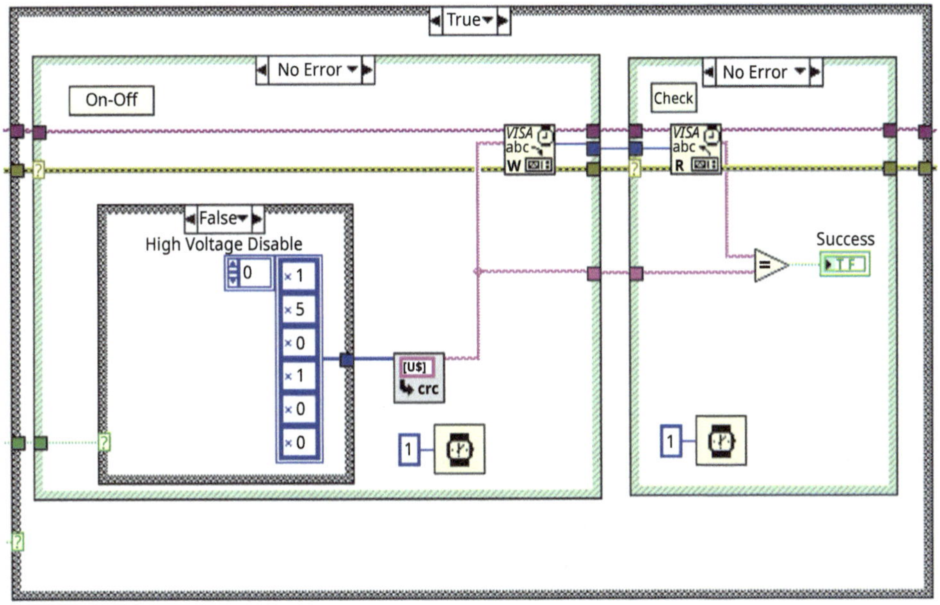

Figure 3.31: Program for the on-off control of high-voltage power supply.

3.2.1.3 Development of data acquisition and conversion module

The software development environment for the data acquisition card used in this system is the LabVIEW platform. The NI-DAQmx driver in LabVIEW provides an easy configuration and programming interface for data acquisition. The date acquisition(DAQ) assistant is used to provide a graphical interface for configuring the settings and parameters for data acquisition. Before performing data acquisition, it is necessary to initialize the measurement type, channel number, range, sampling frequency, and sampling method of the data acquisition card. Using the DAQ Assistant eliminates the need for programming. Instead, users can interactively configure channels and timing, making it easy and convenient to set up. This greatly reduces development time for users. The voltage signals collected by the data acquisition card are transmitted to the computer via the DAQ Assistant. The acquisition path is shown in Figure 3.34.

The routine for data acquisition and conversion is illustrated in Figure 3.35. The signals collected by DAQ assistant are transmitted in the form of array. To monitor

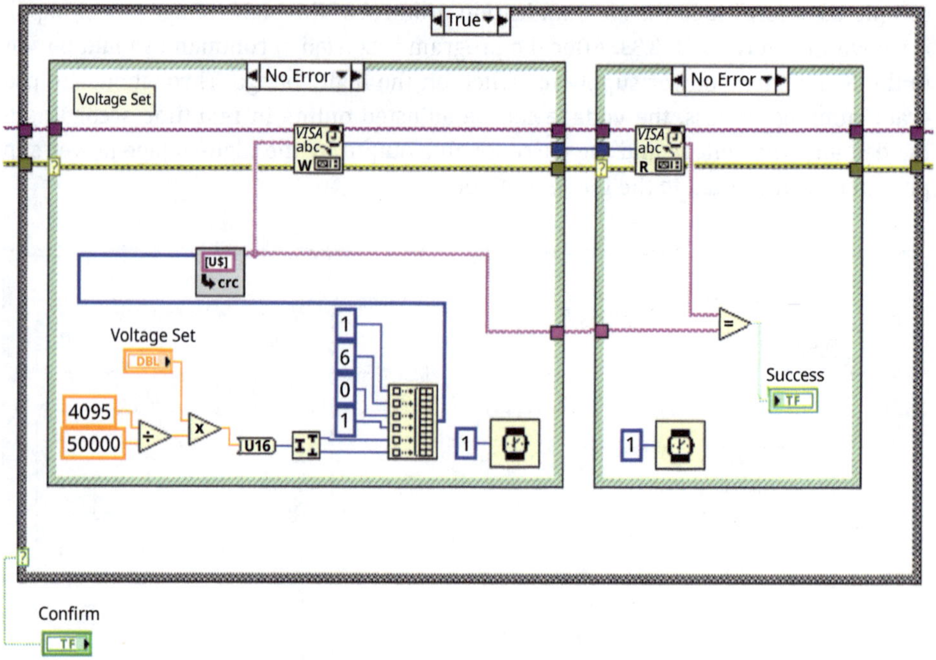

Figure 3.32: Voltage setting program for the high-voltage power supply.

current variation more intuitively, the collected voltage signals are converted back to corresponding current signals according to different amplification factors set in the design of the amplification circuit. In addition, to facilitate the operator's monitoring of the operation status of the amplified circuit and timely handle the abnormal situations, both the collected voltage signal and the actual current signal are simultaneously displayed in the waveform chart on the front panel.

3.2.1.4 Development of the main program for the EDW current monitoring system

The voltage control subprogram and the data acquisition and conversion subroutines are integrated into the main program. The serial port configuration is completed according to the communication protocol of high-voltage DC power supply. The output voltage and current values of the high-voltage power supply, as well as the voltage and current signals from the the data acquisition and conversion subprogram, are connected to a waveform diagram. The variations in all acquired signals are displayed in real time on the front panel. All the collected data are recorded in a file according to measurement needs and are saved for later data processing. An error reporting module is added to the operation interface on the front panel for user maintenance. The diagram of the main program of the EDW current monitoring system is shown in Figure 3.36.

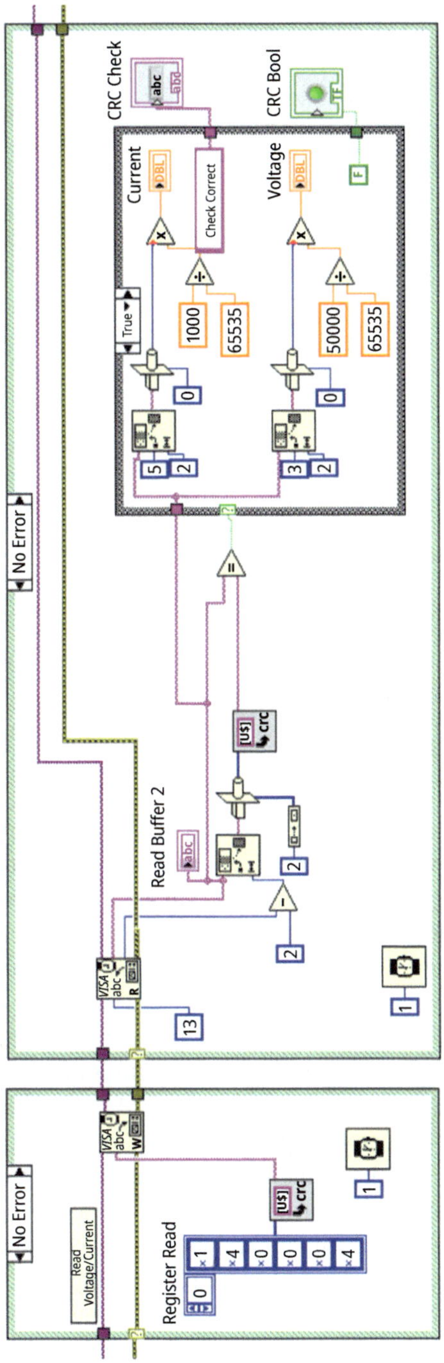

Figure 3.33: Program for reading the voltage and current output of the high-voltage power supply.

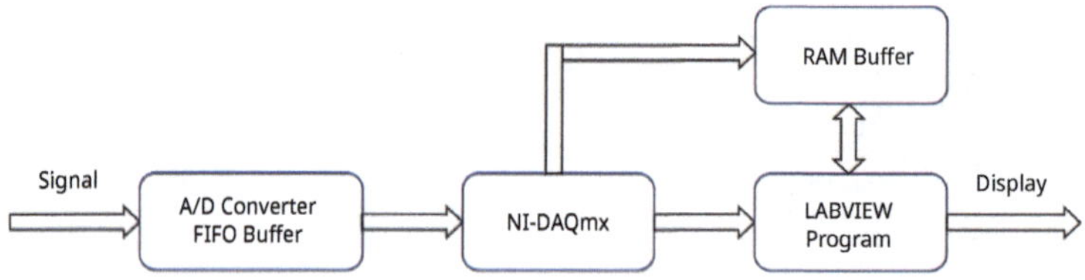

Figure 3.34: DAQ signal acquisition path.

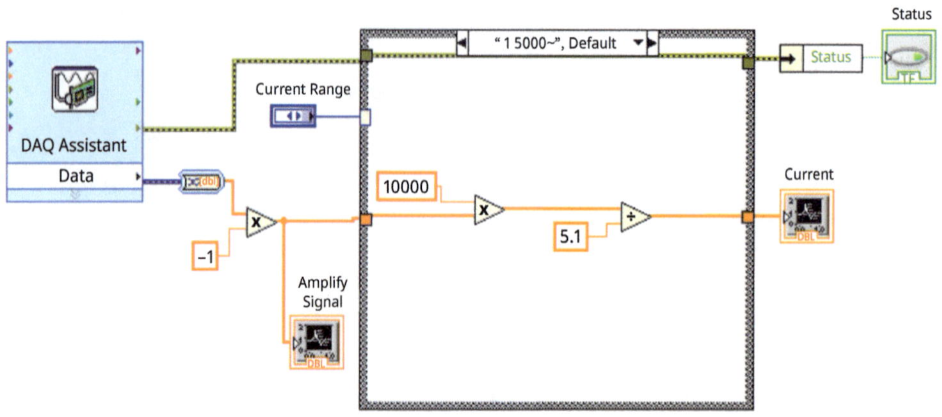

Figure 3.35: Subroutine for data acquisition and conversion.

The control interface of the EDW current monitoring system is shown in Figure 3.37, including four basic functional modules: system parameter setting panel, voltage power supply control panel, monitoring result display interface, and intelligent error reporting panel. The system parameter setting panel allows for the setting of system parameters, including the data storage path in the computer, the VISA interface name, and the current level according to the amplification factor used in the micro-current detection system (there are seven levels available). On the voltage source control panel, the voltage source related parameters, including starting high voltage (the indicator light is on if the onset is successful), setting the desired voltage (the indicator light is on if the setting is successful after confirming the setting), and stopping the program. The monitoring result display interface shows the real-time changes in voltage and current values during the EDW process. The "voltage source output current" waveform displays the real-time current output by the high-voltage DC power supply, which is mainly used to record the impulse current in the EDW process. The waveform of "the voltage source supply voltage" displays the output voltage of the high-voltage DC power supply in real-time, and is used to determine whether the high-voltage power supply has been turned on, reaches the preset voltage and works normally; The waveform of "micro-current detection system current" displays the current value collected by the micro-current detection system,

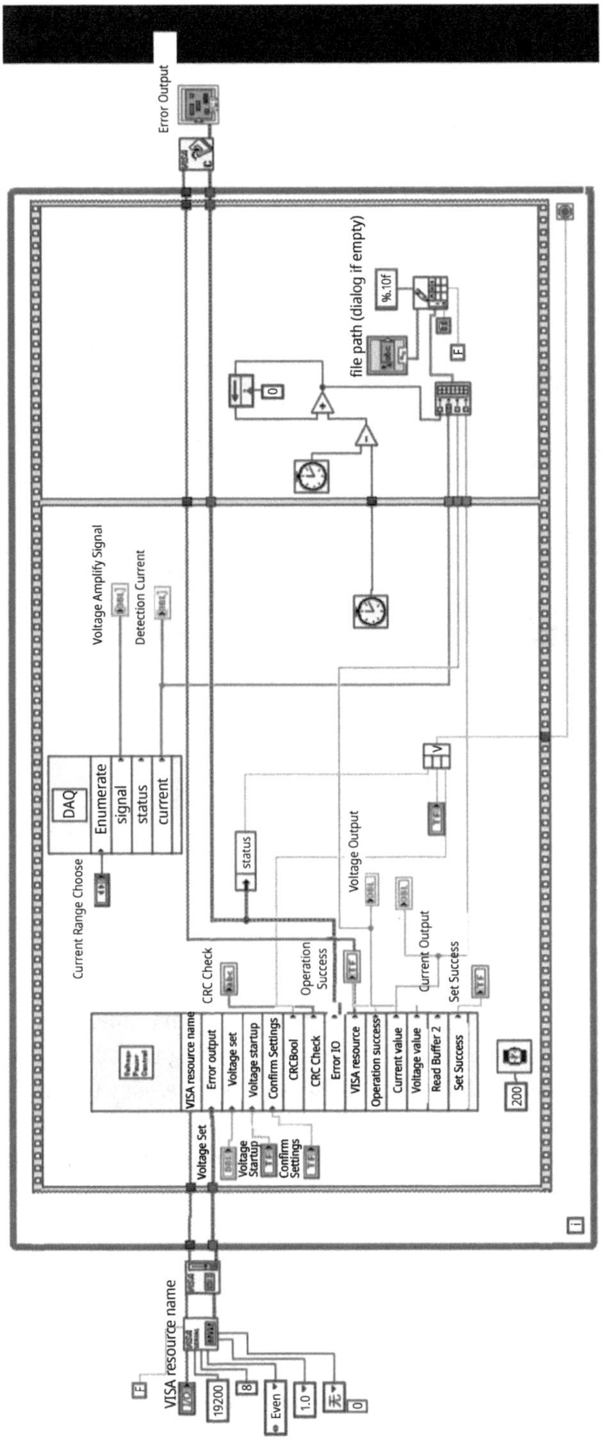

Figure 3.36: Diagram of the main program of the EDW current monitoring system.

which is used for the analysis of EDW current characteristics. The waveform of “voltage amplification signal” displays the voltage signal directly collected by the micro-current detection system, which is used to determine whether the micro-current detection system is working normally. If the system fails to operate normally due to some error, the error reporting panel will display the corresponding error code and the program source, explain the current error, guiding the administrator to deal with the problem in time.

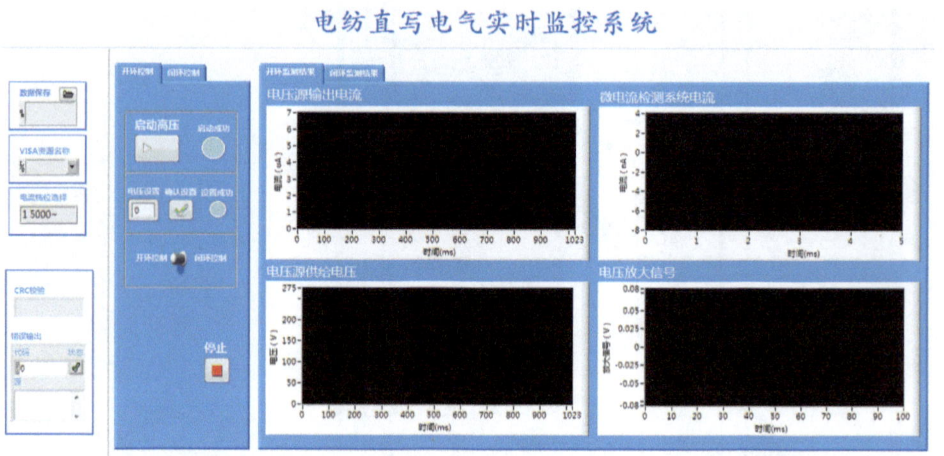

Figure 3.37: Control interface of EDW current monitoring system.

3.2.1.5 Micro-current detection system

3.2.1.5.1 Circuit amplification

The micro-current detection system is composed of an amplifying circuit and a data acquisition part. The data acquisition part is generally achieved by the data acquisition card and software platform. The amplitude of EDW current is very small, generally ranging from tens of nanoampere to a few microamperes. During the acquisition process, the current signal is easily overwhelmed by noise, so it is necessary to reduce the influence of noise while amplifying the current signal. The data acquisition card can only be used for voltage acquisition. Therefore, to collect micro-current signals in EDW system, facilitate signal transmission, and reduce interference, the current signal needs to be converted to voltage signal. The schematic diagram of the amplifier circuit is shown in Figure 3.38. The current-voltage conversion and amplification method is adopted, where a conversion resistor is connected to the inverting input and output of the amplifier. The method has high sensitivity, simple design, and little influence on the circuit to be measured. The inverting input end of that amplifier is connected to the collector, and the output is connected to the data acquisition card.

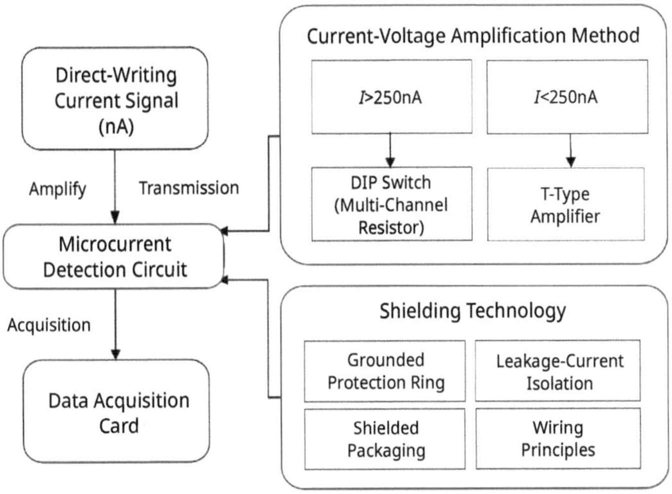

Figure 3.38: Schematic diagram of amplifier circuit.

The amplitude of EDW current is very small, but the span is quite large. The difference between the maximum and minimum values is nearly 100 times. Therefore, constant amplification factor cannot meet the input requirements of the data acquisition card (range of −10 to 10 V). If the gain is set too high, the current with large amplitude may exceed the range of the data acquisition card. Conversely, if the gain is set too small, currents with small amplitudes may not be detected as it does not meet the accuracy requirements. Therefore, two I-V amplification circuits are designed for current amplification, one for amplifying the currents below 250 nA and the other one for currents above 250 nA respectively, both using single-stage amplification to reduce the number of devices and minimize potential noise interference.

The same amplification factor is selected to amplify currents below 250 nA to the voltage range that can be recognized by the data acquisition card. To avoid an increase in the input bias current of the amplifier caused by high-resistance feedback resistors, which can cause amplifier distortion, and to avoid a long response time caused by distributed capacitance, T-type network resistors are selected to improve the response speed of the system. For currents above 250 nA, the current amplitude range is relatively large. To meet the sensitivity requirements, the feedback resistors with different resistances are selected, and a six-bit Dual In-line Package (DIP) switch is used to select different amplification factors to meet different current amplification requirements.

The leakage current on the printed board and the noise generated by interference between components may obscure actual current values, so noise reduction is essential during the circuit board production process. Components are selected based on the principle of high precision and low noise. Grounding protection ring technology and air isolation technology are used in the printing of circuit board to isolate leakage current and to reduce the impact on current signals. The principle of low noise is strictly followed during wiring process, with the circuit extensively covered by copper. The circuit is enclosed in a shielding box to shield electromagnetic interference and is all grounded. Low-noise shielded cables are adopted as the connection lines from the collector to the input end of the detection circuit, and the output end to the data acquisition card. The physical diagram of the amplification circuit is shown in Figure 3.39.

Figure 3.39: Physical diagram of amplification circuit.

3.2.1.5.2 Isolation circuit

When using the above system for testing, applying high voltage on EDW platform can generate a significant impulse current, which is damaging to the computer. In industrial measurement, to reduce various interferences from the outside world on the signals, it is necessary to electrically isolate the detection system from the computer control system. The commonly applied methods of electrical isolation include transformer isolation, optical coupler isolation, and so on. Among these, optical coupler isolation has advantages such as simple circuitry, strong anti-interference capability, complete elimination of electrical interference between stages, and small size, making it widely used in practical applications.

The HCNR201 linear optocoupler is utilized as the isolation circuit in this system, which includes a light-emitting diode (LED) and two photodiodes with identical performance parameters. When current passes through the LED, it emits light, and a current is formed in the photodiodes. The internal packaging of HCNR201 ensures that the two photodiodes receive approximately the same amount of light. With the same performance parameters, equivalent current proportional to the light intensity will be generated in the two diodes, through which the electrical isolation between the input module and the output module is achieved, ensuring high stability.

The schematic diagram of the isolation circuit is shown in Figure 3.40. The input terminal of the operational amplifier OP1 receives the output signal from the amplifier circuit. The amplified current signal passes through the first-stage positive proportional amplifier circuit to increase the input impedance and enhance the load capacity. The input of the optocoupler isolation section includes a LED, an operational amplifier OP2, and a photodiode pd1. When the voltage signal of the previous stage is input, the output of OP1 causes a current to flow through the LED, thereby driving the LED to emit light and generate the induced current i_1 in pd1. The current generated by the voltage signal flows through R_2 and pd1 into the ground, thereby providing negative feedback. As the ideal operational amplifier is "virtual short" and "virtual break," there is no current flowing through the input terminal of OP1, so the current flowing through pd1 is

$$i_1 = \frac{V_1}{R_2 + R_1} \tag{3.25}$$

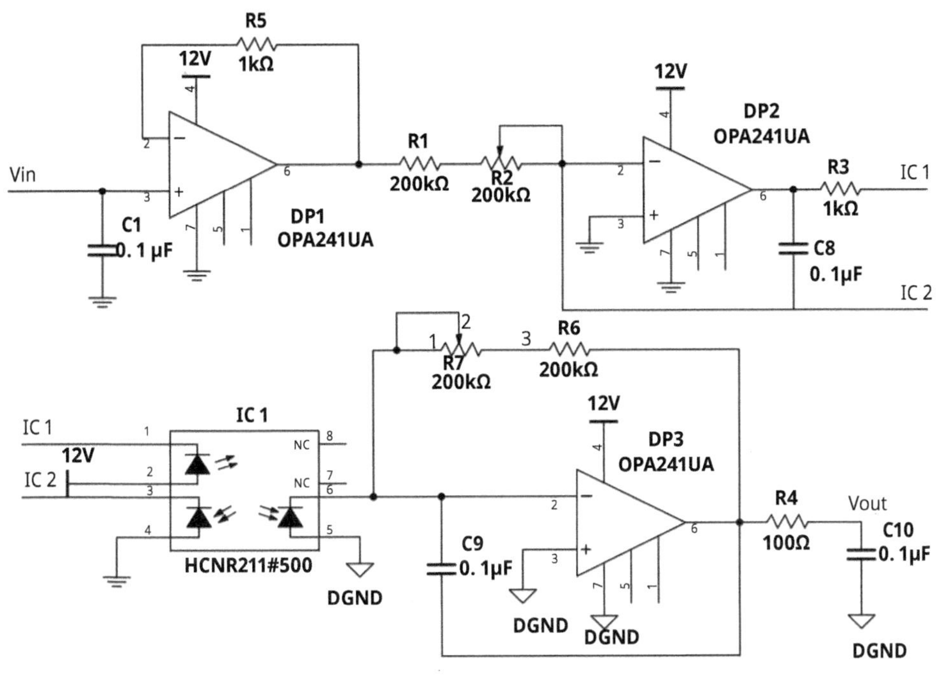

Figure 3.40: Schematic diagram of isolation circuit.

Equation (3.25) shows that the magnitude of i_1 is determined only by V_1, R_1, and R_2. The output of the isolation circuit includes a photodiode pd2 and an operational amplifier OP3. As there is no current flowing through the output terminal of OP3, with the current flowing through pd2 defined as i_2, the output voltage V_6 can be given as

$$V_6 = i_2(R_7 + R_6) \tag{3.26}$$

Ideally, the current passes through pd2 should be equal to the current passing pd1, that is, $i_i = i_2$, and the relationship between the input and output voltages can be obtained by using Eqs. (3.25) and (3.26):

$$\frac{V_6}{V_1} = \frac{R_7 + R_6}{R_2 + R_1} \tag{3.27}$$

Therefore, taking $R_7 = R_2$ will result in the same input and output voltage.

Connecting the output end of the circuit with the date acquisition card, the schematic diagram and physical diagram of the isolation circuit are shown in Figure 3.41. For this circuit, independent power supply is required to supply power to the pre- and post-stage operational amplifiers to achieve the isolation of the applied power supply and ensure the complete isolation of the signal.

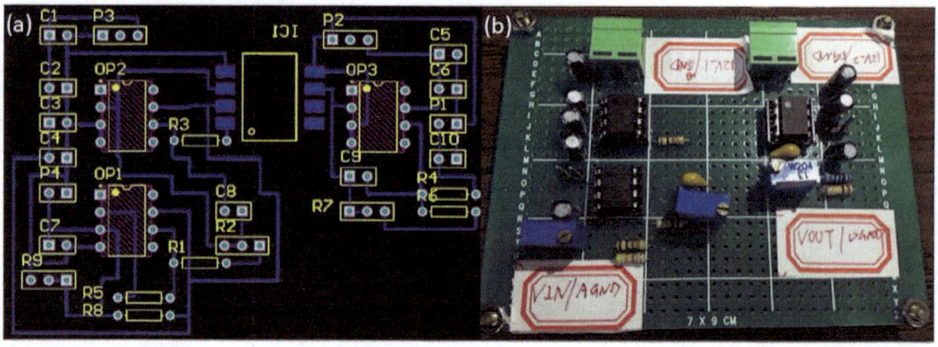

Figure 3.41: The schematic diagram and physical diagram of the isolation circuit: (a) schematic diagram and (b) physical diagram.

3.2.1.6 Test of the micro-current detection system

To test whether the noise of micro-current detection system meets the accuracy requirements for EDW current detection, the no-load current is tested, that is, the output signal by the micro-current detection system is collected under the condition of no liquid supply and only high voltage supply. Since the micro-current detection system consists of two circuit boards, it is necessary to analyze the noise of the two circuits separately. The no-load currents are shown in Figures 3.42 and 3.43, respectively.

In the noise waveform, the moment when the voltage is applied is defined as the zero-moment. Figure 3.42 and Figure 3.43 show that there is no significant change in the noise signal before and after the voltage is applied. For detection circuits with currents above 250 nA, as there are six different amplification resistors, different amplification factors are selected by manipulating the dial switch. It is found that the noise amplitude under different amplification factors shows no much difference. The noise range of the detection circuit designed for currents below 250 nA is −8 to 8 nA, while that for currents above 250 nA is −1.2 to 1.2 nA. Compared with the EDW current signal, the noise amplitude is very small, which will not interfere with the acquisition of the target current signal.

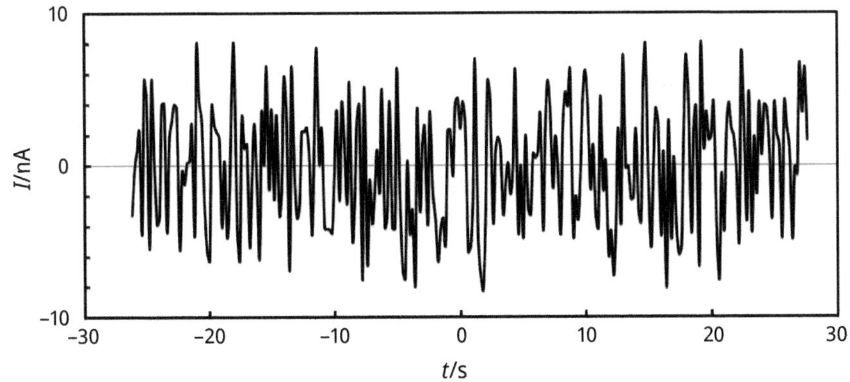

Figure 3.42: Noise waveform of current detection circuit for currents below 250 nA (spinneret-to -collector distance is 6 mm and applied voltage is 3.0 kV).

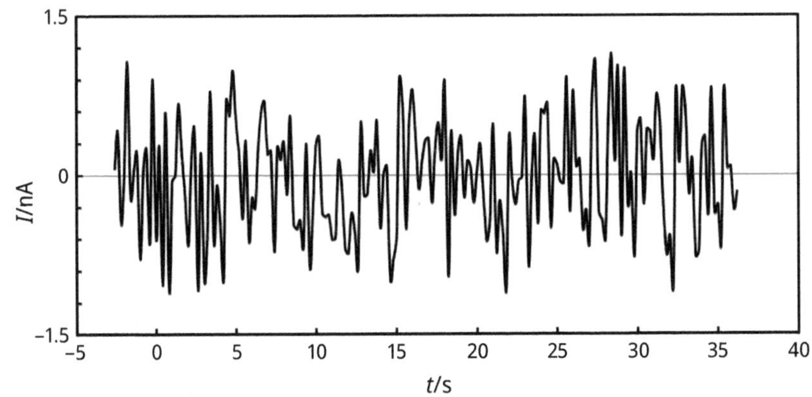

Figure 3.43: Noise waveform of current detection circuit for current above 250 nA (spinneret-to -collector distance is 6 mm and applied voltage is 3.0 kV).

3.2.2 Test of electrical characteristics in EDW Jet

3.2.2.1 Real-time variation of EDW current

A hollow spinneret is used for continuous liquid supply to carry out the EDW experiment, and the real-time current variation in the electrospinning process is recorded. The evolution of the current with time is shown in Figure 3.44. After applying the voltage, the current signal fluctuates due to the strong electric field. The moment when the voltage is applied is defined as time zero. Under the strong space electric field, a large amount of free charges accumulates on the polymer droplet hanging at the nozzle. After a certain ejection delay time, the solution squirts out, with the front end carrying a large amount of free charges. When the jet deposits on the collector, the free charge carried by the jet front is transferred to the ground through the collector, resulting in a large impulse current, which is shown as the peak current in the current signal. As time goes on, the electrospinning steps into the stable ejection stage, where the jet behavior gradually stabilize, and the current tends to be stable. Therefore, the EDW process is divided into the initial ejection stage and the stable ejection stage. In the initial ejection stage, the front end of the jet shows a bead structure with larger diameter as the jet is not fully stretched, and the solvent is not fully volatilized. Moreover, in this stage the jet carries more free charge. This can generate peak current when the jet deposits on the collector. After reaching the collector, the jet quickly enters the stable ejection stage where the jet is fully stretched, the solvent volatilizes faster. The charge density on the jet surface is much smaller than that at the jet front in the initial ejection stage. So the current in the stable jet stage is significantly smaller than the peak current in the initial ejection stage. The average value of the current during the stable ejection phase is 500 nA, with a fluctuation range of 480–540 nA.

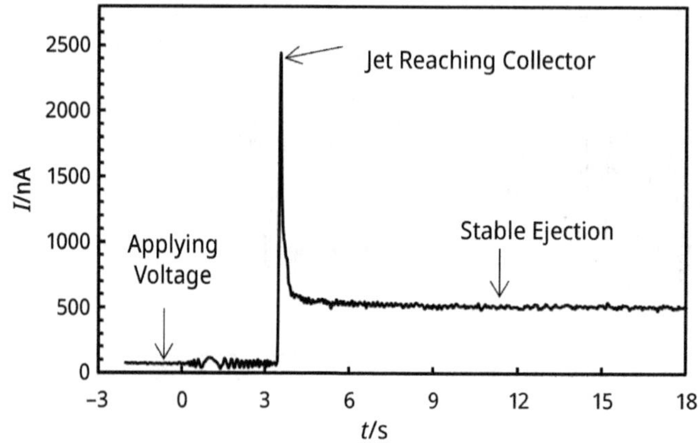

Figure 3.44: Variation of EDW current with time (the solution concentration is 10%, the applied voltage is 3.0 kV, the spinneret-to-collector distance is 6 mm, and the solution supply rate is 100 μL/h).

After applying the voltage, the electric charges accumulate at the tip of the spinneret and drive the solution deformation and jet formation. Figure 3.44 shows that there is a certain time delay between the applying of voltage and the jet ejection. This is because after providing an external electric field, it takes time to accumulate sufficient charge to produce an electric field force that is strong enough to overcome the surface tension and viscoelastic force of the solution. There is also a time delay between the induction of the external force and the deformation of the solution, which reflects the viscous characteristics of PEO and other solutions. The evolution of current over time under different voltages is shown in Figure 3.45. With the increase of voltage, the electric field strength and the charge migration speed increase, the charge accumulation time and the current delay time decrease, and the time for the jet to step into the sta-

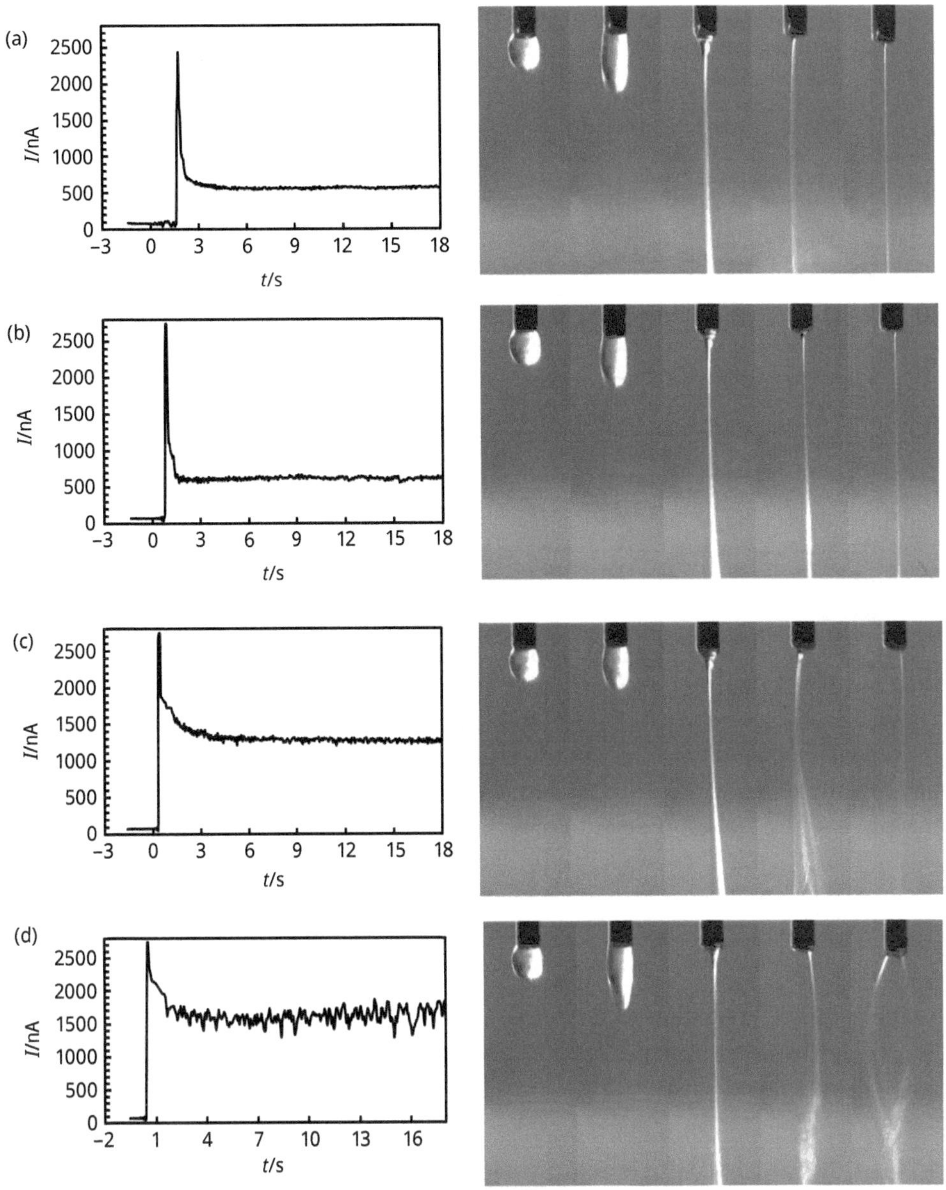

Figure 3.45: Current waveform and jet morphology under different applied voltages: (a) 3.3 kV; (b) 3.6 kV; (c) 3.9 kV; and (d) 4.2 kV (solution concentration is 10%, spinneret-to-collector distance is 6 mm, and solution supply speed is 100 μL/h).

ble ejection and to form a stable current is also shortened. The current delay time and the time required to reach.

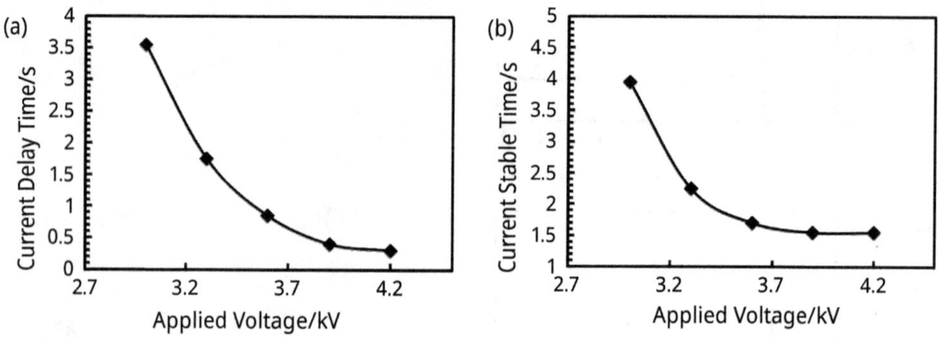

Figure 3.46: Variation in the current delay time and stable ejection time with the applied voltage: (a) plots of current delay time versus the applied voltage and (b) stable ejection time versus the applied voltage (solution concentration is 10%, spinneret-to-collector distance is 6 mm, and the solution feeding rate is 100 μL/h).

Figure 3.46 shows that when the applied voltage increases from 3.0 kV to 3.7 kV, the time for the current to enter the stable ejection stage decreases significantly. This is because when the voltage is small, the rise of the voltage will lead to the increase of the jet velocity, and the effect of the electric field force is more pronounced than the disordered motion caused by the charge repulsion force. As the voltage continues to increase, the oscillation time of the jet front is already very short, and the increased voltage cannot significantly increase the jet motion speed, so the oscillation time shows no obvious change with the applied voltage.

3.2.2.2 Impact of process parameters on average current

To further understand the electrical characteristics of the jet, the effects of the applied voltage, the spinneret-to-collector distance, and the solution concentration on the average current are studied.

3.2.2.2.1 The applied voltage and the average current

The average current for PEO solution electrospun at different applied voltages is shown in Figure 3.47. The average current increases with the applied voltage. As the voltage increases, the space electric field strength, the charge density on the jet surface, and the jet speed increase. Accordingly, both the Ohmic bulk conduction current and the convection current increase, resulting in an increase in the average current.

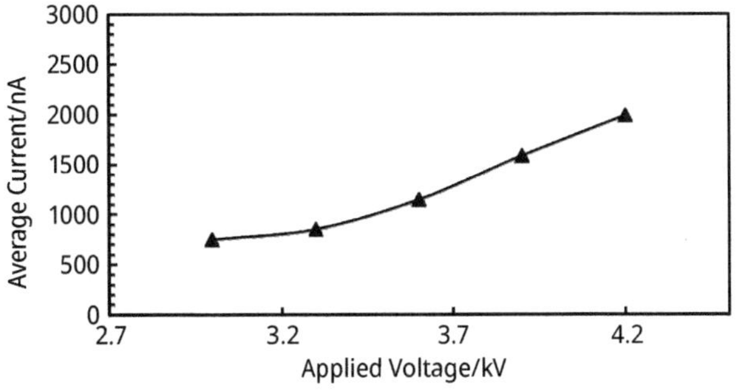

Figure 3.47: Variation of the average current with the applied voltage (solution concentration is 12%, spinneret-to-collector distance is 6 mm, and solution supply speed is 100 μL/h).

Figure 3.48 shows the jet morphology under different voltage conditions. The charge migration speed increases with the increase of the electric field strength. Meanwhile, the charge repulsive force carried by the jet increases, causing the jet to be stretched to a greater extent and its diameter to decrease. When the voltage is higher than 3.6 kV the jet transitions from a stable cone jet mode to an unstable mode. The jet swings irregularly along the edge of the spinneret, resulting in unstable jet deposition, which increases the difficulty of fiber precise positioning. The magnitude of the voltage is the most important and easily adjustable parameter in EDW technology, and it has the greatest impact on the jet ejection behavior and the current. The voltage is the most important and easily adjustable parameter in EDW technology, and it has the greatest impact on the jet ejection behavior and current. Adjusting the applied voltage is the most effective way to control the EDW current and jet behavior.

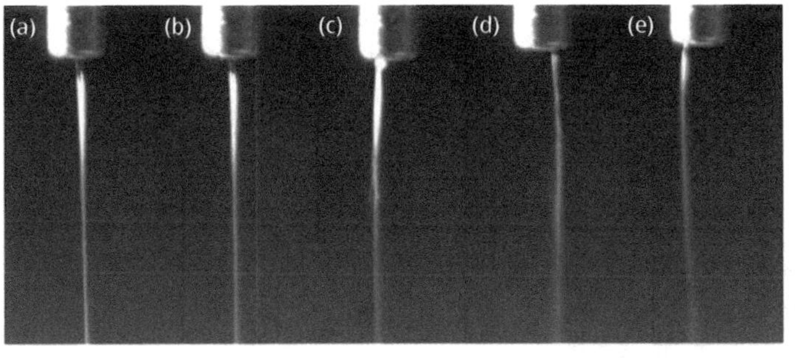

Figure 3.48: Jet morphology under different voltages (a) 3.0 kV; (b) 3.3 kV; (c) 3.6 kV; (d) 3.9 kV; and (e) 4.2 kV (solution concentration is 12%, spinneret-to-collector distance is 6 mm, and solution supply speed is 100 μL/h).

3.2.2.2.2 Relation between spinneret-to-collector distance and the average current

Figure 3.49 shows the average current obtained when the PEO solution is spun at different spinneret-to-collector distances. As the spinneret-to-collector distance increases, the electric field strength declines and the Ohmic conduction current significantly decreases. Therefore, the average current decreases with the increase of the spinneret-to-collector distance, and the influence of the applied voltage on the average current is more significant for shorter distance.

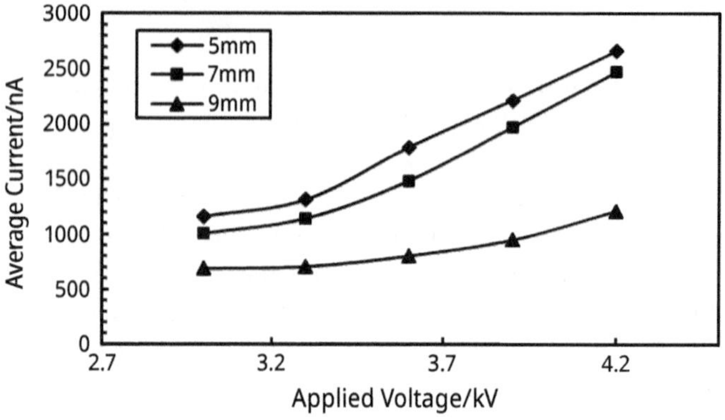

Figure 3.49: Variation of the average current with the spinneret-to-collector distance (the concentration of the solution is 10% and the solution feeding rate is 100 μL/h).

Figure 3.50 shows the jet morphology at different spinneret-to-collector distances. As the distance increases, the electric field strength decreases and the constraining effect of the electric field force on the jet weakens, resulting in more violent jet whipping and reduced fiber deposition accuracy. At a spinneret-to-collector distance of 5 mm, the high-voltage electric field has a significant influence on the jet, causing it to be stretched

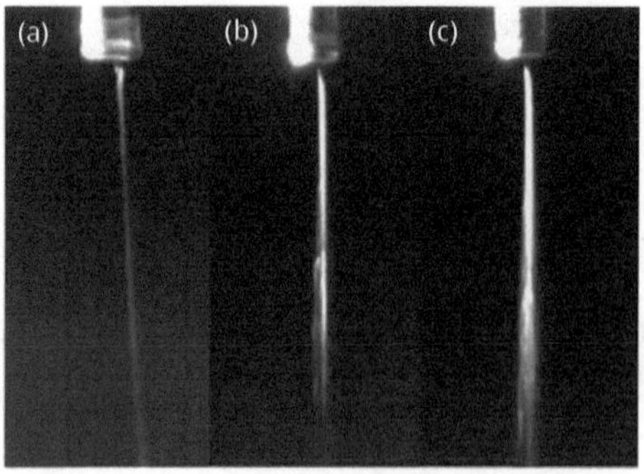

Figure 3.50: Jet morphology at different spinneret-to-collector distances: (a) 5 mm; (b) 7 mm; and (c) 9 mm (solution concentration is 10%, applied voltage is 3.0 kV, and solution flow rate is 100 μL/h).

to a greater extent. Figure 3.50 (a) shows that both the jet diameter and the jet whipping range are smaller compared to those for larger spinneret-to-collector distances. As the distance from the spinneret to the collector grows, the electric field constraint on the jet decreases, and the jet whipping becomes more violent, making it difficult to achieve precise positioning deposition.

3.2.2.2.3 The relation between PEO solution concentration and average current

The average current obtained by spinning PEO solution of different concentrations is shown in Figure 3.51. Increasing the concentration of the solution increases its viscosity and viscoelastic force, which slows down the jet velocity and reduces the current. However, it also hinders the refinement of the jet, resulting in an increase in the jet diameter, which in turn reduces the jet resistance and increases the current. Besides, increasing the solution concentration improves the charge carrying capacity of the jet, thus increasing the charge density. These factors jointly determine the value of the current. In this experiment, the average current increases with the solution concentration, indicating that the reduction in jet resistance and the increase in charge density have a greater impact on the total current. At a solution concentration of 8%, as there are fewer free charges in the solution, the voltage shows little effect on the average current.

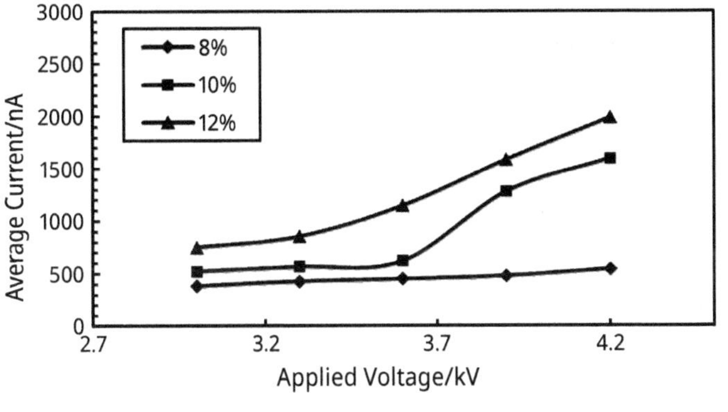

Figure 3.51: Variation of average current with solution concentration (the spinneret-to-collector distance is 6 mm and the solution supply rate is 100 μL/h).

The jet morphology under different PEO solution concentrations is shown in Figure 3.52. At a solution concentration of 8%, the free charge in the solution is small in quantity, and the applied voltage has weak constraining force on the jet, resulting in severe jet whipping and low deposition accuracy. For a solution concentration of 10%, the jet deposits on the collector before the solvent can fully evaporate, leading to the formation of liquid droplets or bead structures on the collector. As the concentration increases to 12%, the solution viscosity increases, and the jet becomes more viscoelastic. The increased viscoelastic force weakens the stretching effect of the applied electric field, resulting in an

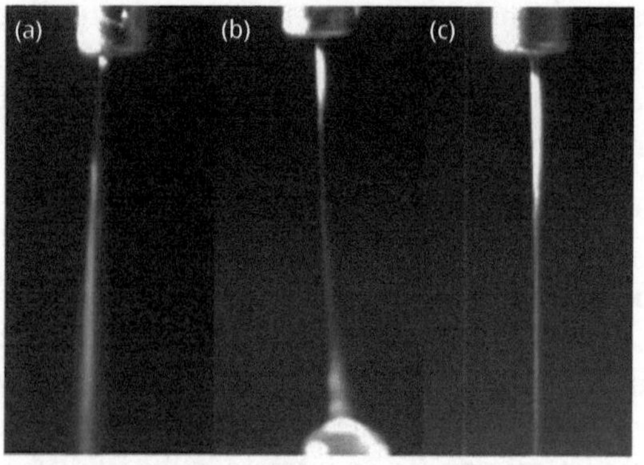

Figure 3.52: Jet morphology at different PEO solution concentrations: (a) 8%; (b) 10%; and (c) 12% (spinneret-to-collector distance is 6 mm, applied voltage is 3.6 kV, and solution feeding rate is 100 μL/h).

increased jet diameter. Before the jet deposits on the collector, the solvent evaporation rate increases, resulting in linear nanofiber structures.

3.3 Closed-loop control system of EDW with current feedback

3.3.1 Development of EDW closed-loop control system based on current feedback

To maintain the stability of the current during the EDW process, we have developed a closed-loop control system based on the current feedback mechanism. In the system, current is used as a real-time feedback detection parameter, and a current feedback controller is built up based on the LabVIEW platform in the host computer. The voltage output by the high-voltage power supply is adjusted according to the change in the current, thereby improving the stability of the jet and the uniformity of the prepared structures.

3.3.1.1 System setup

The EDW system with current feedback control, as shown in Figure 3.53(a), includes a precision syringe pump, a spinneret, a precision current detection module, a host computer control system, and a numerical control high-voltage DC power supply. The precision current detection module, and the host computer control system, as shown in Figure 3.53(b), adopt a current-voltage conversion circuit, through which the input current signal can be converted into an output voltage signal after signal amplification. The resolution is 1 nA, and the system's sampling frequency range is 1–2 kHz.

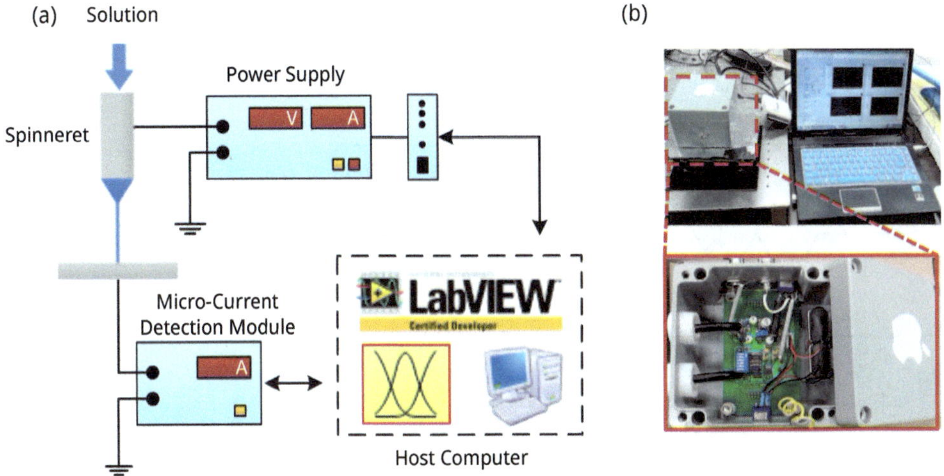

Figure 3.53: EDW system with current feedback control: (a) schematic diagram; and (b) host computer and current detection modules.

3.3.1.2 System control model

It is difficult to establish an accurate mathematical model for EDW because of the jet's high motion speed, low current and various influencing factors. Based on the characteristics of the EDW current, a fuzzy controller is used to control the system. The diagram of the control system is shown in Figure 3.54, which includes a fuzzy controller $G_f(s)$, a power supply module $G_v(s)$, an EDW jet ejection module $G(s)$, and a current detection module $G_c(s)$. Where, I^* is the given target current, ΔI is the current deviation, ΔU is the adjustment of the output voltage, U^* is the given output voltage of the controller, U is the actual output voltage of the power supply, I is the current formed during jet ejection process, and I' is the feedback current after transformation. ΔI is the current deviation defined by $\Delta I = I^* - I$. According to the requirements of fuzzy control, the membership function of ΔI is defined, as shown in Figure 3.55(a). The current deviation is categorized into seven levels to describe the degree of deviation.

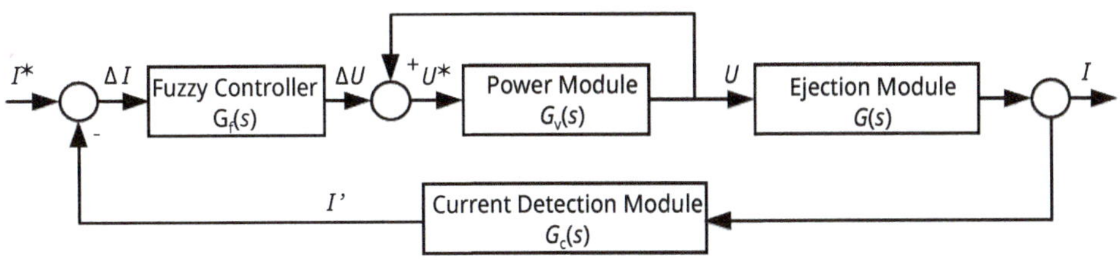

Figure 3.54: The closed-loop control system with current feedback.

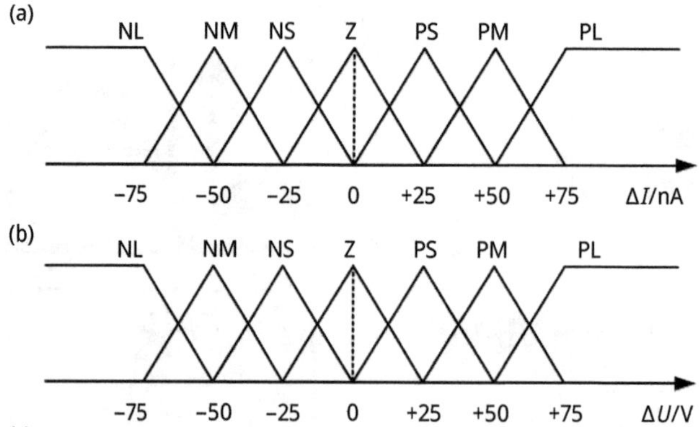

(c)

ΔI	NL	NM	NS	Z	PS	PM	PL
ΔU	PL	PM	PS	Z	NS	NM	NL

Figure 3.55: Current feedback control rule: (a) the membership function of the current deviation; (b) the membership function of voltage adjustment; and (c) basic control rules for voltage adjustment.

The output voltage of the power supply is adjusted according to the current deviation. The adjustment amount ΔU, is also divided into seven levels in accordance with the current deviation ΔI, and the membership function is shown in Figure 3.55(b). Therefore, the given output voltage of the power supply can be described as $U^* = U + \Delta U$. U^* is input into the high-voltage DC power supply to achieve voltage regulation. According to the influence principle of the applied voltage on the current in EDW, the corresponding table of basic control rules for voltage adjustment as shown in Figure 3.55(c) is created to achieve the closed-loop feedback control based on the EDW current.

3.3.2 EDW experiment with current feedback

3.3.2.1 Verification of closed-loop control system with current feedback

The voltage is finely tuned based on current variation to maintain current stability, thereby attaining a stable jet ejection state. Figure 3.56 depicts the current waveform without/with current feedback during EDW process. It shows that when EDW is carried out at a fixed voltage, the jet is disturbed, and the current fluctuates greatly. With the introduction of current feedback, the fluctuation of current can be effectively suppressed by adjusting the voltage, ensuring the stability of jet ejection and the uniformity of deposited structure.

The motion platform moves according to the designed trajectory as shown in Figure 3.57(a) to induce the orderly deposition along the collector's motion trajectory for patterned structure. A capillary spinneret with an inner diameter of 0.21 mm, and the PEO aqueous solution with a concentration of 10% are used to conduct comparative experi-

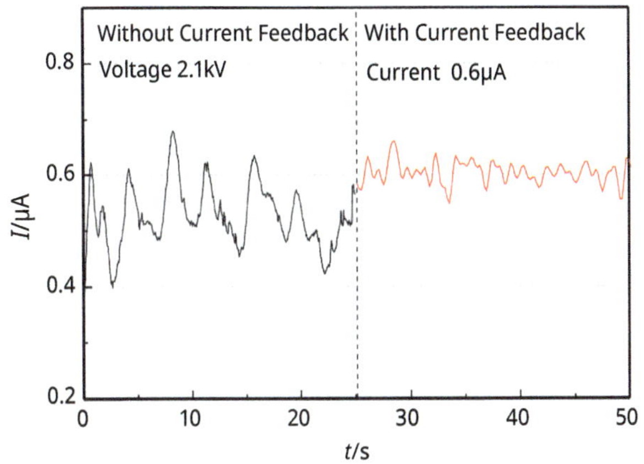

Figure 3.56: Current waveform without/with current feedback in EDW process (the solution concentration is 10%, the inner diameter of the spinneret is 0.21 mm, the spinneret-to-collector distance is 2 mm, the solution supply speed is 100 μL/h, and the collector speed is 100 mm/s).

ments on the EDW system without and with current feedback. The experimental results and data statistics are shown in Figure 3.57(b) and (c). When the voltage is fixed at 2.3 kV, the prepared liquid film structure shows a wide diameter distribution (34.50–54.86 μm), with a standard deviation of 4.73 μm, and a variation coefficient of 0.107. By introducing current closed-loop feedback, and setting the target current value to 0.7 μA, the uniformity of the liquid film diameter is improved, and the distribution range is reduced to 39.86–49.01 μm, the standard deviation and the coefficient of variation are reduced to 2.67 μm and 0.060, respectively.

If the surface of the collector is uneven, the distance from the spinneret to the collector will change, which will affect the space electric field strength and lead to the changes in the jet ejection state (speed, diameter, etc.). After introducing the current feedback, the stability of the jet, and the uniformity of the diameter of the deposition structure can be improved by stabilizing the current. Figure 3.58 compares the deposition of nanofibers at different spinneret-to-collector distances without/with current feedback. In the experiment, a hollow needle with an inner diameter of 0.06 mm is used. The solution is 14% PEO solution, and the solvent is a mixture of deionized water and alcohol in a volume ratio of 3:1. Due to the small inner diameter of the spinneret and the fast evaporation of the solvent, the jet has solidified into solid nanofibers before reaching the collector, and the effect of the collector speed on the nanofiber diameter is negligible. At a voltage of 1.6 kV, the average diameter of nanofibers obtained at different spinneret-to-collector distances is quite different. When the distance is shortened from 2.0 mm to 1.5 mm, the average diameter increases from 0.696 μm to 0.878 μm. In addition, the uniformity of fiber diameter is poor, with variation coefficients of 0.279 and 0.155, respectively. After introducing current feedback, the difference in the average diameter of the nanofibers deposited at different spinneret-to-collector distance re-

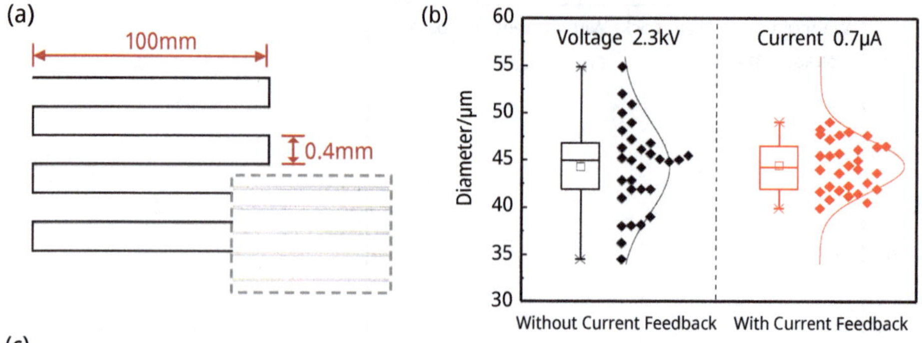

(c)

Control Strategy	Average Diameter/μm	Maximum Diameter/μm	Minimum Diameter/μm	Standard Deviation/μm	Coefficient of Variation
Without Current Feedback (Voltage 2.3kV)	44.26	54.86	34.50	4.73	0.107
With Current Feedback (Voltage 0.7μA)	44.36	49.01	39.86	2.67	0.060

Figure 3.57: EDW liquid film: (a) the collector's motion trajectory; (b) diameter distribution without/with current feedback control; and (c) statistics results of the diameter (solution concentration is 10%, spinneret inner diameter is 0.21 mm, spinneret-to-collector distance is 2 mm, liquid supply rate is 100 μL/h, and collector motion speed is 100 mm/s).

duces when the target current is 0.2 μA. At spinneret-to-collector distances of 2.0 mm and 1.5 mm, the average diameters are 0.734 μm and 0.757 μm, respectively, and the uniformity of fiber diameter is also improved, with the coefficient of variation reducing to 0.060 and 0.135, respectively

The closed-loop control system with current feedback can also ensure the repeatability for direct-writing of nanofibers. As illustrated in Figure 3.59(a), the collector's motion follows a meticulously designed trajectory, enabling the direct-writing of a single nanofiber for multiple times. Post-electrospinning of each nanofiber, the power supply is momentarily disconnected to ensure precision. Experimental results show that current feedback can effectively improve the repeatability of electrospun nanofibers, as shown in Figure 3.59(b) and (c). In the absence of current feedback at a voltage of 1.6 kV, nanofiber diameters exhibit a broad distribution from 0.161 to 2.010 μm, with a standard deviation of 0.518 μm and a coefficient of variation standing at 0.596. Conversely, implementing current feedback with a target current set at 0.2 μA markedly narrows the diameter distribution to a range of 0.670–1.124 μm. This adjustment yields a standard deviation of merely 0.126 μm and reduces the coefficient of variation to a mere 0.139.

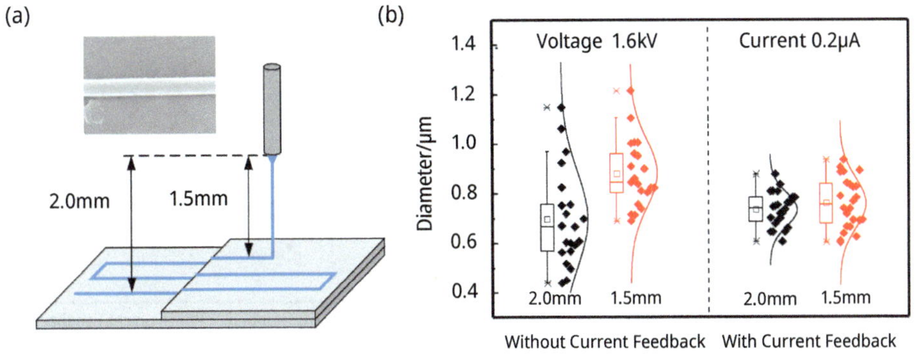

(c)

Control Strategy	Spinneret-to-Collector Distance	Average Diameter/μm	Maximum Diameter/μm	Minimum Diameter/μm	Standard Deviation/μm	Coefficient of Variation
Without Current Feedback (Voltage 1.6kV)	2mm	0.696	1.149	0.439	0.195	0.279
	1.5mm	0.878	1.213	0.688	0.136	0.155
With Current Feedback (Current 0.2μA)	2mm	0.734	0.876	0.605	0.068	0.060
	1.5mm	0.757	0.932	0.599	0.102	0.135

Figure 3.58: Nanofibers spun at different spinneret-to-collector distances: (a) collector structure and its motion trajectory; (b) diameter distribution without/with current feedback control; and (c) statistics results of fiber diameter (solution concentration is 14%, spinneret inner diameter is 0.06 mm, liquid supply rate is 100 μL/h, and collector motion speed is 100 mm/s).

3.3.2.2 Control of structure deposition by EDW with current feedback

3.3.2.2.1 Closed-loop control for electrohydrodynamic direct-writing liquid film structures with current feedback

The experiments are conducted using 8% PVP solution with different control currents, solution supply rates and collector motion speeds. Figure 3.60 shows the diameter distribution of the direct-written liquid films at different currents. Since the collector moving speed is less than the jet deposition speed, the variation in current value has little effect on the diameter. With the use of current feedback, the diameter distribution range is 80–95 μm, which shows certain improvement compared to 75–100 μm of the voltage control mode (Figure 2.36). Figure 3.61 exhibits the diameter distribution at different solution supply rates. With current feedback, the diameter fluctuation range is ±15 μm, while it is ±20 μm when there is no current feedback (Figure 2.37). Figure 3.62 shows the variation in diameter as the collector speed is increased from 5 to 20 mm/s. It clear that with current feedback, the diameter uniformity is improved compared to the constant voltage control (Figure 2.39).

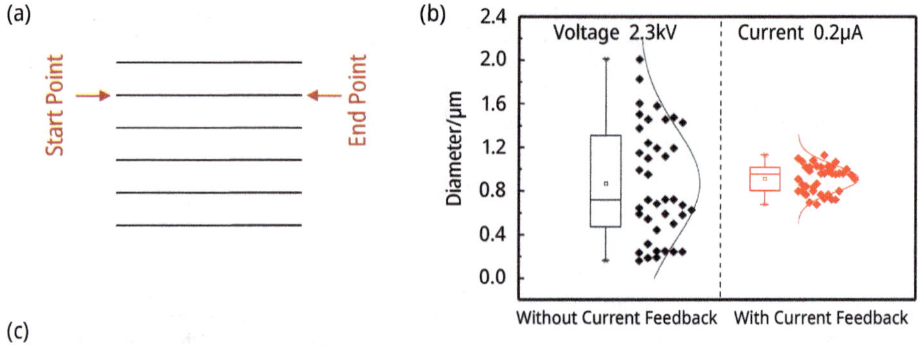

(c)

Control Strategy	Average Diameter/μm	Maximum Diameter/μm	Minimum Diameter/μm	Standard Deviation/μm	Coefficient of Variation
Without Current Feedback (Voltage 1.6kV)	0.869	2.010	0.161	0.518	0.596
With Current Feedback (Current 0.2μA)	0.905	1.124	0.670	0.126	0.139

Figure 3.59: Repeatedly direct-written nanofiber: (a) designed motion trajectory of the collector; (b) diameter distribution without/with current feedback control; and (c) diameter statistics (solution concentration is 14%, inner diameter of the spinneret is 0.06 mm, and spinneret-to-collector distance is 2.0 mm).

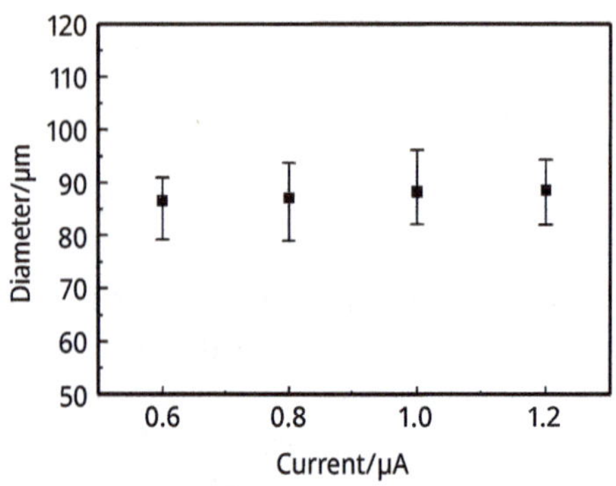

Figure 3.60: Effect of current on the diameter of electrohydrodynamic direct-written structures (an 8% PVP solution was used with an inner spinneret diameter of 0.06 mm, a liquid supply rate of 70 μL/h, a collector speed of 20 mm/s, and a distance of 1.5 mm between the spinneret and the collector).

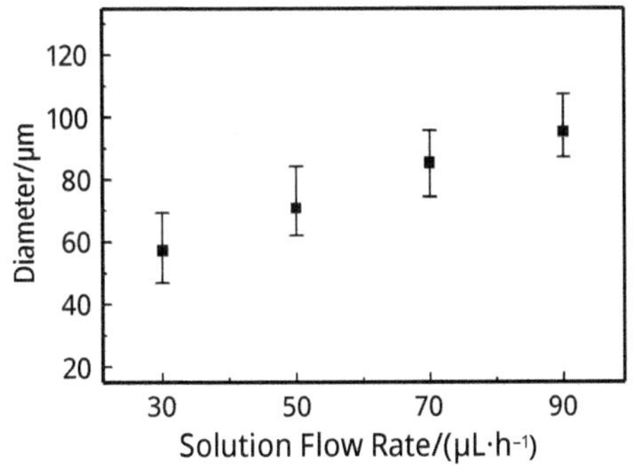

Figure 3.61: Influence of solution supply rate on the diameter of electrohydrodynamic direct-written structure (8% PVP solution, spinneret inner diameter is 0.06 mm, target current 0.8 µA, collector speed is 20 mm/s, and spinneret-to-collector distance is 1.5 mm).

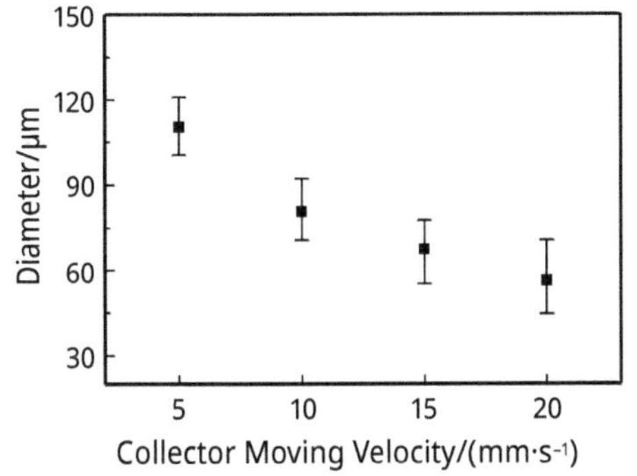

Figure 3.62: Influence of collector moving velocity on the diameter of the electrospun structure (8% PVP solution, spinneret inner diameter is 0.06 mm, target current is 0.8 µA, solution supply speed is 30 µL/h, and spinneret-to-collector distance is 1.5 mm).

3.3.2.2.2 The closed-loop control for electrohydrodynamic direct-writing of nanofibers with current feedback

Figure 3.63 shows the distribution of nanofiber diameters with different currents at a spinneret-to-collector distance of 2.0 mm. During the EDW process, the degree of jet stretching is small, and the diameter of nanofibers mainly depends on the initial jet diameter. Increasing the voltage increases the electric field force acting on the solution, causing the initial diameter of the jet to increase, which eventually leads to an increase in the diameter of the nanofibers.

In the closed-loop control system with current feedback, to maintain a constant current, the voltage must increase with the spinneret-to-collector distance. On the one hand, increasing the spinneret-to-collector distance can improve the degree of jet stretching and refinement, reducing the nanofiber diameter. On the other hand, the initial diameters of the jet, and the diameter of the nanofibers increase with the applied voltage. Figure 3.64 shows the influence of spinneret-to-collector distance on the diameter of the nanofibers at a current of 0.2 μA. The results indicate that the spinneret-to-collector distance has a smaller effect on the diameter of the nanofibers under a constant current.

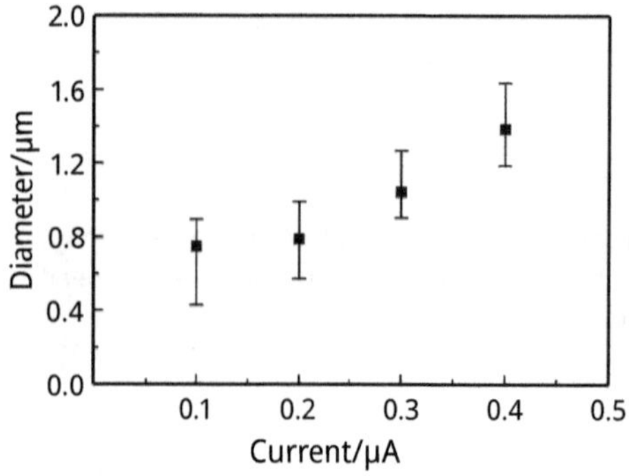

Figure 3.63: Influence of current on nanofiber diameter (14% PEO solution, spinneret inner diameter is 0.06 mm, and spinneret-to-collector distance is 2.0 mm).

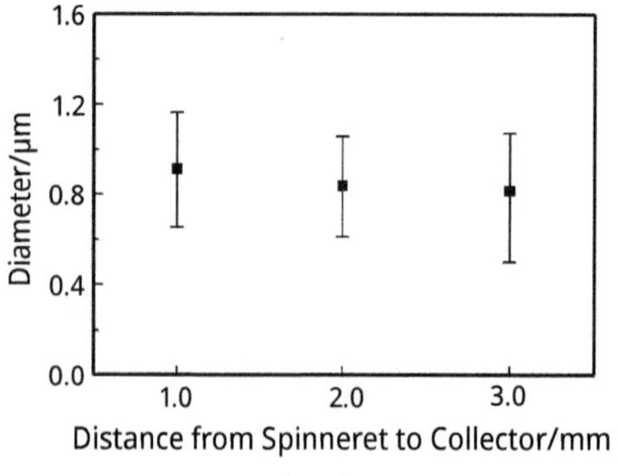

Figure 3.64: Influence of spinneret-to-collector distance on nanofiber diameter (14% PEO solution, spinneret inner diameter is 0.06 mm, target current is 0.2 μA).

References

[1] Bober D B, Chen C H. Pulsating electrohydrodynamic cone-jets: From choked jet to oscillating cone. J Fluid Mech 2011, 689, 552–563.

[2] Yarin A L, Koombhongse S, Reneker D H. Bending instability in electrospinning of nanofibers. J Appl Phys 2001, 89(5), 3018–3026.

[3] Melcher J R, Taylor G I. Electrohydrodynamics: A review of the role of interfacial shear stresses. Annu Rev Fluid Mech 1969, 1(1), 111–146.
[4] Saville D A. Electrohydrodynamics: The Taylor-melcher leaky dielectric model. Annu Rev Fluid Mech 1997, 29(1), 27–64.
[5] Yarin A L, Koombhongse S, Reneker D H. Taylor cone and jetting from liquid droplets in electrospinning of nanofibers. J Appl Phys 2001, 90(9), 4836–4846.
[6] Theron S A, Zussman E, Yarin A L. Experimental investigation of the governing parameters in the electrospinning of polymer solutions. Polymer 2004, 45(6), 2017–2030.
[7] Deitzel J M, Kleinmeyer J D, Harris D E A, Tan N C B. The effect of processing variables on the morphology of electrospun nanofibers and textiles. Polymer 2001, 42(1), 261–272.
[8] Lee C K, Kim S I, Kim S J. The influence of added ionic salt on nanofiber uniformity for electrospinning of electrolyte polymer. Synthetic Met 2005, 154(1–3), 209–212.
[9] Kadomae Y, Amagasa M, Sugimoto M, Taniguchi T, Koyama K. Effect of electric current on beads formation in electrospinning of poly (vinyl alcohol). Int Polym Proc J the Polym Proc Soc 2008, 23(4), 377–384.
[10] Lin K, Chua K N, Christopherson G T, Lim S, Mao H Q. Reducing electrospun nanofiber diameter and variability using cationic amphiphiles. Polymer 2007, 48(21), 6384–6394.

Chapter 4
Precision positioning and patterning deposition of micro-nanostructures

4.1 Study on electrohydrodynamic direct-writing of patterned microstructures

4.1.1 Development of graphic compilation software for electrohydrodynamic direct-writing of patterned microstructures

The graphic compilation software's primary role is to interpret the DXF file created by the CAD software, adjust settings, such as movement, speed, and acceleration, transform 2D graphics into a control program compatible with the precision motion platform, and manage the motion trajectory of the platform.

Parker ACR8020 eight-axis control card is used as the controller of the precision EDW platform. ACRGView control software is used in the host computer, and the recognizable control code is AcroBasic advanced programming language.

4.1.1.1 Graphic compilation software interface

The human-machine program interface of this software is depicted in Figure 4.1, which mainly includes the following components:

(1) Menu bar: Contains function menu items of the software.
(2) Toolbar: Provides shortcuts to software tools.
(3) Graphic display area: Displays graphics given in the DXF documents, allows for setting of graphic starting point and motion direction, and simulates motion path through control codes.
(4) Program box: Displays the AcroBasic program codes automatically generated by the graphic compilation software.
(5) Simple graphic area: For simple graphics like lines, circles, and arcs, graphic parameters can be inputted directly in this area. The graphics will then be displayed in the graphic display area, and control code will be generated automatically.
(6) Prompt message bar: Displays information such as the date and the coordinates of the mouse in the graphic display area.

4.1.1.2 Functions of the graphic compilation software

(1) Importing the DXF file: Importing the designed two-dimensional graphics and displaying them in the graphic display area, as depicted in Figure 4.2.
(2) Setting up the starting point for the platform movement and the motion direction for the graph, as denoted by the small circle in Figure 4.2.

https://doi.org/10.1515/9783110797879-004

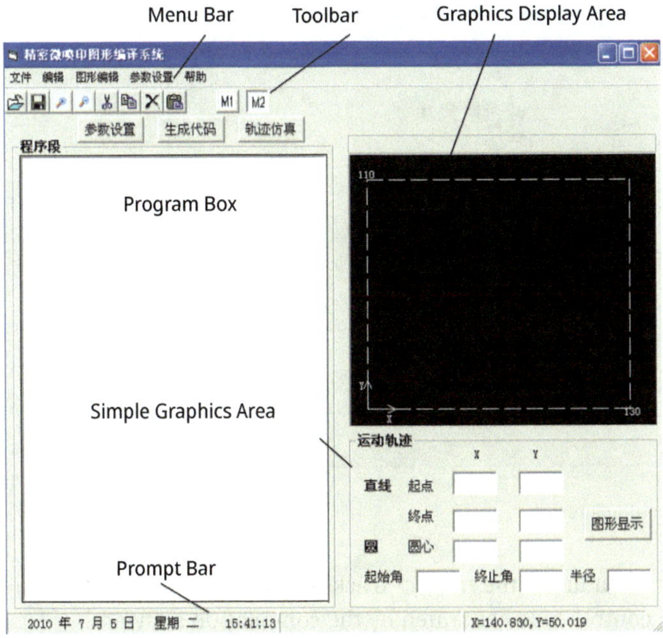

Figure 4.1: User interface of the compilation software.

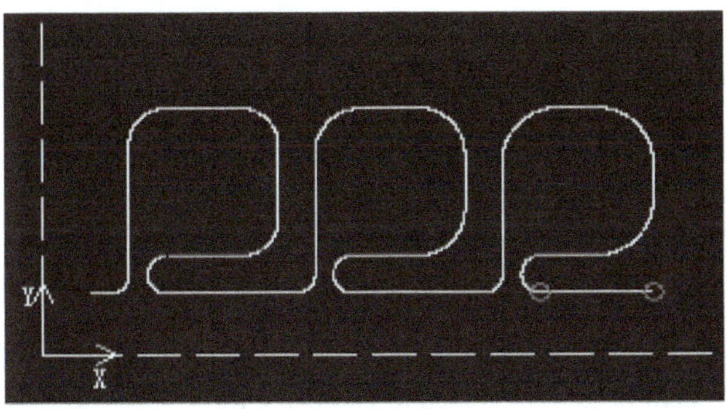

Figure 4.2: Display the designed graphic and determine the starting point and direction of the platform movement.

(3) Adjusting motion parameters setting up parameters such as the movement, speed, and acceleration of the platform, and the spinneret-to-collector distance. The parameter setting interface is shown in Figure 4.3.

(4) Code generation: to automatically generate the codes that can be recognized by the controller according to the designed DXF file and the set motion parameters, display them in the program box.

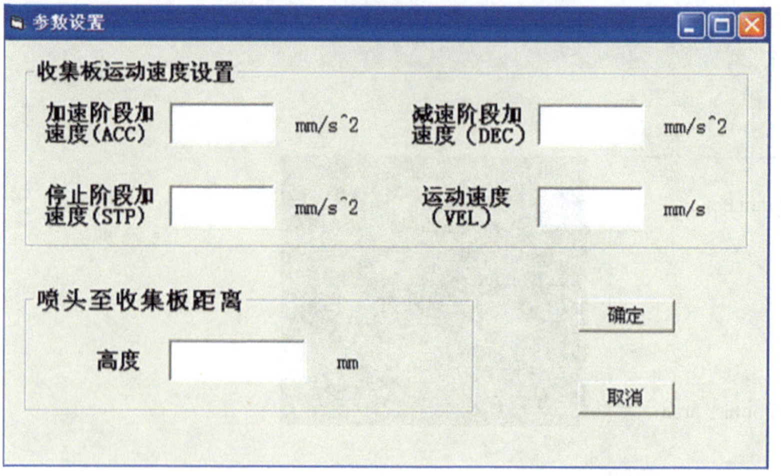

Figure 4.3: Motion parameter setting interface.

(5) Program verification: to simulate the motion track of the collector and verifying the correctness of the control code generated by the compilation software.
(6) Code transmission: transmitting the generated control code to the ACR-View control program to drive the motion platform to move according to the preset track.

4.1.2 Creating patterned microstructures using EDW

Create a design as shown in Figure 4.4(a) and use graphic compilation software to convert it into a control code that can be understood by the Parker controller. Transfer this control code to the controller using ACR-View software, which will then command the motion platform to follow the designed graphic path. This will guide the electrospun nanofibers to deposit in accordance with the motion trajectory of the collector. The collector will pause for 0.3 s at the corners of the pattern. The resulting rectangular micro-nanocoil structures on the silicon substrate are shown in Figure 4.4(b).

The slower movement speed of the collector allows spiral structures to persist in the micro-pattern. By strategically pausing the collector at the corners, distortion in the directly written micro-pattern can be overcome, enabling the preparation of complex patterned micro-nanostructures. Additionally, more nanofibers are deposited at the corners of the pattern in Figure 4.4(b).

Create a square-waveform pattern with varying amplitudes, as illustrated in Figure 4.5. Use this pattern to guide the collector in moving along a specific trajectory to direct the nanofibers in depositing and forming a micro-pattern, as shown in Figure 4.6. The collector should move at a speed of 0.35 m/s, which is faster than the matching speed, to prevent a helical structure from forming in the straight-line segments of the

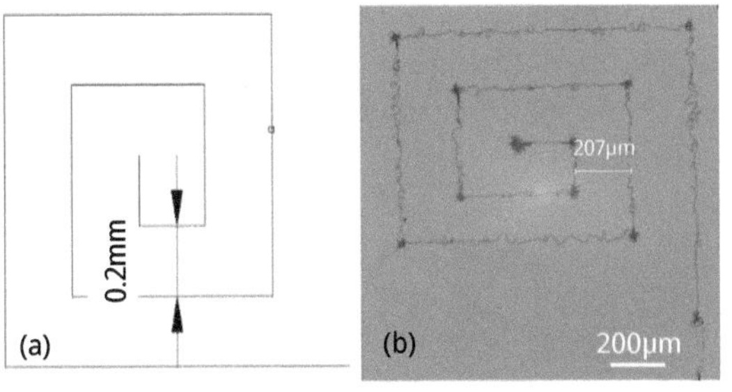

Figure 4.4: Electrohydrodynamic direct-writing of rectangular coil micro-nanostructures: (a) designed rectangular coil pattern and (b) electrohydrodynamic direct-written pattern (spinneret diameter, solution concentration, applied voltage, collector speed, and spinneret-to-collector distance are 40 μm, 18%, 1.3 kV, 0.20 m/s, and 1 mm, respectively, silicon substrate).

micro-pattern. During the EDW process, the collector follows the designated trajectory, resulting in micro-nanostructure patterns that are symmetrical to the designed pattern.

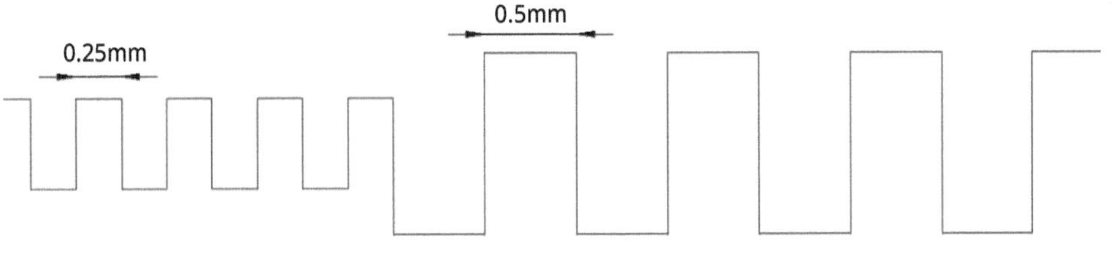

Figure 4.5: Designed square-waveform pattern with varying amplitudes.

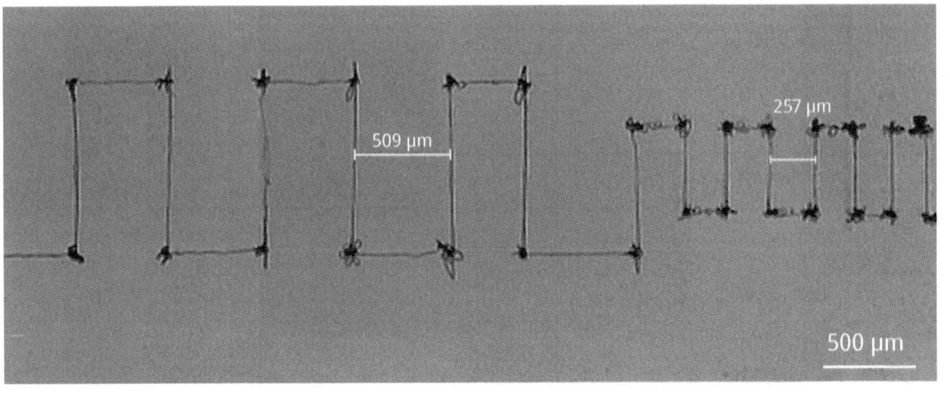

Figure 4.6: Electrohydrodynamic direct-written square-wave micro-nanostructures with different amplitudes (spinneret diameter is 40 μm, solution concentration is 18%, applied voltage is 1.3 kV, and spinneret-to-collector distance is 1 mm, silicon substrate).

Reducing the positioning error of direct-written micro-patterned structures in electrospinning is a crucial factor in controlling of EDW. Using the square-wave pattern displayed in Figure 4.6 as the study case, the impact of collector moving speed, spinneret-to-collector distance, applied voltage, and collector moving distance on the positioning error of electrospun patterns is examined. The results are illustrated in Figures 4.7–4.10.

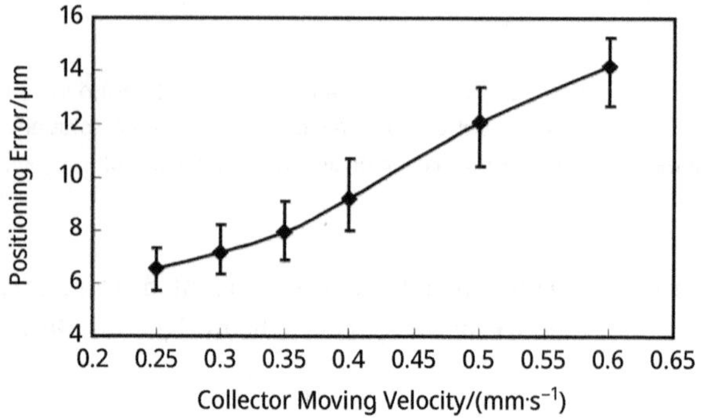

Figure 4.7: Impact of the collector moving velocity on the positioning error (the spinneret diameter, solution concentration, applied voltage, spinneret-to-collector distance, and moving distance of the collector are 40 μm, 18%, 1.3 kV, 0.5 mm, and 500 μm, respectively, silicon substrate).

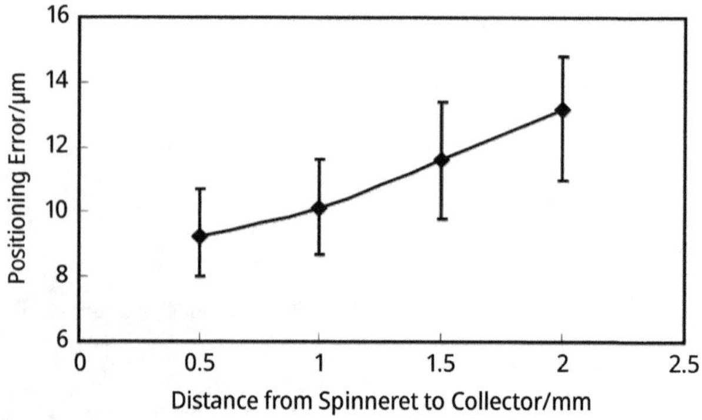

Figure 4.8: Effect of the spinneret-to-collector distance on the positioning error (the spinneret diameter, solution concentration, applied voltage, collector speed, and moving distance of the collector are 40 μm, 18%, 1.3 kV, 0.20 m/s, and 500 μm, respectively, silicon substrate).

The collector moving speed is the most crucial factor affecting the positioning accuracy of the micro-nanostructures fabricated by EDW. It primarily impacts the movement precision of the platform and the viscoelastic force within the jet flow. As depicted in Figure 4.7, the positioning error increases with the collector moving speed. Additionally, widening the gap between the spinneret and the collector amplifies the

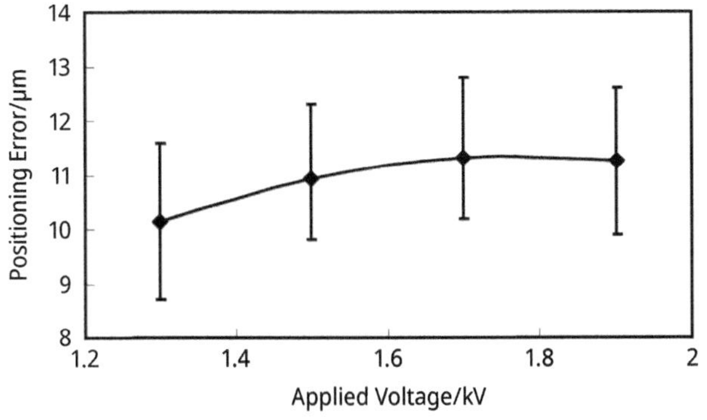

Figure 4.9: Effect of the applied voltage on the positioning error (the spinneret diameter, solution concentration, spinneret-to-collector distance, collector speed, and moving distance of the collector are 40 μm, 18%, 1 mm, 0.40 m/s, and 500 μm, respectively, silicon substrate).

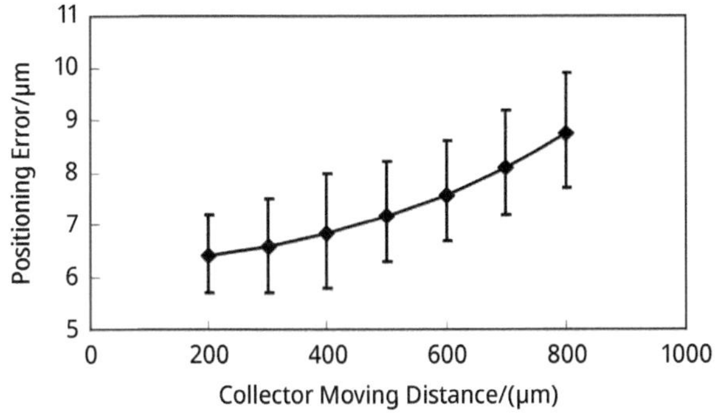

Figure 4.10: Impact of the collector moving distance on the positioning error (the spinneret diameter, solution concentration, spinneret-to-collector distance, collector velocity, and applied voltage are 40 μm, 18%, 0.5 mm, 0.30 m/s, and 1.3 kV, respectively, silicon substrate).

impact of unstable factors like the charge repulsion force, leading to a greater extent of jet spiral whipping and ultimately decreasing the positioning accuracy. This trend is illustrated in Figure 4.8, where the positioning error increases with the distance between the spinneret and the collector.

As the applied voltage increases, the charge density on the jet surface, the repulsion force of the charged jet during movement and deposition, and the amplitude of the jet spiral whip all increase, leading to reduced positioning accuracy. Figure 4.9 illustrates how positioning error rises with higher applied voltage. Additionally, the accumulation of error causes positioning error in micro-nanostructures to increase with the collector moving distance, as depicted in Figure 4.10.

Using graphic compilation software, the designed pattern is automatically converted into control code recognizable by the controller. This drives the collector to move along the predetermined trajectory, guiding the direct-written nanofibers to deposit according to the designed pattern, thereby achieving the precise fabrication of patterned electrospun direct-written micro-nanostructures.

4.1.3 Electrohydrodynamic direct-writing patterns on collectors with three-dimensional microstructures

Three-dimensional (3D) micro-patterned structures are fabricated on single-crystal silicon substrates via wet etching. These structured substrates are used as collectors to investigate fiber deposition behavior on microstructured surfaces and to explore positioning control strategies in electrostatic direct writing (EDW) assisted by microstructure guidance.

4.1.3.1 Influence of microstructures on nanofibers

The space distribution of the electric field close to the collector surface greatly influences how the charged jet and micro-nanostructures are deposited [1]. When a silicon substrate with microstructures is used as the collector, it changes the electric field distribution, causing the nanofibers to be deposited in particular areas due to variations in electric field strength. Simulation results shown in Figure 4.11 indicate that the electric field strength is strongest at the surface of the three-dimensional microstructure on the collector, especially at the edge where the electric field lines converge toward the surface.

When there are no microstructures present on the silicon substrate, the electric field force guides the jet to deposit at the point where the spinneret axis extension intersects the collector surface. If the electric field force is not uniform in certain areas, the charged jet may bend and change its deposition trajectory. Figure 4.12 exhibits that the convex structure on the collector alters the electric field distribution, causing the charged jet to deflect toward the convex portion. Experimental findings suggest that the convex microstructure influences the deposition of electrospun nanofibers.

(a) (b)

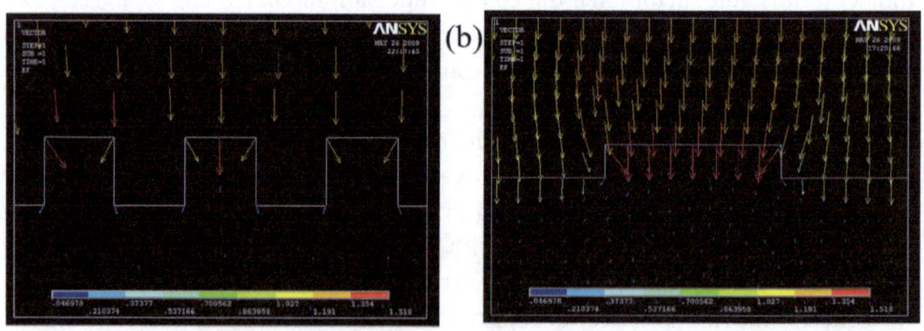

Figure 4.11: Distribution of the space electric field above the silicon microstructure: (a) electric field distribution over the microstructure array and (b) local electric field distribution over a single microstructure.

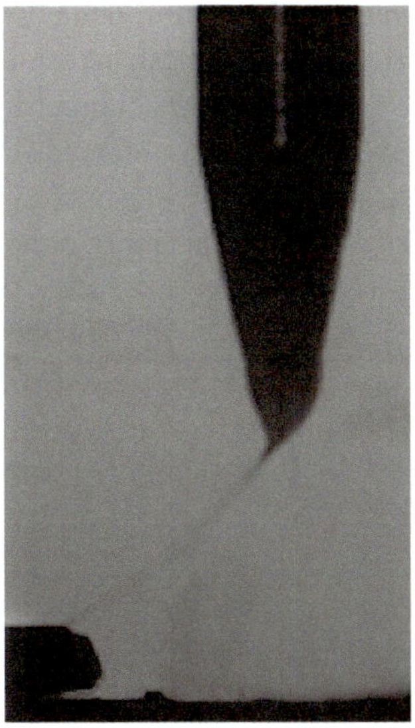

Figure 4.12: The charged jet is induced to bend by the convex structure on the collector (spinneret-to-collector distance is 1 mm, applied voltage is 1.7 kV, solution concentration is 16%, spinneret diameter is 40 μm, and height of the convex structure is 250 μm).

At a low collector moving speed, the jet moves smoothly, and the nanofibers gather on the silicon substrate's microstructure surface under the influence of the local electric field. In Figure 4.13, the strip microstructure's width ranges from 3 to 20 μm. With a 20% solution concentration, the nanofibers assemble on the opposite side of the collector's motion direction due to the jet's viscoelastic force, as depicted in Figure 4.13(a) and (b). At a 16% solution concentration, the nanofibers on the microstructure surface form solid fiber clusters, as shown in Figure 4.13(c). In all three scenarios, the electrospun nanofibers span the strip-shaped microstructures in a suspended, tensioned state, forming near-linear bridges.

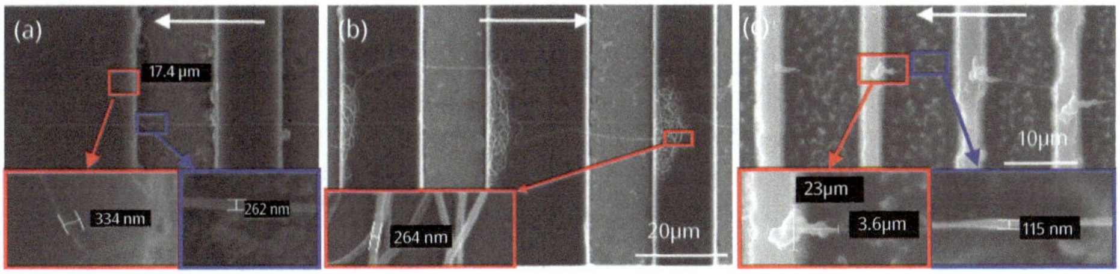

Figure 4.13: Nanofibers deposited on the surface of strip-shaped microstructure at low collector speeds: (a) collector speed is 0.20 m/s and solution concentration is 20%; (b) collector speed is 0.15 m/s and solution concentration is 20%; and (c) collector speed is 0.15 m/s and solution concentration is 16% (spinneret diameter is 70 μm, applied voltage is 1.2 kV, spinneret-to-collector is 1 mm, and microstructure height is 5 μm. The arrows at the top of the figure indicate the direction in which the collector is moving).

Larger microstructures are employed to further investigate the deposition behavior of direct-written nanofibers. In Figure 4.14, the width of the silicon microstructure is 200 μm. At a collector moving speed of 0.25 m/s, straight nanofibers are obtained in regions outside the microstructure, but straight nanofibers could not be obtained on the upper surface of the microstructure, as shown in Figure 4.14(b). The edges of the silicon microstructure exhibit higher electric field intensity, which guides more nanofibers to accumulate in these areas, as illustrated in Figure 4.14(c).

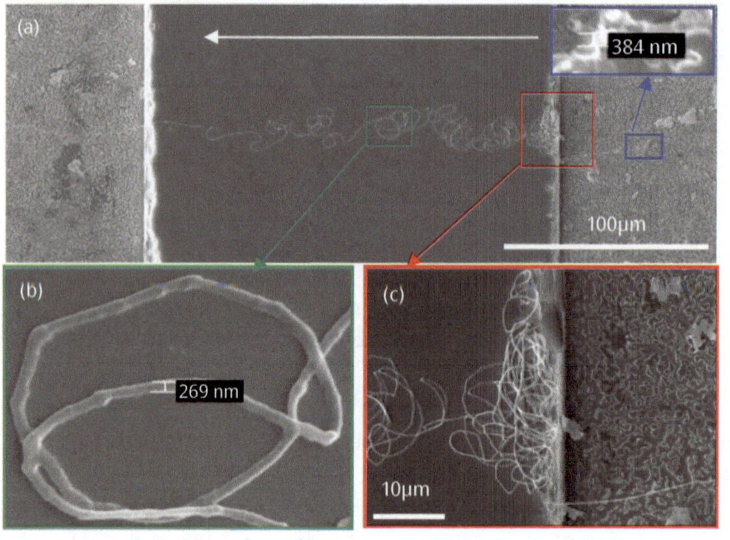

Figure 4.14: Nanofibers deposited on a silicon substrate with microstructures: (a) nanofibers deposited on a 200-μm-wide microstructure at a lower collector speed; (b) nanofibers deposited on the surface of the microstructures; and (c) nanofibers stacked at the edge of the microstructure (solution concentration is 20%, spinneret diameter is 40 μm, applied voltage is 1.7 kV, collector speed is 0.25 m/s, spinneret-to-collector distance is 1 mm, and microstructure height is 5 μm. The arrow in the figure indicates the direction of collector movement).

We further explore how nanofibers deposit on the silicon square microstructure with 5- μm height by adjusting the speed of the collector, as illustrated in Figure 4.15. When the collector moves at a speed of 0.10 m/s, we observe that straight nanofibers are unable to form on the surface of the microstructure or other areas of the silicon substrate (Figure 4.15(a)). By increasing the collector moving speed to 0.45 m/s, we are able to achieve straight nanofibers across the entire collector. When the collector moving speed is set at 0.25 m/s, straight nanofibers can be obtained in the area outside the microstructure , whereas aligned deposition is not observed on the microstructure surface (Figure 4.15(b)). This discrepancy is due to the higher electric field strength above the microstructure compared to other areas, resulting in faster movement of nanofibers during deposition and an increased number of nanofibers deposited on the collector surface per unit time. Consequently, a higher collector moving speed is necessary to achieve straight nanofibers on the microstructure surface.

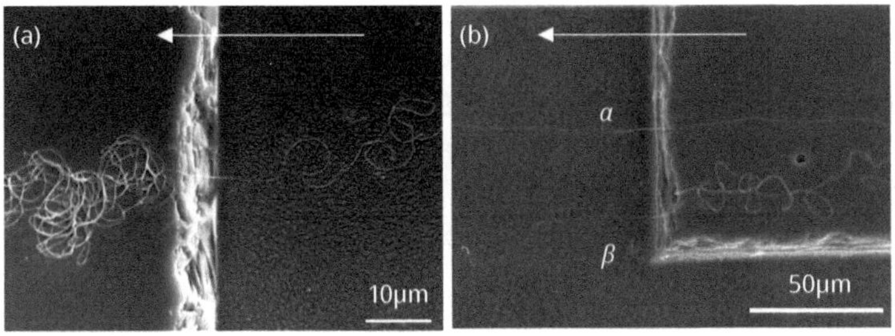

Figure 4.15: Nanofibers deposited on microstructured substrate at different collector movement speeds: (a) the collector moves at a speed of 0.10 m/s; (b) direct-written nanofibers α and β at collector speeds of 0.45 and 0.25 m/s, respectively (solution concentration is 18%, spinneret diameter is 40 μm, applied voltage is 1.7 kV, spinneret-to-collector distance is 1 mm, and microstructure height is 5 μm. The arrow indicates the direction of collector movement).

The microstructure on the silicon substrate alters the distribution of the electric field in the surrounding space above the collector, boosting the strength of the electric field on the surface of the microstructure. This shift in electric field distribution impacts not only where electrospun nanofibers are deposited but also influences the behavior of the charged jet and nanofibers during deposition. It enhances the swirling motion of the charged jet above the microstructure and increases the matching speed of collector.

4.1.3.2 Positioning deposition of nanofibers on the microstructure array

As shown in Figure 4.16(a), a circular microstructure array with a diameter of 5 μm is fabricated on a silicon substrate. By carefully adjusting the positions of the spinneret and collector under a microscope, we can achieve precise positioning and alignment of the spinneret and microarray. This allows to adjust the deposition position and trajectory of the jet on the collector. The motion of the jet and deposition processes is illustrated in Figure 4.16(b) and (c).

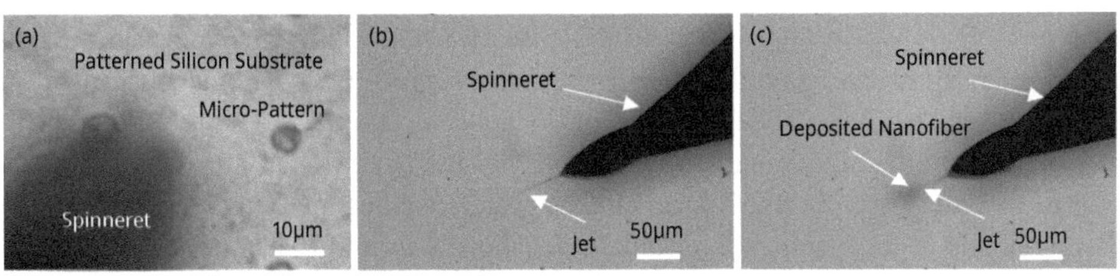

Figure 4.16: Positioning of electrospinning nozzle and jet ejection process: (a) alignment and positioning of the nozzle on a silicon substrate with microstructures. The diameter of the circular micro-structure is 5 μm; (b) jet ejection; and (c) jet with deposited nanofibers.

To investigate the influence of the microstructure array on the deposition path of EDW nanofibers, experiments were conducted in which two nanofibers were deposited along the same trajectory at different collector speeds. The results, shown in Figure 4.17, indicate that the two nanofibers, labeled δ and γ, follow different paths. Nanofiber δ is deposited with the collector moving at 0.15 m/s, while nanofiber γ is deposited at 0.35 m/s. At a low collector speed, nanofiber δ remains in a relaxed state and is easily influenced by the local electric field near the microstructure surface, resulting in a deviated deposition path. In contrast, at a higher collector speed, the nanofiber maintains a straight trajectory, remaining unaffected by external disturbances.

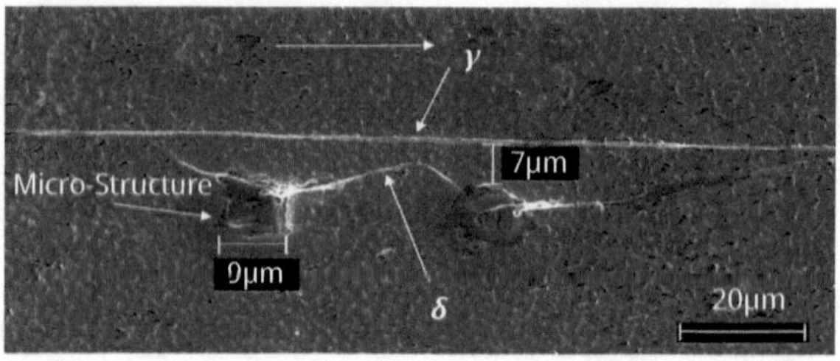

Figure 4.17: The impact of microstructures on the deposition trajectory of nanofiber. The collection has the same movement trajectory. The speeds of the collection are 0.15 (nanofiber δ) and 0.35 m/s (nanofiber γ), respectively. The microstructure measured 9 μm in width and 3 μm in height. The arrow at the top of the figure indicates the movement direction of the collector.

Figure 4.18 shows the direct-written nanofibers precisely deposited on micro-pillar structures. The microstructures in Figure 4.18(a) and (c) are 5 μm tall, while the one in Figure 4.18(b) is 2 μm tall. The nanofibers in Figure 4.18(a) and (c) are positioned between two micro-pillar structures that are approximately 10 μm in diameter and spaced 27 μm apart. As for Figure 4.18(b), the nanofiber is bridged to a micro-stage structure array with a diameter of just 1.6 μm, demonstrating an impressive positioning accuracy of less than 2 μm.

At low collector speeds, the jet is easily influenced by the external electric field force and its own viscoelastic force, causing nanofibers to accumulate on the microstructure, as shown in Figure 4.19. In the case of a 10-μm diameter and 5-μm height micro-pillar structure, nanofibers gather on the surface under the electric field, as shown in Figure 4.19(a). The low speed of the collector causes the nanofibers to bend and deposit on the silicon substrate rather than being stretched and suspended between the two micro-pillars. Figure 4.19(b) and (c) displays the outcomes for a 1.5-μm diameter and 2-μm height pillar structure, where nanofibers cluster on the microstructure and its surroundings.

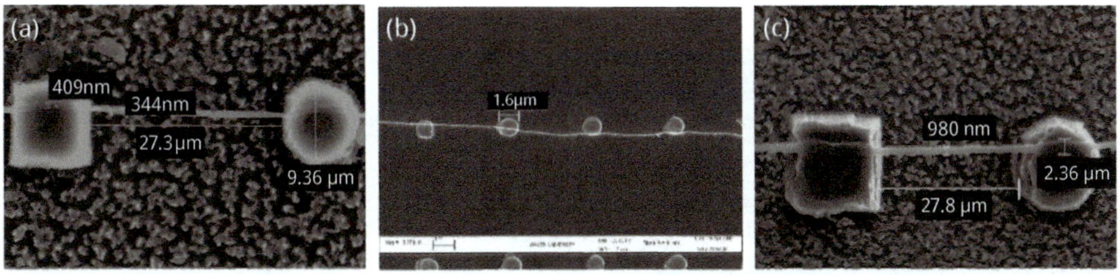

Figure 4.18: Electrospun nanofibers precisely deposited on the micro-pillar structures in three different configurations: (a) and (c) nanofibers positioned between two micro-pillar structures and (b) a nanofiber covering the micro-pillar array. The parameters used for this process include a collector movement speed of 0.35 m/s, a spinneret diameter of 40 μm, a spinneret-to-collector distance of 1 mm, a solution concentration of 18%, and an applied voltage of 1.7 kV for configurations (a) and (b). For configuration (c), the parameters are a collector movement speed of 0.50 m/s, a spinneret diameter of 232 μm, a spinneret-to-collector distance of 2 mm, a solution concentration of 22%, and an applied voltage of 2.8 kV.

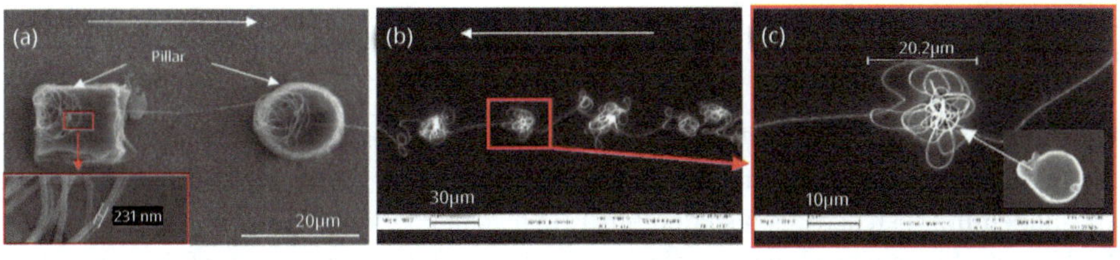

Figure 4.19: Electrospun nanofibers bridge micro-pillars and accumulate at low collector speeds: (a) diameter of the micro-pillar structure is 10 μm; (b) diameter of the micro-pillar structure is 1.5 μm; and (c) local amplification of a nanofiber gathered on the micro-pillar structure and its vicinity. The arrow on the top of the figure indicates the direction in which the collector moves (collector speed is 0.15 m/s, spinneret diameter is 40 μm, spinneret-to-collector distance is 1 mm, solution concentration is 18%, and applied voltage is 1.7 kV).

4.1.3.3 The guiding effect of strip microstructure on the deposition of nanofibers

In the experiment, a silicon substrate with strip-shaped microstructures is used as the collector to study the guiding effect of these microstructures on the deposition of electrospun direct-written nanofibers, thereby enabling controlled, microstructure-assisted deposition.

The electric field distribution above the collector is altered by the silicon microstructure. EDW nanofibers being deposited on the upper surface of the strip-shaped microstructure is exhibited in Figure 4.20. The collector moving speeds in Figure 4.20(a) and (b) are 0.20 and 0.45 m/s, respectively. The width of the strip microstructure in Figure 4.20(a) is 5 μm. The nanofibers only deposit on the upper surface of the strip microstructure due to the space electric field constraint, resulting in a much smaller deposition area compared to the silicon substrate without microstructures. Straight nanofibers can also be achieved on the upper surface of the strip microstructure in Figure 4.20(b) when the collector moving speed exceeds the matching speed.

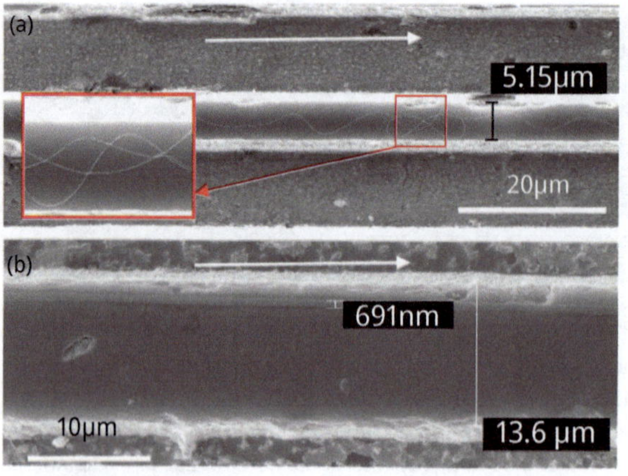

Figure 4.20: Nanofibers deposited on strip microstructures, with the collector movement speeds of (a) 0.20 and (b) 0.45 m/s (spinneret diameter is 40 μm, spinneret-to-collector distance is 1 mm, solution concentration is 18%, applied voltage is 1.7 kV, and microstructure height is 5 μm. The arrows in that figure indicate the motion direction of the collector).

We use a silicon substrate with arc-shaped microstructures as the collector to verify the inducing effect of strip-shaped microstructures on the deposition of nanofibers. The arc-shaped microstructures also demonstrate a strong guiding effect on the nanofibers. As the collector moves along a straight path, the deposition trajectory of the nanofibers adapts to the curvature of the microstructure, as depicted in Figure 4.21. As illustrated in Figure 4.22, the arc-shaped microstructures with large chord length induce positioning deposition of nanofibers. In Figure 4.22(a) and (b), the chord lengths of the arc-shaped microstructures are 64 and 86 μm, respectively, with widths of 5 and 1 μm. In both cases, nanofibers accurately adhere to the microstructure surfaces. Additionally, the strip microstructure on the silicon substrate played a significant role in the EDW process of micron film deposition. Figure 4.23 displays the micron film deposited on the upper surface of the arc microstructure, with chord lengths of 184 and 107 μm and a width of 5 μm.

To enhance the precision of EDW micro-nanostructures, incorporating a stripe microstructure can achieve positioning accuracy of less than 1 μm. Results from experiments demonstrate that EDW technology is compatible with micro-silicon manufacturing processes, paving the way for its integrated use in micro-silicon systems.

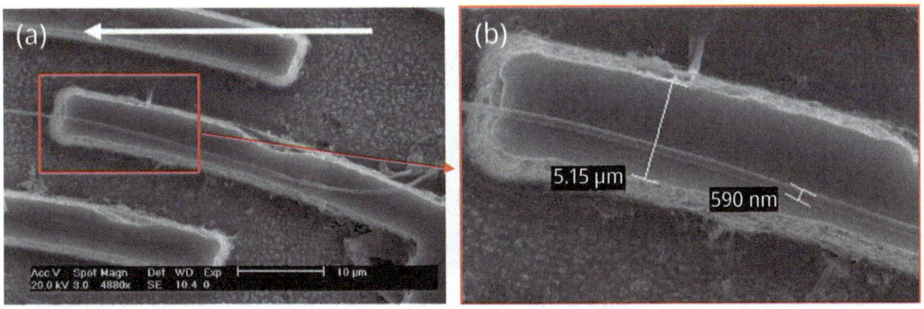

Figure 4.21: Deposition of nanofibers induced by arc-shaped microstructure: (a) nanofibers deposited on the surface of the arc-shaped microstructure and (b) local magnification of nanofibers deposited on the surface of the microstructure (collector speed is 0.30 m/s, spinneret diameter is 40 μm, spinneret-to-collector distance is 1 mm, solution concentration is 18%, applied voltage is 1.7 kV, and microstructure height is 5 μm. The arrow indicates the collector movement direction).

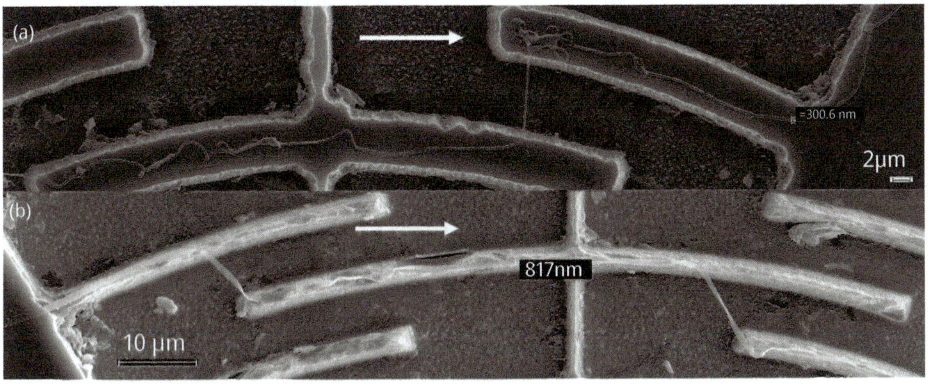

Figure 4.22: Deposition of nanofibers induced by large chord-length arc-shaped microstructure: (a) microstructure width is 5 μm and collector speed is 0.25 m/s and (b) microstructure width is 1 μm and collector speed is 0.30 m/s (at spinneret diameter of 40 μm, spinneret-to-collector distance of 1 mm, solution concentration of 18%, applied voltage of 1.7 kV, and microstructure height of 5 μm. The arrow indicates the movement direction of the collector).

4.2 Deposition and control of stretchable jet

4.2.1 Motion model of stretchable jet

Figure 4.24 shows the jet ejection morphology at varying collector moving speeds. When the speed is slow (5 and 10 mm/s), the jet travels vertically or nearly vertically, with spiral whipping occurring as it approaches the collector, as depicted in Figure 4.24(a) and (b). As the collector moving speed increases, the transverse tensile force between the jet and the collector also increases. When the collector moves at 20 mm/s, the electric field

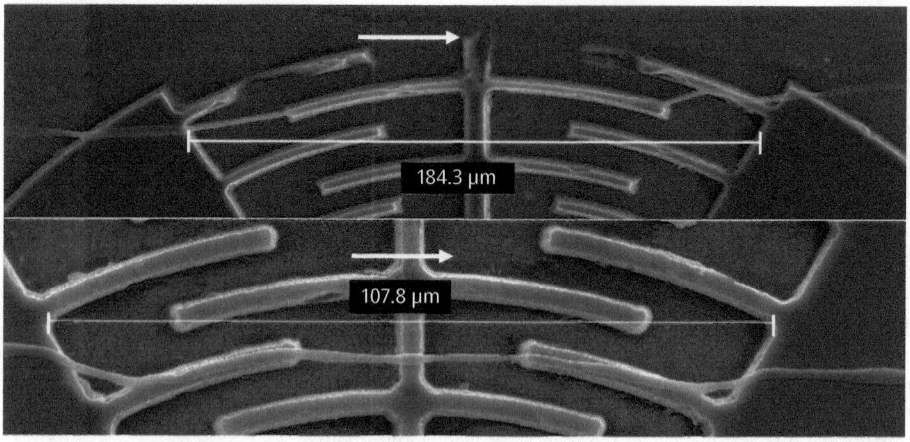

Figure 4.23: Deposition of micron thin films induced by large chord-length arc-shaped microstructure: (a) chord length of 184 μm and (b) chord length of 107 μm (spinneret inner diameter, applied voltage, collector movement speed, solution feeding rate, spinneret-to-collector distance, and microstructure height are 232 μm, 2.8 kV, 1.20 m/s, 5 μL/h, 2 mm, and 5 μm, respectively. The arrow shows the direction of the collector movement).

force and gravity on the jet balance out with the transverse tensile force, causing the jet to deposit directly downward on the collector, as shown in Figure 4.24(c). At a collector moving speed of 40 mm/s, the deposition point of the jet lags behind the position directly beneath the spinneret, resulting in a deflection angle relative to the spinneret's axis, as shown in Figure 4.24(d). As the collector moving speed increases further, the deflection angle of the jet increases, promoting the stretching and thinning of the fiber, as illustrated in Figure 4.24(e).

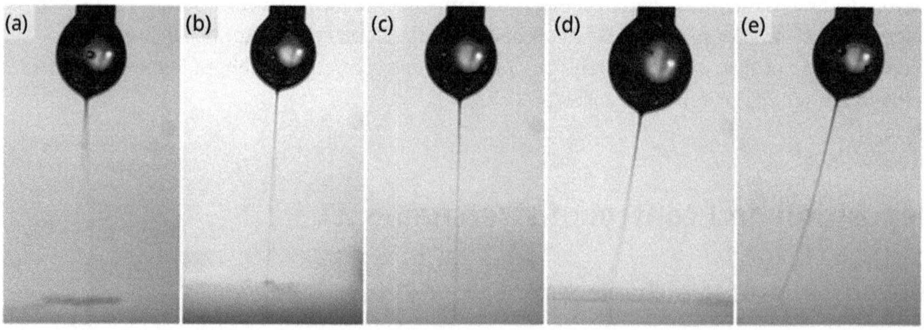

Figure 4.24: Jet ejection morphology at collector speeds of (a) 5; (b) 10; (c) 20; (d) 40; and (e) 60 mm/s (10% solution concentration, spinneret-to-collector distance, spinneret inner diameter, applied voltage, and solution feeding rate are 2 mm, 0.11 mm, 1 kV, and 30 μL/h, respectively).

The jet deposition speed v_j and the collector moving speed v_s jointly determine the ejection and deposition morphology of the jet. Figure 4.25 illustrates how the jet morphology is influenced by the speed of the collector. When $v_s < v_j$, the charged jet moves freely and is not heavily impacted by the collector moving speed. The deposition point of the jet is below the spinneret and forms a curved wave-like fiber structure along the direction of the collector's movement, resulting in a wide deposition range. When the collector moving speed matches the jet deposition speed ($v_s = v_j$), the charged jet deposits straight downward, creating a stable straight fiber. However, when $v_s > v_j$, the jet is stretched, the deflection angle increases, and the jet deposition point is affected by the stretching collector. The jet deposits in the direction of the collector's movement, resulting in a straight fiber structure, but deposition accuracy may be lost at the corners. At the spinneret-to-collector distance of 2 mm and the applied voltage of 1.5 kV, the deposition velocity of the jet is about 20 mm/s.

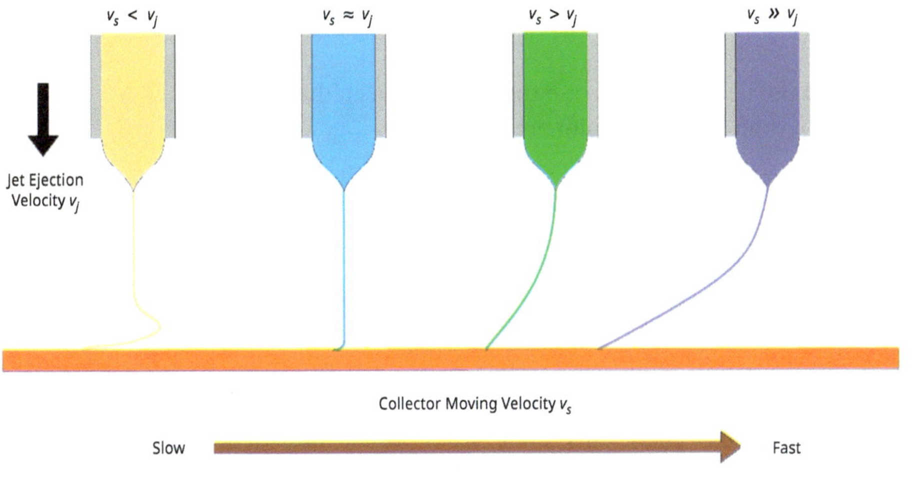

Figure 4.25: Illustration of relationship between jet ejection morphology and collector movement speed.

To better explore the motion and deposition behaviors of the jet when the collector moving speed exceeds the jet deposition speed, the model shown in Figure 4.26 is employed. Let v_s be the collector moving speed, v_j be the deposition speed of the charged jet, v be the velocity of the charged jet, φ be the deflection angle of the jet, Δx be the deviation of the jet, and H be the distance from the spinneret to the collector.

Then, the speed of charged jet can be calculated by

$$v = \sqrt{(v_s - v_j)^2 + v_j^2} \tag{4.1}$$

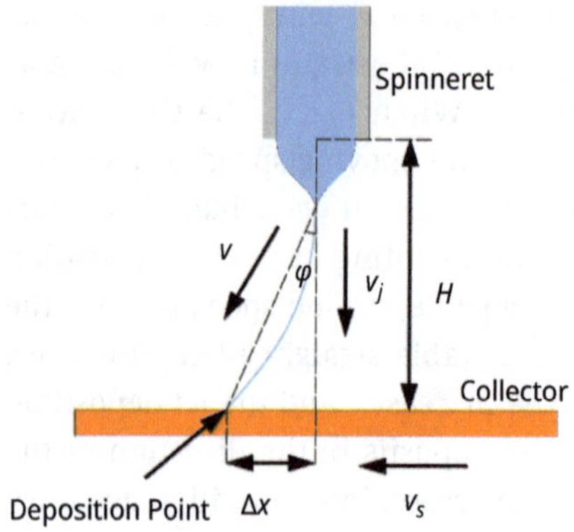

Figure 4.26: Illustration for the stretching effect of the moving collector on the charged jet.

And the angle φ can be given as

$$\tan\varphi = \frac{v_s - v_j}{v_j} \tag{4.2}$$

As depicted in Figure 4.26, φ is also the angle between the straight trajectory of the jet and the extension line of the spinneret. So

$$\tan\varphi = \frac{\Delta x}{H} \tag{4.3}$$

Therefore, the relationship between the deviation of the charged jet and the velocity of the collector can be defined as

$$\frac{\Delta x}{H} = \frac{v_s - v_j}{v_j} \tag{4.4}$$

That is, the deviation of the charged jet Δx is proportional to the velocity difference $v_s - v_j$. In the experiments, when the process parameters remain constant, the jet deposition speed is fixed, and the jet deviation can be reduced by adjusting the collector moving speed.

When the collector's moving speed exceeds the jet's deposition rate, the jet undergoes additional stretching, resulting in a lag in the deposition process. This lag significantly affects the precision of jet placement, especially at pattern corners. The duration of this lag can be represented by the equation:

$$t = \frac{\Delta x}{v_0} \tag{4.5}$$

It follows from Eqs. (4.4) and (4.5) that

$$t = \frac{(v_s - v_j)H}{v_j^2} \tag{4.6}$$

Provided that the jet deposition speed remains unchanged, an increase in the collector's movement speed leads to a prolonged time lag. This delay also extends the duration for the charged jet to revert to its vertical ejection state. Consequently, investigating the synchronization between the collector's movement speed and the jet deposition speed is crucial for enhancing the positioning precision of jet deposition.

The research findings highlight that the behavior of the jet varies with the moving speed of the collector. At lower speeds, the jet exhibits whipping, whereas at higher speeds, a hysteresis effect emerges. This effect leads to positional inaccuracies in the deposition of nanofibers, particularly at the corners of complex patterns. To achieve precision deposition in the electrohydrodynamic direct-writing of complex patterns, it is imperative to fine-tune the collector's movement speed in relation to the jet deposition speed.

4.2.2 Controlling of stretchable jet deposition

Upon ejection from the apex of the Taylor cone, the jet initially moves a brief linear path before entering the whipping instability phase. Figure 4.27(a) illustrates that at a spinneret-to-collector distance of 30 mm, the jet advances freely toward the collector, exhibiting significant whipping amplitude. Consequently, the unstable jet results in nanofiber belts with a random distribution along the collector's trajectory, as depicted in Figure 4.27(b). The whipping amplitude diminishes as the spinneret-to-collector distance decreases. At a distance of 10 mm, the jet's motion stability enhances, yielding a more condensed fiber band, as observed in Figure 4.27(c) and (d). Further reduction of the spinneret-to-collector distance to 3 mm, as presented in Figure 4.27(e), enables the jet to suppress the whipping instability, culminating in the formation of structured fiber arrangements as shown in Figure 4.27(f).

As the speed of the collector increases, the jet undergoes further stretching, transitioning from a free motion state to a stretched state, which effectively mitigates the jet's whipping instability. The distance between the jet deposition point and the Taylor cone's apex expands in tandem with the collector's speed, as illustrated in Figure 4.28(a). At lower velocities of the collector, the jet is in a state of free motion, resulting in a random distribution of fibers on the collector (Figure 4.28(b)). However, when the collector's speed exceeds the jet deposition rate, the jet exists in a stretched state, successfully minimizing the disruption from charge repulsion. This allows for the deposition of an ordered fiber structure on the collector, as evidenced in Figure 4.28(c) and (d).

The impact of the process parameters on the diameter of stretched fibers is further examined, as depicted in Figures 4.29–4.32. The increase of the collector moving speed induces stretching in the jet, significantly enhancing fiber uniformity

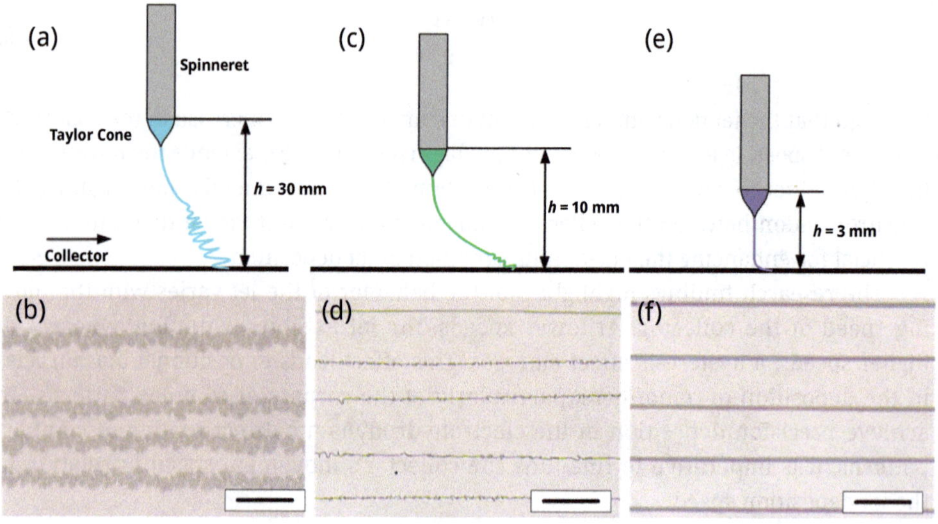

Figure 4.27: The deposition behaviors of charged jet with different distances between the spinneret and the collector. The distance between spinneret and collector: (a, b) 30 mm; (c, d) 10 mm; and (e, f) 3 mm. The solution concentration, flow rate, and collector velocity are 12 wt%, 50 μL/h, and 500 mm/s, respectively. Scale bar is 100 μm.

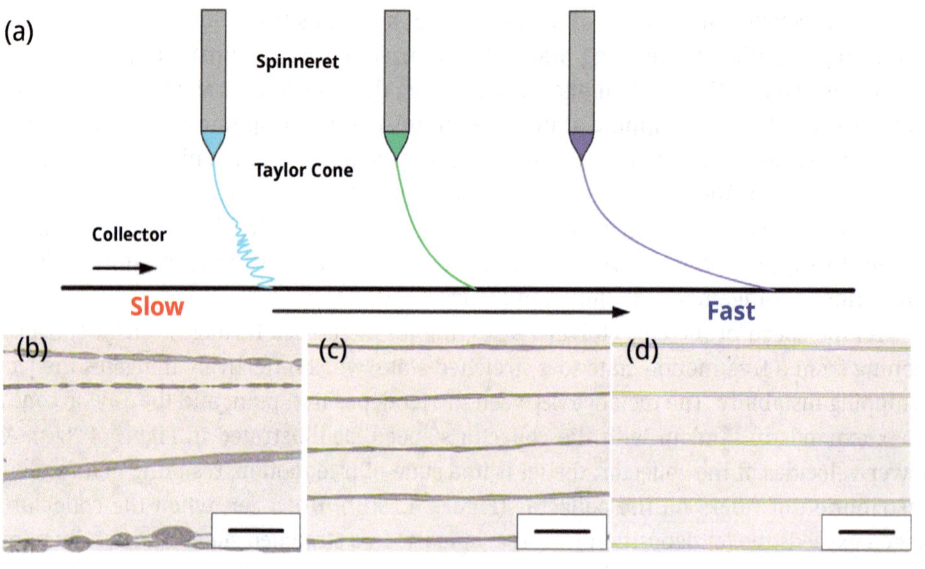

Figure 4.28: Deposition behaviors of charged jets at varying collector velocities: (a) schematic diagram of jet deposition; the collector velocity is 50, 100, and 300 mm/s for (a), (b), and (c), respectively. The applied voltage, PEO solution concentration, flow rate, and distance between spinneret and collector are 1.5 kV, 8 wt%, 50 μL/h, and 2 mm, respectively. Scale bar is 500 μm.

and diminishing fiber diameter (Figure 4.29). As the collector moving speed increases from 10 to 1,000 mm/s, the average fiber diameter decreases from 18.89 μm to 0.89 μm, with the distribution range contracting from 34.9 μm to 1.57 μm. An increased spinneret-to-collector distance allows for adequate solvent evaporation, leading to a reduction in fiber diameter. As the distance increases from 2 to 10 mm, the average diameter of the deposited nanofibers in the stretched state decreases from 4.36 μm to 2.75 μm, as shown in Figure 4.30. Increasing the applied voltage strengthens electric field's confinement effect on the jet, which facilitates a reduction in fiber diameter. When the applied voltage increases from 1.1 to 1.9 kV, the average diameter of the deposited fiber in the stretched state decreases from 5.12 μm to 3.98 μm, as indicated in Figure 4.31. High solution concentration reduces the flowability of the solution, consequently decreasing the fiber diameter. As the concentration of PEO solution increases from 8 to 16 wt%, the average diameter of the fibers decreases from 4.36 μm to 3.21 μm (Figure 4.32).

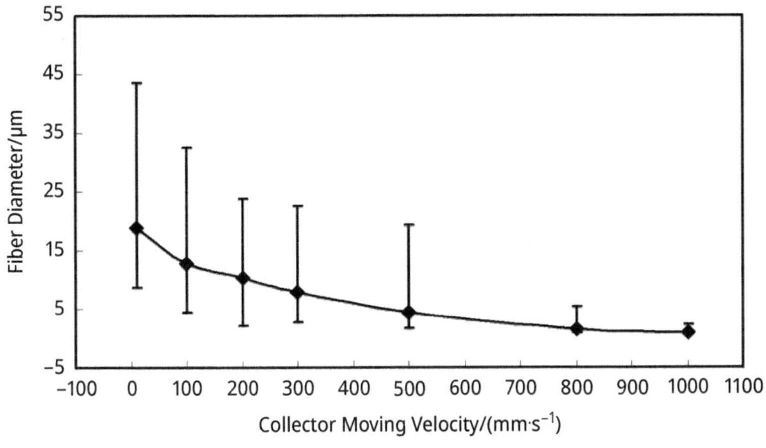

Figure 4.29: The relationship between diameter of fiber and collector velocity. The applied voltage, PEO solution concentration, flow rate, and distance between spinneret and collector are 1.5 kV, 8 wt%, 50 μL/h, and 2 mm, respectively.

4.3 External field-controlled positioning deposition of micro-nanostructures

4.3.1 Cored spinneret-induced EDW

4.3.1.1 Simulation of cored-spinneret-induced space electric field

A cored spinneret is a spinneret that has been modified by adding a core, typically a needle or wire electrode, within the hollow portion of the spinneret. Firstly, the distribution

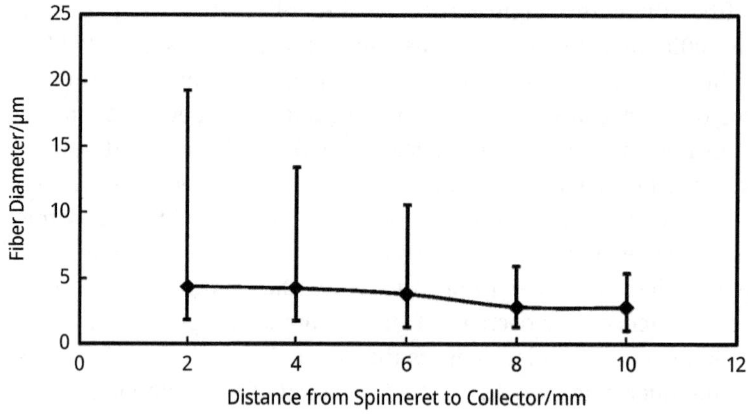

Figure 4.30: The relationship between diameter of fiber and distance between spinneret and collector. The applied voltage, PEO solution concentration, flow rate, and collector velocity are 1.5 kV, 8 wt%, 50 µL/h, and 500 mm/s, respectively.

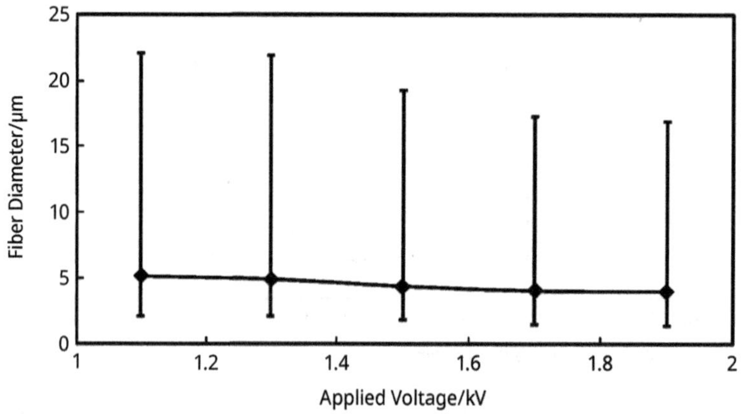

Figure 4.31: The relationship between diameter of fiber and applied voltage. The PEO solution concentration, flow rate, distance between spinneret and collector, and collector velocity are 8 wt%, 50 µL/h, 2 mm, and 500 mm/s, respectively.

of the space electric fields around the spinneret with and without needle is simulated. The inner and outer diameters of the spinneret tube are 0.5 and 0.7 mm. For the cored spinneret, the distance from the core needle tip to the collector is 1 mm and the core protrudes 0.1 mm out of the spinneret tube, while the distance from the original coreless spinneret to the collector is 1.1 mm as depicted in Figure 4.33. A voltage of 1 kV is applied to the spinnerets. In Figure 4.33, the dashed line represents the path along which the electric field strength values are selected, extending from the tip of the needle located on the central line to the collector. As shown in Figure 4.33(a), the

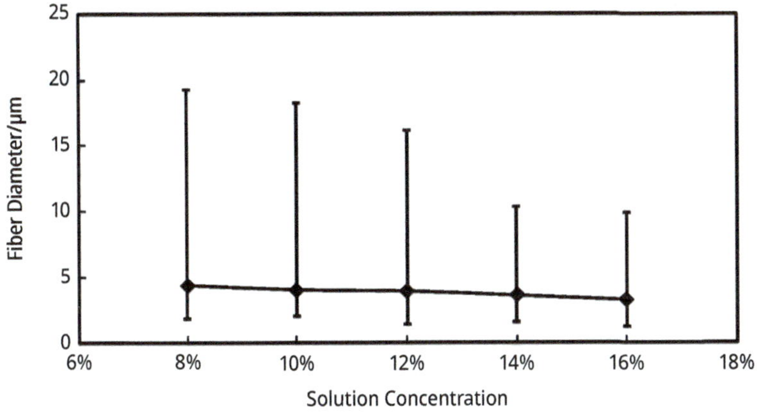

Figure 4.32: The relationship between diameter of fiber and solution concentration. The applied voltage, flow rate, distance between spinneret and collector, and collector velocity are 1.5 kV, 50 μL/h, 2 mm, and 500 mm/s, respectively.

maximum electric field strength of the coreless spinneret-1,943.93 V/mm-is concentrated at the tube edge, while the field along the centerline is significantly weaker, with some regions lacking any electric field. The field strength gradually decreases from the tube edge toward the collector. In contrast, Figure 4.33(b) shows that the cored spinneret generates a peak electric field strength of 2,917.87 V/mm at the tip of the core needle. The core tip and spinneret edge together form an arc-shaped electric field envelope that aligns with the surface of the solution droplet. This region experiences a strong tangential electric field force, which, according to dynamic analysis, facilitates jet initiation by lowering the energy barrier for jet formation.

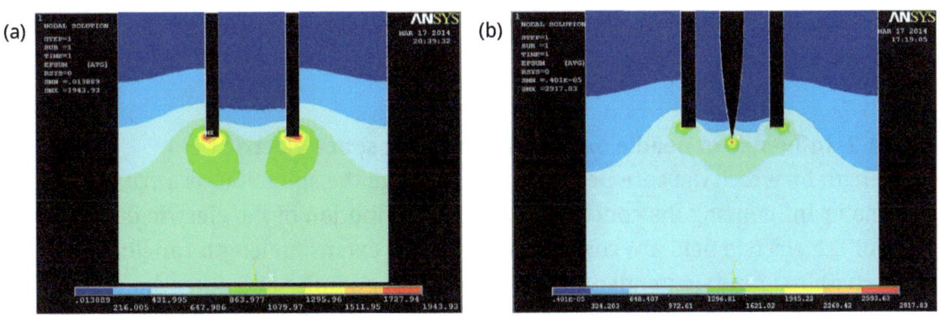

Figure 4.33: Distribution of electric field around the spinnerets: (a) a coreless spinneret and (b) cored spinneret.

Dashed lines aligned with the central axis of each core needle are drawn in Figure 4.33 to indicate the path of electric field analysis. The corresponding electric field strength distribution along the lines are shown in Figure 4.34, clearly demonstrating that the core needle significantly enhances the electric field intensity along the central axis. The overall electric field strength between the coreless spinneret and the collector is relatively low, with only slight changes, indicating a low concentration of electric field between the spinneret and the collector. In contrast, the electric field strength around the cored spinneret is larger than that around the coreless spinneret, and it decreases rapidly from the core to collector. As shown in Figure 4.34(b), the electric field strength changes by a staggering 1,700 V/mm over a distance of just 0.1 mm, indicating a high concentration of electric field around the core and a pronounced tip effect.

As the electric field force grows with increasing electric field strength, the Rayleigh equilibrium can be disrupted at a lower applied voltage. This indicates that the critical voltage needed for the cored spinneret is lower than that of the coreless spinneret. The presence of the core enhances the local electric field strength at the tip of the needle, which plays a crucial role in stabilizing the conical jet during ejection.

4.3.1.1.1 Influence of core needle placement on space electric field

The positioning of the core needle within the spinneret tube significantly influences the surrounding electric field distribution. When the core needle is aligned at the center of the spinneret, the electric field exhibits a symmetrical pattern, and the high-intensity field region near the needle tip closely resembles the shape of a pendant droplet, as shown in Figure 4.35(a). In contrast, when the core needle is offset by 0.1 mm to the left for simulation purposes, as depicted in Figure 4.35(b), the strong electric field region becomes discontinuous, and the field strength on the right side is markedly reduced. The field lines also lose their uniform downward orientation and instead tilt toward the left, concentrating the electric field in the left half of the spinneret. This asymmetry alters the force distribution acting on the emerging jet, ultimately causing it to deflect toward the left.

4.3.1.1.2 Influence of needle extension length on space electric field

The length by which the core needle protrudes from the spinneret- is a critical process parameter influencing the concentration and distribution of the electric field. Simulations of the electric field are conducted for needle extension length ranging from −0.1 (retracted) to 0.5 mm (extended). The spinneret tube's inner and outer diameters measure 0.26 and 0.51 mm, respectively. The distance between the needle tip and the collector remains at a constant 3 mm. A voltage of 1 kV is applied to the core needle tip and the spinneret tube, while no voltage is applied to the collector. The outcomes of these simulations are illustrated in Figure 4.36. In Figure 4.36(a) and (b), where the needle does not protrude out of the spinneret tube, the interior of the spinneret tube being a sealed domain devoid of electric field distribution results in no significant en-

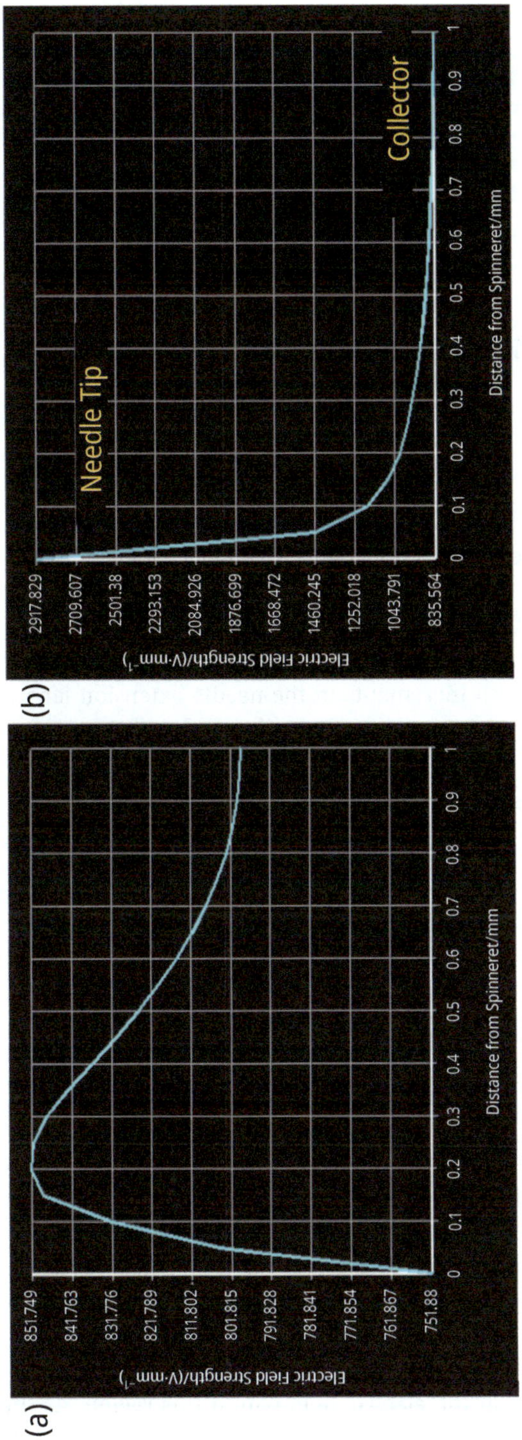

Figure 4.34: Distribution of electric field strength on the center line of the spinneret: (a) spinneret without a core and (b) spinneret with a core.

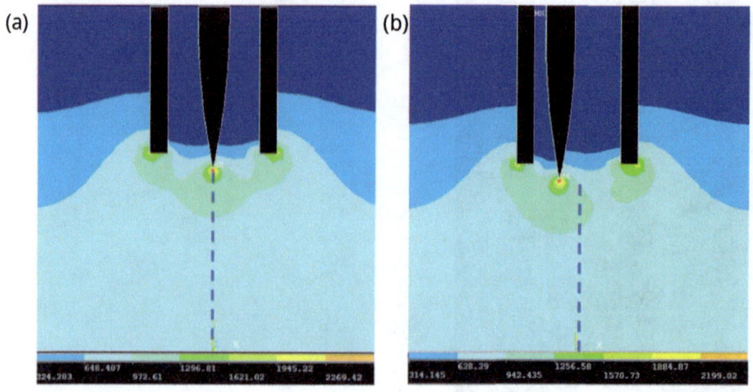

Figure 4.35: Influence of core needle placement on electric field distribution: (a) distribution of the electric field with the core located in the center of the needle tube and (b) distribution of the electric field with the core shifted to the left by 0.1 mm.

hancement of the electric field. As a result, regions of intensified electric field are primarily concentrated around the edge of the spinneret tube. However, when the core needle extends out of the spinneret tube, as depicted in Figure 4.36(c)–(g), a concentrated electric field emerges at the tip of the protruding needle. The electric field strength at the needle tip intensifies with increments in the needle extension length, as demonstrated in Figure 4.37.

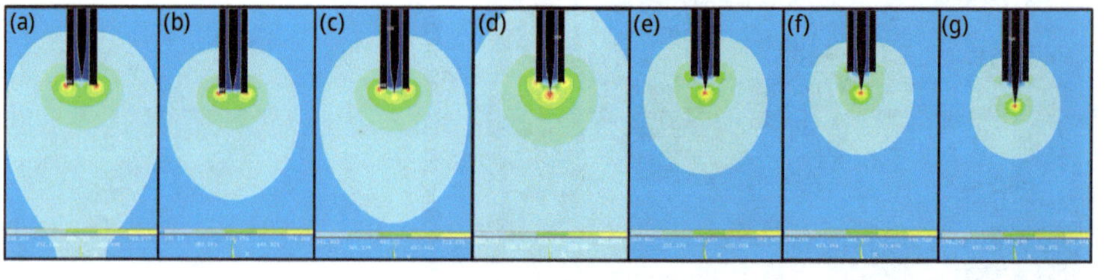

Figure 4.36: Electric field for varying core extension lengths of (a) −0.1; (b) 0; (c) 0.1; (d) 0.2; (e) 0.3; (f) 0.4; and (g) 0.5 mm.

To further illustrate the concentration effect of the needle tip on the electric field, the electric field strength is evaluated at the horizontal plane of the needle tip. The corresponding electric field strength distribution curve is shown in Figure 4.38, where the x-axis represents the horizontal distance to the core needle. The three curves correspond to needle extension lengths of 0, 0.1, and 0.5 mm, respectively. When the needle does not extend beyond the spinneret tube, the peak electric field strength at the needle tip is only 600 V/mm. Additionally, the nearby electric field gradient is relatively gentle, indicating a low rate of change in the electric field within this region. As the length of the needle extension increases, the peak electric field strength at the needle

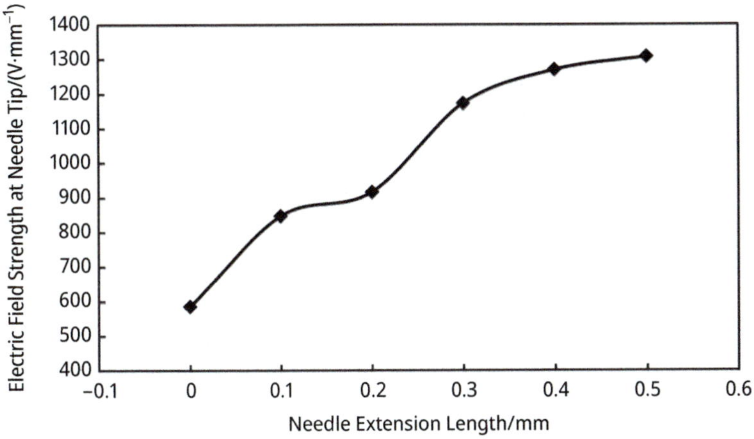

Figure 4.37: Plot of the electric field strength at the needle tip versus needle extension length.

tip rapidly increases, culminating in a maximum electric field strength of 1,307.5 V/mm when the needle extension length is 0.5 mm. The augmentation in the peak electric field strength is accompanied by an increment in the curve's gradient, which bespeaks a substantial augmentation in the electric field strength encompassing the needle tip.

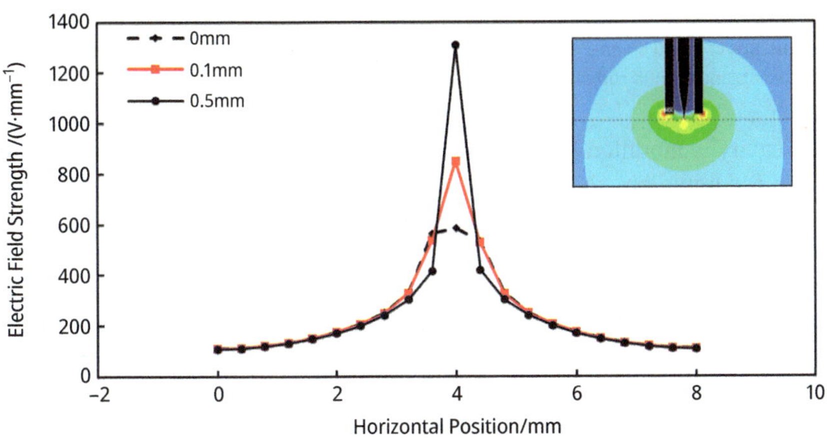

Figure 4.38: Electric field strength curves on the horizontal plane where the needle tip is located at needle extension lengths of 0, 0.1, and 0.5 mm (the needle tip is at the horizontal position of 4 mm).

4.3.1.2 Design and construction of experimental equipment for cored spinneret-induced EDW

As shown in Figure 4.39, the structure of cored spinneret includes solution cavity, clamp, pull rod, screw nut, and the main body. The clamp is designed to securely grasp core needles that range from 0.1 to 1.0 mm in diameter. Based on the principle of the elastic clamp, the design of the core needle clamping device is illustrated in Figure 4.39(a). This elastic clamp is composed of four movable petals with uniform elastic deformation capabilities, ensuring excellent concentricity under identical radial forces and thereby preventing core needle misalignment. The base of the clamp is connected to the pull rod. Turning the rotating nut clockwise raises the pull rod, which in turn lifts the collet. The four movable petals are then squeezed by the inclined plane, effectively securing the core needle in place.

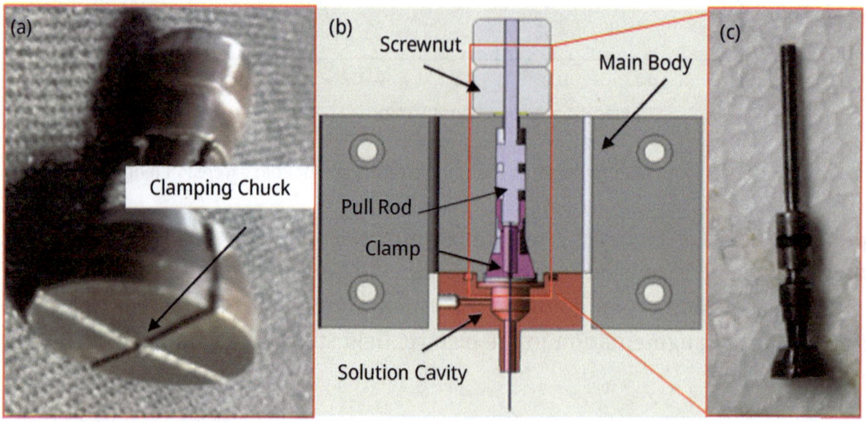

Figure 4.39: Structure of cored spinneret and its core clamping component: (a) the clamp; (b) schematic diagram of spinneret; and (c) pull rod and clamp installed together.

The solution cavity is mounted beneath the main body of the spinneret, with a solution inlet hole for liquid supply. By using sheet molding compound (SMC) joint, this solution inlet hole is connected to the syringe through a hose, as shown in Figure 4.40 (a) and (b). The interface between the solution cavity and the spinneret's main body is sealed with a seal ring to prevent the escape of the solution from the cavity. A dispensing needle is screwed onto the base of the solution cavity to serve as the spinneret tube, with the core needle protruding from the spinneret tube. The needle extension length can be modified by adjusting the position of the dispensing needle, as illustrated in Figure 4.40(c) and (d).

The entire spinneret is installed on a precision motion platform, with the wire being clamped by two screw nuts, and the high voltage is applied to the core needle through the pull rod. The experimental setup for cored spinneret-induced EDW is shown in Figure 4.41.

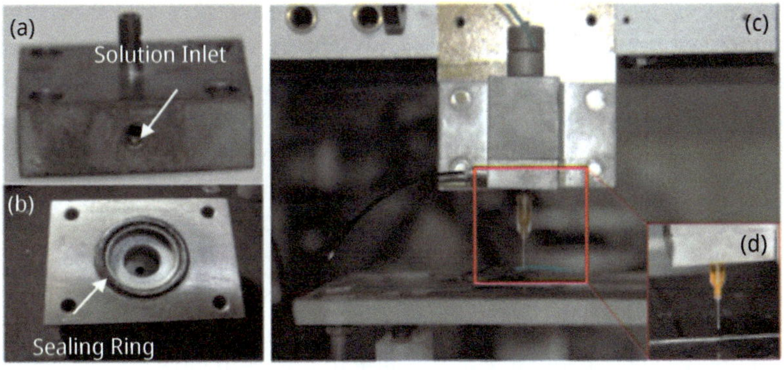

Figure 4.40: Physical map of the cored spinneret: (a) outer of the solution cavity; (b) the interior of the solution cavity; (c) overall drawing of the spinneret (installed on the precision motion platform); and (d) local magnification of the dispensing needle.

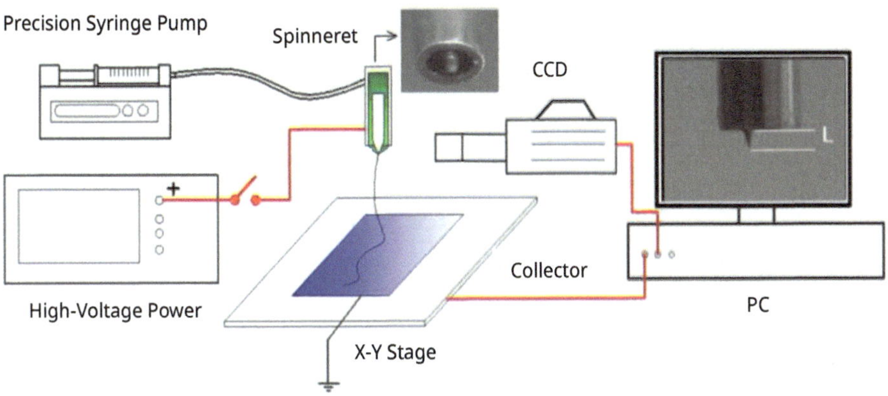

Figure 4.41: Experimental setup for cored spinneret-induced EDW.

4.3.1.3 Ejection behaviors of cored spinneret-induced EDW jet

4.3.1.3.1 Jet formation process of cored spinneret-induced EDW

After the voltage is applied, the suspended droplet goes through three stages before a stable jet is formed. The first stage is the deformation of the suspended droplet, where the polymer solution deforms under the force of electric field. This stage ends with the formation of a Taylor cone, as shown in Figure 4.42(a)–(c). The second stage is the unstable ejection stage, as shown in Figure 4.42(d) and (e), where a jet forms at the tip of the Taylor cone and charges are rapidly released. This causes the Taylor cone to recede, while the jet becomes larger and thicker, and the volume of the suspended droplet's solution decreases rapidly. It is a process where the consumption and replenishment of the solution reach an equilibrium, and until they do, the size of the jet and the height of the Taylor cone continuously change. Figure 4.43 shows the video time sequence images between Figure 4.42(e) and (f), with a time interval of 0.68 s. It can be observed that the height of the Taylor cone gradually decreases from 1.332 mm at the start of the ejection to 0.325 mm. When the jet ejection speed and the solution supply rate reach equilibrium, the height of the Taylor cone will no longer change, and this marks the beginning of the third stage – the stable cone ejection stage, as shown in Figure 4.42(f) and (g), where the height of the Taylor cone remains constant.

The precise jet printing process for micro-nanostructures via EDW occurs during the stable cone jet stage. Observations from Figure 4.42 indicate that in the jet ejection with a coreless spinneret, the Taylor cone at the base of the jet follows the collector's movement to the right. Once the process enters the stable ejection stage, the base of the jet exhibits intense oscillations, alternating between left and right positions, as depicted in Figure 4.42(f) and (g). It is evident that the jet emanating from the coreless spinneret is susceptible to disturbances caused by the motion of the collector.

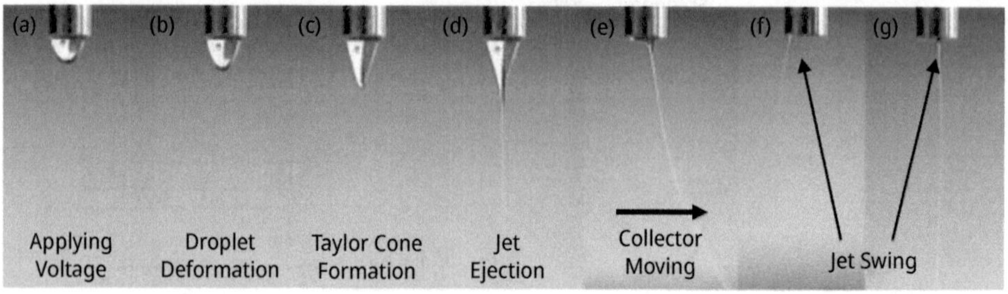

Figure 4.42: Formation process of EDW Jet with coreless spinneret: (a) starting to apply a voltage; (b) deformation of the hanging droplet; (c) formation of a Taylor cone; (d) formation of a jet; (e) the collector begins to move to the right (the arrow indicates the movement direction of the collector); (f) the jet swings to the left; and (g) the collector stops moving (the solution concentration is 8%, the spinneret-to-collector distance is 5 mm, the voltage is 3.3 kV, and the liquid supply speed is 60 μL/h).

Upon embedding a core needle into the hollow spinneret, the jet formation process still adheres to the three aforementioned stages, as shown in Figure 4.44. However, when the jet reaches the stable stage, the Taylor cone remains anchored at the needle tip's apex, unaffected by the collector's rightward movement, as demonstrated in Figure 4.44(f) and (g). The jet is in a bent state. This demonstrates that the presence of the core needle imposes constraints on the flow of the solution within the spinneret, thereby governing the fluid dynamics of the Taylor cone and mitigating its oscillations.

4.3.1.3.2 Oscillation law of the cored spinneret-induced jet

Incorporating a core needle into the hollow spinneret not only reduces the starting voltage but also modifies the flow state of the solution at the spinneret, effectively

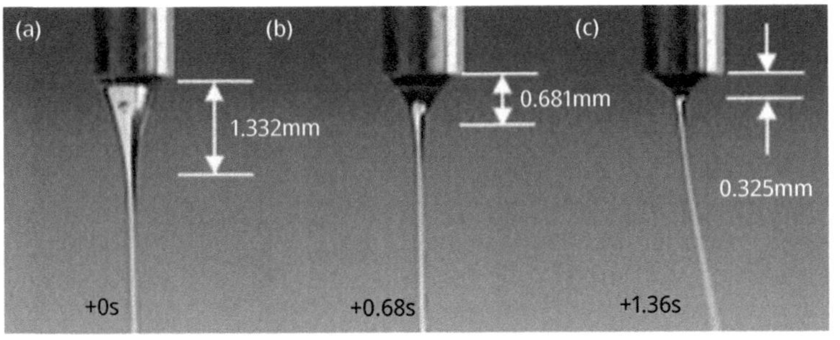

Figure 4.43: Variation of Taylor cone height: (a) at the beginning of the jet formation, the height of the Taylor cone is 1.332 mm; (b) after 0.68 s, the height of the Taylor cone is 0.681 mm; and (c) after 1.36 s, the height of the Taylor cone decreases to 0.325 mm (solution concentration is 8%, spinneret-to-collector distance is 5 mm, applied voltage is 3.3 kV, and liquid supply speed is 60 μL/h).

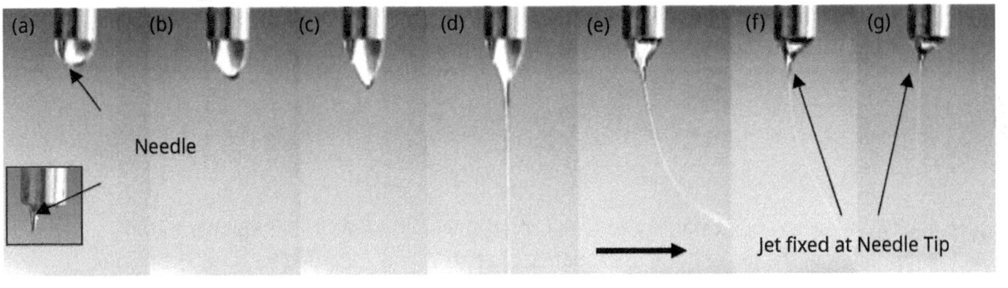

Figure 4.44: Formation process of jet with a cored spinneret: (a) starting to apply a voltage; (b) deformation of the droplet; (c) formation of the Taylor cone; (d) formation of a jet; (e) the collector begins to move to the right (the arrow indicates the direction of movement); (f) the jet bends; and (g) the collector stops moving (the jet remains straight down, with the base still at the tip of the core needle, core needle extends by 1 mm, solution concentration is 8%, spinneret-to-collector distance is 3 mm, applied voltage is 3.3 kV, and liquid supply velocity is 60 μL/h).

inhibiting the backflow of the solution. For coreless spinneret, after the conical jet is formed, the flow of the solution inside the hanging drop does not remain constant; both the flow velocity and direction are continuously changing, and the root of the jet often swings. Conversely, after adding the core needle inside the syringe, the swing of the jet root below the syringe is suppressed, and the amplitude of the swing is greatly reduced compared to that of the coreless spinneret jet.

To investigate the suppressive effect of the core needle on the oscillation of the jet at the root, the oscillation angles of the jets formed by cored spinneret and coreless spinneret under the identical voltage are observed. As illustrated in Figure 4.45, the oscillation angles within 45 s are contrasted. The left side of the image is considered positive and the right side is considered negative for recording the angular deviation. The graph elucidates that the jet ejecting from the coreless spinneret exhibits significant instability, with an oscillation angle spanning from −15° to +15°, culminating in a total oscillation range of approximately 30°. On the other hand, the jet generated by the cored spinneret shows a slight bias toward the left, yet its maximum oscillation range does not exceed 10°, which is merely one-third of the oscillation range experienced by the coreless spinneret.

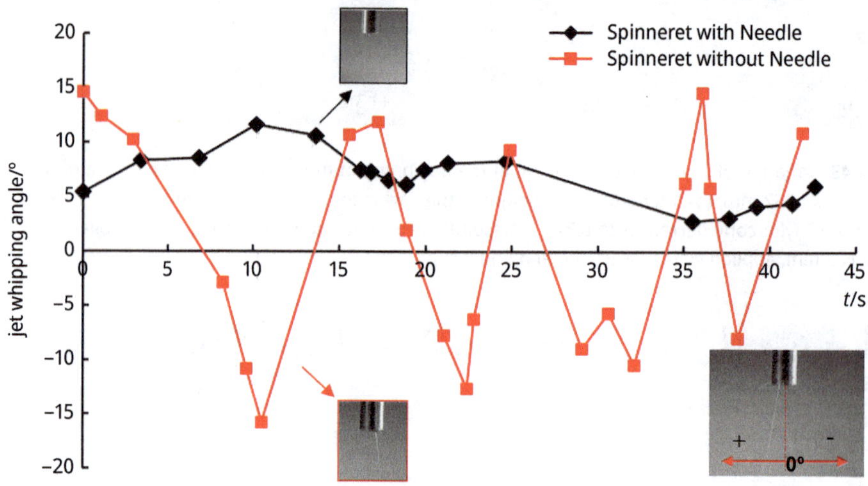

Figure 4.45: Comparison of jet stability between cored spinneret and coreless spinneret (timing starts from the formation of the jet, and the oscillation angle of the jet is recorded for 45 s. The needle extension length is 0.1 mm, the solution concentration is 8%, the spinneret-to-collector distance is 5 mm, the applied voltage is 3.3 kV, and the liquid supply speed is 60 μL/h).

Experimental investigations have revealed that the extension length of the core needle substantially affects the stability of the jet. An excessively long extension length can reduce the stability of the jet. Figure 4.46 illustrates the jet morphology at different needle extension lengths. In Figure 4.46(a), a single stable jet is formed at an extension length of 0.1 mm. Influenced by the core needle, the jet exhibits a tilt angle of 5°–8° and a swing span limited to 1°–3°. However, in Figure 4.46(b), augmenting the extension length to 0.4 mm results in the formation of multiple unstable jets. In Figure 4.46(c), the coreless spinneret displays a jet oscillation range of 15°. It can be concluded that the core needle exerts a stabilizing effect on the jet, provided that the extension length of the needle should match with other parameters.

4.3.1.3.3 Critical voltage for jet ejection of cored spinneret-induced EDW

Electrohydrodynamic printing is a process that relies on the synergistic interaction between electric field forces and the surface tension of a solution. The term "critical voltage" denotes the threshold voltage necessary for the solution to deform into a Taylor cone and subsequently produce a jet. This critical voltage can be calculated using the following formula [2]:

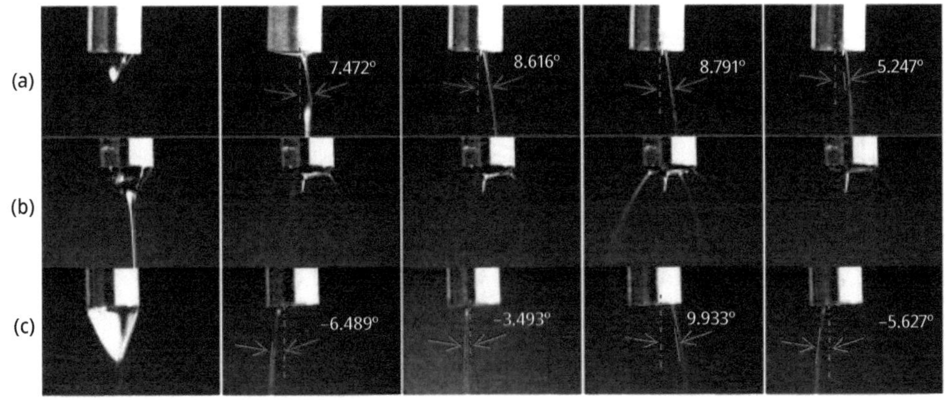

Figure 4.46: Comparison of jet flow at different needle extension lengths: (a) extension length of 0.1 mm; (b) extension length of 0.4 mm; and (c) coreless spinneret (applied voltage is 4 kV, solution concentration is 8%, and spinneret-to-collector distance is 5 mm).

$$V_{\text{onset}} = \sqrt{\frac{\gamma d_N}{2\varepsilon}} \ln\left(\frac{8L}{d_N}\right) \tag{4.7}$$

where V_{onset} is the critical voltage, γ is the surface tension of the solution, d_N is the effective flow diameter of the spinneret, and L is the distance between the spinneret and the collector. Since the solution surface tension γ is inherently constant, a decrease in the effective flow diameter d_N results in a diminished critical voltage. Consequently, it is theoretically predicted that the critical voltage for a spinneret equipped with a core will be lower compared to a spinneret without a core.

As shown in Figure 4.47, the critical voltage increases with the spinneret-to-collector distance. It is lower for the cored spinneret than that for the coreless spinneret as the addition of the core needle lower effective flow diameter of the entire cored spinneret.

4.3.1.3.4 Maintaining voltage for jet ejection induced by cored spinneret

The surface tension coefficient of polymer solution directly influences the jet formation process. Changes in solution concentration will alter the surface tension of the suspended droplet at the needle tube: a higher concentration leads to greater surface tension, requiring a stronger electric field force to maintain jet ejection, thus increasing the maintaining voltage. Figure 4.48 demonstrates that for PEO solutions ranging from 4% to 12%, a spinneret with a core needle consistently requires less voltage than one without a core needle, regardless of the solution concentration. As solution concentration increases, solution viscosity and the surface tension of the suspended droplet rise. To overcome this surface tension for sustained ejection, a higher electric field force (and thus a higher maintaining voltage) is needed.

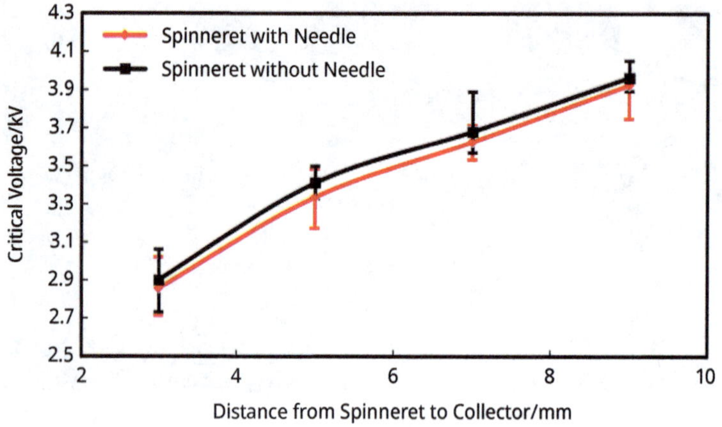

Figure 4.47: Variation of critical voltage with the spinneret-to-collector distance (needle extension length is 0.1 mm, solution supply rate is 100 µm/L, and solution concentration is 10%).

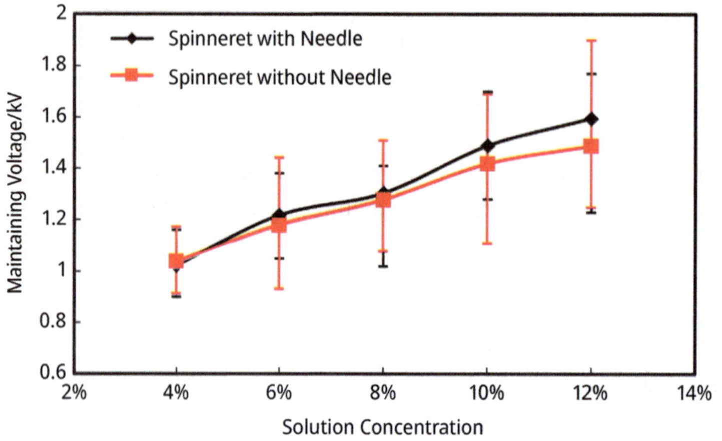

Figure 4.48: Relationship between maintaining voltage and solution concentration. The extension length of the core needle is 0.1 mm, the distance between the spinneret and the collector is 5 mm, and the liquid supply rate is 100 µL/h.

4.3.1.3.5 Ejection delay time of cored spinneret-induced EDW

Adding a core needle strengthens the local electric field, facilitating the formation of a jet from polymer solution droplets at lower voltages. Moreover, the inclusion of the core needle modifies the internal flow dynamics of the spinneret tube, leading to increased flow velocities and a reduced effective cross-sectional area within the cored spinneret, compared to a conventional design under identical feed rates. As a result, this leads to higher internal pressures within the pendant droplets in the cored spinneret.

The time required for jet formation after applying the voltage is influenced by various factors such as the applied voltage, solution concentration, and ambient conditions like temperature and humidity. Figure 4.49 illustrates the relationship between ejection delay time and the applied voltage. A stronger electric field at the tip of the core needle increases the tangential component of the electric field on the droplet surface, which accelerates the stretching of the polymer solution, reducing the time needed to form the Taylor cone. Consequently, a spinneret with a core needle forms a cone jet more quickly than one without a core needle, particularly at lower applied voltage (3.0 kV). On average, a cored spinneret takes less than 4 s to form a cone jet, whereas a coreless spinneret requires about 12 s from voltage application to cone jet formation. Additionally, the stretched hanging droplets exhibit distinct morphologies between the two types of spinnerets, as depicted in Figure 4.50.

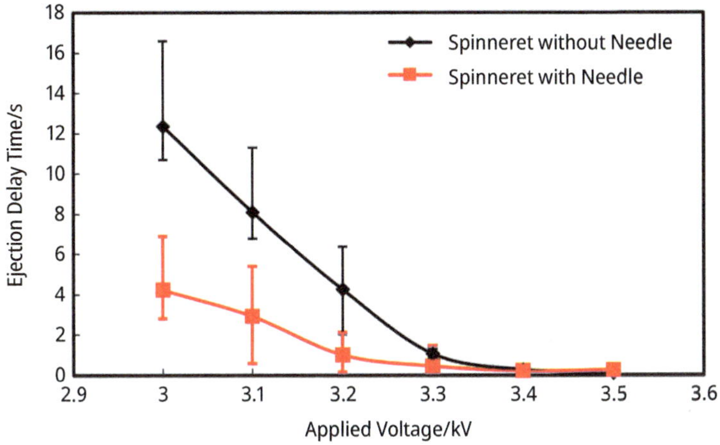

Figure 4.49: Variation of ejection delay time with the applied voltage (needle extension length is 0.1 mm, solution concentration is 8%, spinneret-to-collector distance is 5 mm, and solution supply rate is 100 μL/h).

For the coreless spinneret, the electric field force experienced by the droplet is relatively small. The hanging droplet is stretched to a length of 1,580 μm over 12 s, with a stretching speed of only 99.5 μm/s. Concurrently, its diameter continuously increases, with a radial growth rate of 27 μm/s, and the volume of the entire droplet gradually increases, as depicted in Figure 4.50(a)–(c). On the other hand, for the cored spinneret, the hanging droplet is rapidly stretched by the electric field force to a length of 1,617 μm in 2.58 s, reaching a stretching speed of 467.8 μm/s. The diameter of the hanging droplet gradually decreases, with a reduction speed of 38.4 μm/s, maintaining a nearly constant volume due to the influence of electric field. The electric field force plays a dominant role in deforming the hanging droplet, which tends to sharpen the droplet's apex, as shown in Figure 4.50(d)–(f).

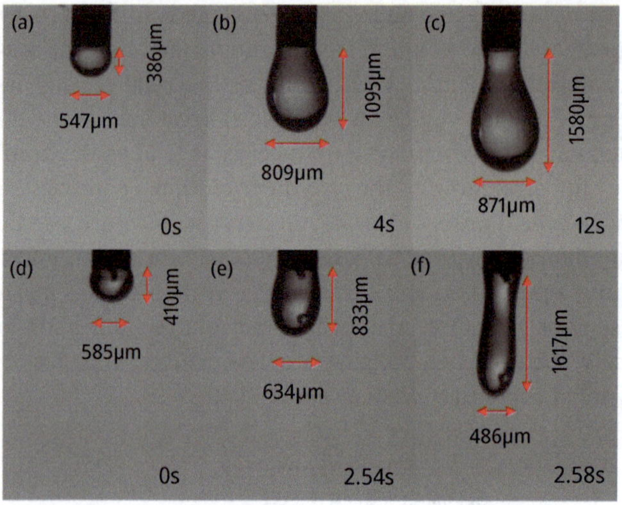

Figure 4.50: The hanging droplets at initial ejection stage, (a)–(c) is for coreless spinneret, while (d)–(f) is for core spinneret: (a) starting to apply voltage; (b) the hanging drop is being stretched; (c) the hanging drop is about to form a jet; (d) starting to apply voltage; (e) the hanging drop is being stretched; and (f) the hanging drop is about to form a jet (extension length of the core needle is 0.1 mm, solution concentration is 8%, spinneret-to-collector distance is 5 m, applied voltage is 3 kV, and liquid supply rate is 100 μL/h).

4.3.1.4 Patterned deposition by cored spinneret-induced EDW

4.3.1.4.1 Cored spinneret-induced electrohydrodynamic direct-writing of micro-nanofiber structures

The core needle within the spinneret intensifies the local electric field and manages solution reflux, enabling stable jet ejection at lower voltages. It also dampens the oscillations generated by the flow of the solution at the jet root, resulting in a more stable jet and smaller structure sizes. Figure 4.51 shows the direct-written micro-nanofiber structures with both cored and coreless spinnerets.

To investigate the performance of the cored spinneret on the direct-writing of single fiber, we directly write 40 parallel fibers at a consistent voltage and measure the diameters using measurement software. Figure 4.52 shows that the fiber diameters produced by a coreless spinneret range from 60 to 120 μm, which is approximately 40 μm larger than those produced by a cored spinneret (20–80 μm). Statistical analysis reveals that the average diameters for fibers directly written with cored and coreless spinnerets at an apply voltage of 3.5 kV are 45 and 88.5 μm, respectively.

In the EDW process, the jet exhibits various forms (such as liquid, solid, and semisolid) before depositing onto the collector, depending on the properties and content of the solvent in the solution. If the jet remains in liquid form before reaching the collector,

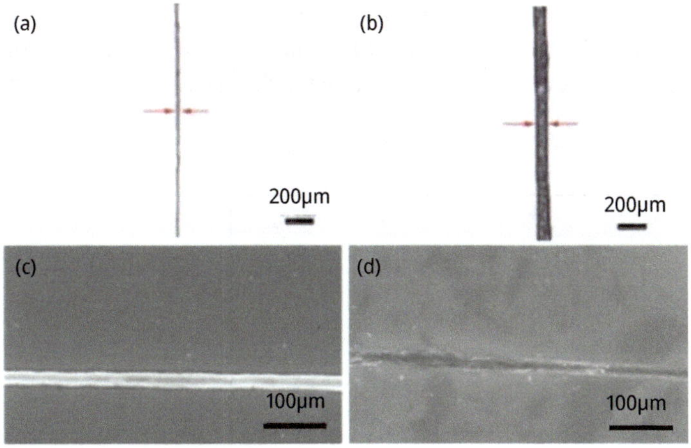

Figure 4.51: Electrohydrodynamic (a) direct-written micro-nanofiber structure scored spinneret at needle extension length of 0.1 mm; (b) coreless spinneret; (c) SEM image of a single fiber prepared by cored spinneret; and (d) SEM image of a single fiber produced by coreless spinneret (applied voltage, spinneret-to-collector distance, liquid supply speed, and collector speed are 3.5 kV, 5 mm, 100 μL/h, and 100 mm/s, respectively).

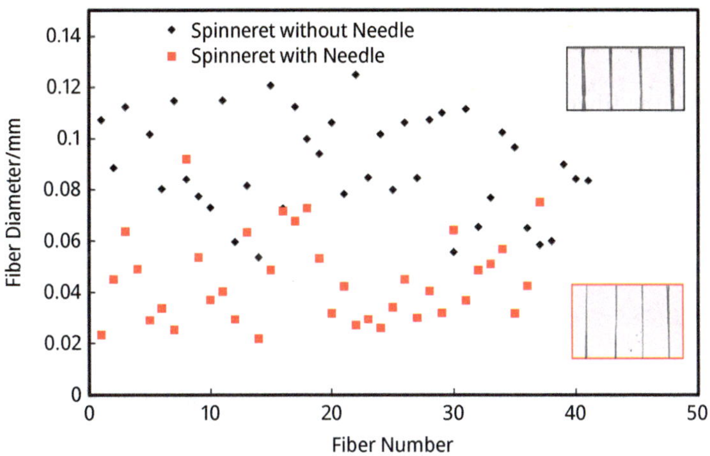

Figure 4.52: Diameters of parallel fibers directly written by cored spinneret and coreless spinneret (the extension length of the core needle is 0.1 mm, the applied voltage is 3.5 kV, the spinneret-to-collector distance is 5 mm, the liquid supply speed is 100 μL/h, and the collector speed is 100 mm/s).

it will be further refined and even atomized by the continuous action of the electric field force in the movement stage. Higher voltages increase the tensile force on the jet, leading to more extensive elongation in a short period. Therefore, theoretically, increasing the voltage further refines this type of polymer solutions. However, if the jet has already solidified or semi-solidified before it reaches the collector, the further refining effect of the electric field is greatly weakened, resulting in thicker fiber diameter. As shown in Figure 4.53, the relationship between the direct-written fiber diameter and the applied voltage for an 18% PEO solution shows that the fiber diameter decreases with increasing voltage.

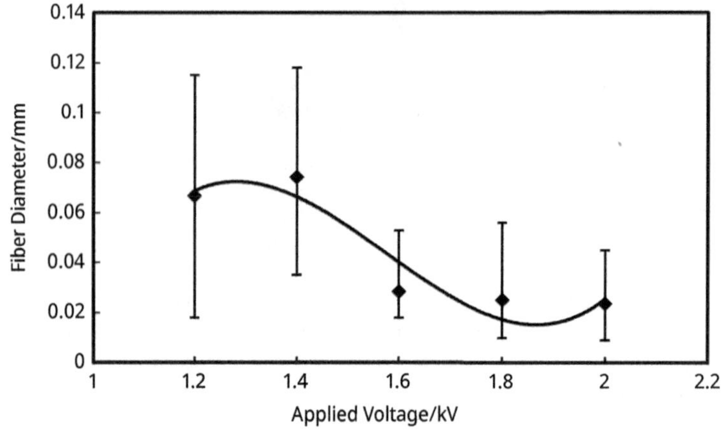

Figure 4.53: Relationship between fiber diameter and applied voltage (the solution concentration is 18%, spinneret-to-collector distance is 1 mm, and the solution supply speed is 50 µL/h).

As shown in Figure 4.54, the fiber morphology of PEO solutions with an 8% concentration varies under different voltages. After the jet reaches the collector, flat fibers form following a period of solvent evaporation. Additionally, as the diameter decreases, serrated defects become apparent on the fiber edges. This suggests that the 8% PEO solution remains in a liquid state before reaching the collector. During this process, the electric field force further refines the jet, resulting in a decrease in the diameter of the formed PEO fibers as the applied voltage increases.

4.3.1.4.2 Direct-writing of pattern structures via cored spinneret-induced EDW

As shown in Figure 4.55(a), the designed square-wave pattern is depicted, while Figure 4.55(b) illustrates the actual pattern achieved through direct-writing. Overall, the square-wave structure clearly separates its peaks from troughs, maintaining a shape that closely resembles the intended design.

The corners of the square wave pose a challenge in EDW. Moreover, jet oscillation during the ejection process introduces further instability that remains a critical issue. In this technology. As the jet approaches the collector, phenomena such as spiral deposition occur frequently due to the combined effects of the electric field force, gravity, charge repulsion from the already deposited fibers, and other environmental disturbances. If the collector's motion is too rapid relative to the jet deposition rate, the jet gets stretched and be pulled to the corner's opposite side before reaching the col-

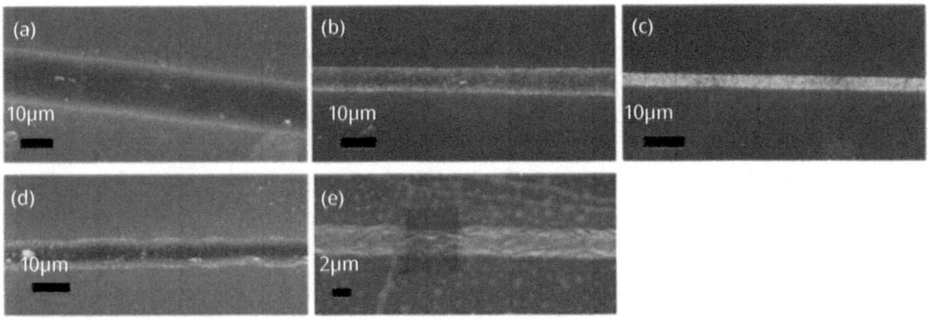

Figure 4.54: Electrohydrodynamic direct-written fibers obtained at voltages of (a) 1.2; (b) 1.4; (c) 1.6; (d) 1.8; and (e) 2.0 kV (solution concentration is 8%, spinneret-to-collector distance is 1 mm, and solution supply speed is 50 μL/h).

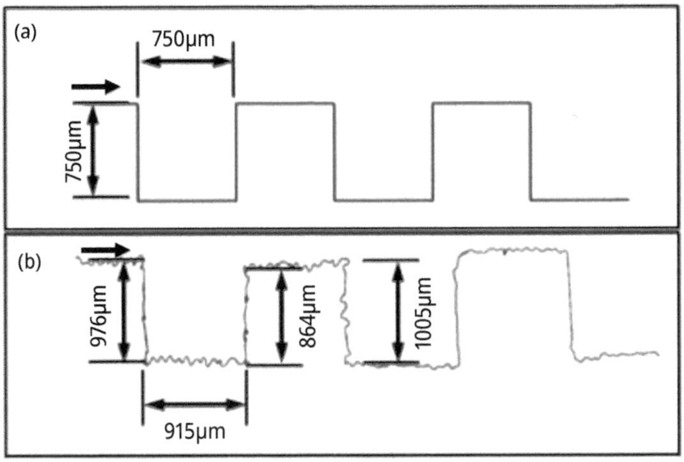

Figure 4.55: Electrohydrodynamic direct-writing of square wave: (a) the design of square-wave pattern and (b) direct-written square-wave pattern (duty ratio is 1:1, height and width of square-wave pattern are 750 μm, and the arrow in the figure represents the direct-writing direction) (solution concentration is 10%, spinneret-to-collector distance is 1 mm, collector speed is 10 mm/s, and applied voltage is 1.5 kV).

lector. Figure 4.56(b) depicts an arc corner formed during the direct-writing process without disrupting the jet flow, exhibiting a curve radius of 32 μm. The fiber distribution width of the straight segment is 65 μm. The appearance of wavy fibers, as shown in Figure 4.56(c), is caused by the mismatch between the collector moving speed and the jet deposition speed.

In the domain of patterned electrohydrodynamic direct-writing, the primary goals are to enhance pattern accuracy and minimize positioning errors. Using the aforementioned square wave as a reference, we analyzed fiber positioning errors at various needle extension lengths. The results are illustrated in Figure 4.57. When the

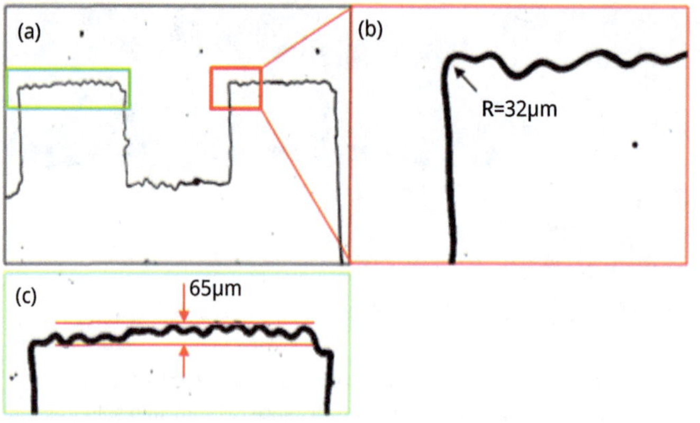

Figure 4.56: Illustration of the direct-written square wave (solution concentration is 10%, collector speed is 5 mm/s, applied voltage is 1.4 kV, and spinneret-to-collector distance is 1 mm).

core needle is not extended beyond the spinneret tube, there is no enhancement of the electric field at the needle tip, as shown in Figure 4.36(b). This leads to a square wave with a positioning error approximately 100 μm greater compared to when the needle extension length is 0.1 mm. As the needle extension length increases, a rapid intensification of the localized electric field at the needle tip occurs. This intensification amplifies the impact of destabilizing factors, which in turn reduces the stability of the ejected stream, resulting in a rapid increase in fiber positioning errors.

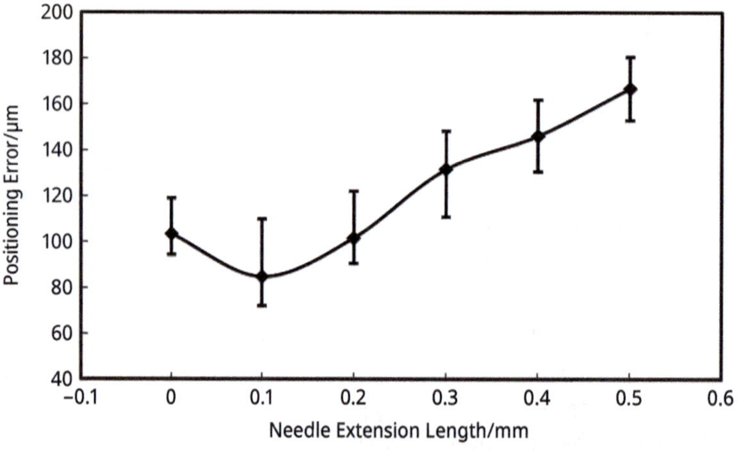

Figure 4.57: Position error of the vertical line of the square-wave pattern (the solution concentration is 10%, the collector speed is 5 mm/s, the applied voltage is 1.4 kV, and the distance from the spinneret to the collector is 1 mm).

In the EDW process, reducing the voltage gradually thins the jet, which aids in decreasing the fiber diameter to some extent. This is exemplified in Figure 4.58(a) and (e), where at a spinneret-to-collector distance of 1 mm, the jet exhibits a tapered form when the voltage is reduced to 0.6 kV. The high speed of the collector results in an incomplete pattern, particularly at the corners, a phenomenon observed in both pentagonal and triangular patterns. Gradually increasing the voltage, the whipping of the jet becomes more violent due to the stronger electric field force, and it is evident that wavy fibers are formed. By decreasing the collector moving speed to 5 mm/s, patterns as shown in Figure 4.58(b) and (f) can be obtained. The overall shape is well maintained, and the edges and corners of the pentagram and triangle are accurately represented. Under these conditions, even after the pattern is directly written twice, it remains intact, as shown in Figure 4.58(g), where the lines from the first and second writings are neatly deposited together without significant positional deviation.

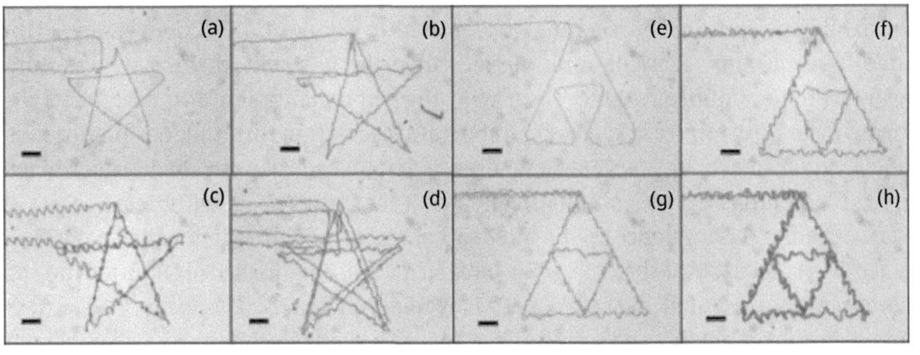

Figure 4.58: Direct-written pentagram and triangle patterns: (a) the applied voltage is 0.6 kV and the collector speed is 10.0 mm/s; (b) the applied voltage is 1.0 kV and the collector speed is 5.0 mm/s; (c) the applied voltage is 1.2 kV, and the collector speed is 5.0 mm/s; (d) the applied voltage is 1.5 kV, and the collector speed is 5.0 mm/s, furcated jets; (e) the applied voltage is 0.6 kV, and the collector speed is 10.0 mm/s; (f) the applied voltage is 1.0 kV, and the collector speed is 5.0 mm/s; (g) the applied voltage is 1.0 kV, and the collector speed is 5.0 mm/s, direct-writing for two times; (h) the applied voltage is 1.5 kV, and the collector speed is 1.0 mm/s. The scale bar is 170 μm.

As illustrated in Figure 4.58(c), when the applied voltage increases to 1.2 kV, the Taylor cone elongates downward due to the electric field force. Consequently, the entire jet begins to thicken, and the fiber exhibits significant fluctuations. The specifics of the prepared pentagram are depicted in Figure 4.59, where each line represents a wavering fiber that is thicker compared to the one produced at lower voltages. Despite the increased thickness and waviness, the overall star shape is preserved, although some curling and adhesion occur at the corners and line intersections.

Upon further elevation of the applied voltage, the electric field force will stretch the Taylor cone downward and break the surface tension at any point of the hanging droplet surface to initiate jet formation. At this point, two distinct scenarios may

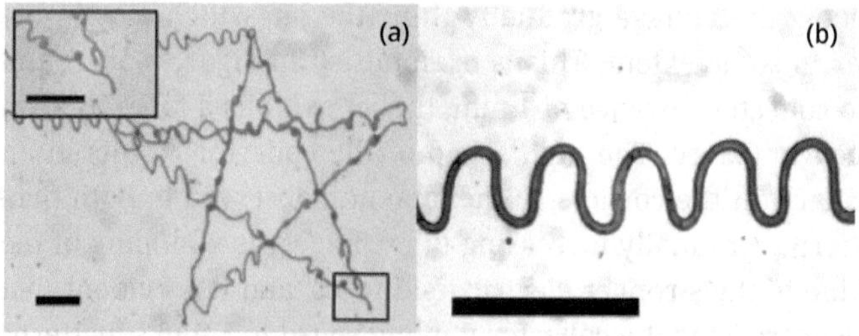

Figure 4.59: Electrospun pentagram and wavy fiber: (a) pentagram pattern and (b) wavy fibers (solution concentration is 10%, applied voltage is 1.2 kV, and collector speed is 5.0 mm/s. The scale bar is 170 μm).

arise: (i) the Taylor cone and the entire jet column are drawn toward the collector, establishing a new equilibrium state or (ii) multiple jets emerge to dissipate excess surface charge. In the former case, the resultant jet and the subsequently deposited fiber exhibit pronounced thickening, as depicted in Figure 4.58(h). Conversely, if the charge is released by forming multiple jets, then the jets split, making the resulting fiber morphology unpredictable. Even with the implementation of a core needle to mitigate instability, bifurcated jets exhibit erratic oscillations around the spinneret orifice, as evidenced in Figure 4.58(d), where the pentagram pattern manifests as overlapping filamentous traces, culminating in a characteristic "ghosting" effect.

To examine the influence of the applied voltage on the precision of fiber positioning, geometric analysis of the triangular base structures (Figure 4.58(e,f)) yields the correlation curve presented in Figure 4.60. Elevated voltages augment the jet's surface charge density, amplifying the susceptibility to electrostatic instability and exacerbating whipping dynamics during its flight. This phenomenon introduces significant positional deviations in the deposited fiber trajectory, thereby compromising spatial resolution.

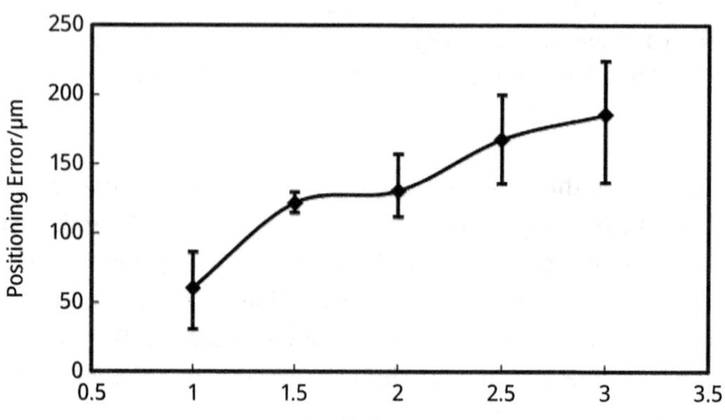

Figure 4.60: Relationship between positioning error and applied voltage (the solution concentration is 10%, and the collector speed is 5.0 mm/s).

4.3.2 Tip-induced electrohydrodynamic direct-writing

4.3.2.1 Simulation of space electric field induced by needle tip

In tip-induced electrohydrodynamic direct-writing (tip-induced EDW), a tip electrode is positioned close to the substrate to manipulate the space electric field. This tip electrode can be a sharp conductive object like a metal needle or a more sophisticated microelectrode.

4.3.2.1.1 Simulation of space electric field with different tip positions

In the process of tip-induced EDW, the structure of the spinneret and the shape of the tip electrode directly determine the distribution of the space electric field, which are important factors affecting the jet formation. To investigate how the inducing tip affects the electric field, we use ANSYS software to simulate the electric field for various spinneret and electrode combinations. The simulation model is established with specific parameters: the inner diameter of the spinneret is 0.26 mm, while the gap between the spinneret and the underlying electrode is set at 1 mm. The relative dielectric constant of the simulation area is fixed at 1. A voltage of 1.0 kV is applied to the spinneret, with all electrodes beneath grounded to 0 kV. Figure 4.61 illustrates four distinct scenarios:

- In Figure 4.61(a), a stainless steel needle tube without a core is used as the spinneret, paired with a planar electrode.
- Figure 4.61(b) features a conductive core needle inserted within the needle tube, while still employing a planar electrode.
- Figure 4.61(c) presents the setup with a coreless spinneret in conjunction with an inducing tip electrode.
- Figure 4.61(d) shows a needle tube with a conductive core needle alongside an inducing tip electrode.

These simulations provide insight into how different spinneret and electrode arrangements affect the distribution of the electric field. Understanding this is essential for controlling and optimizing the jet formation in EDW processes.

The electric field distribution of the coreless spinneret and planar electrode combination is depicted in Figure 4.62(a). A concentrated electric field is formed at the edge of the needle tube. It declines as it approaches the collector. The curve in Figure 4.62(b) shows the electric field strength along the centerline of the needle tube, extending from the spinneret toward the flat electrode. The electric field strength in the center of the spinneret is approximately 0 V/mm and peaks just beneath the spinneret. The electric field strength slowly decreases with the decrease of the distance to the spinneret. When a core needle is introduced into the needle tube, the electric field at the tip of the core needle increases rapidly. As shown in Figure 4.63, the electric field strength at the tip of the core needle reaches the maximum and rapidly declines with decreasing distance

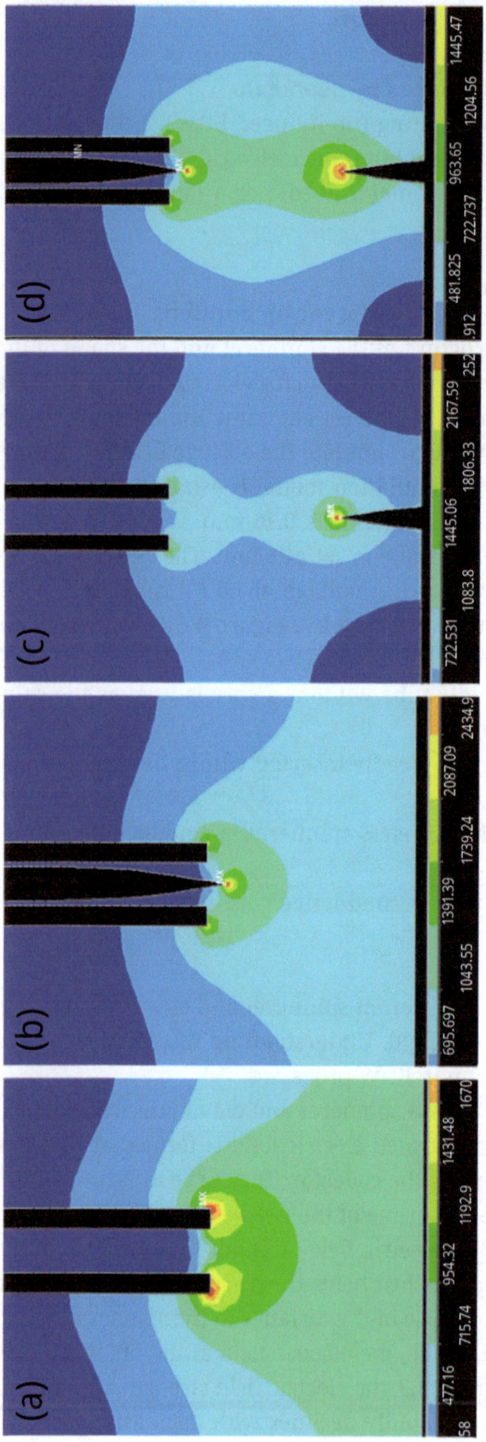

Figure 4.61: Simulation of electric field for different spinneret and electrode combinations: (a) coreless spinneret and flat electrode; (b) cored spinneret and flat electrode; (c) coreless spinneret and inducing tip electrode; and (d) cored spinneret and inducing tip electrode (the spinneret-to-collector distance is 1 mm, and the applied voltage is 1.0 kV).

from the spinneret. The steep gradient of the electric field strength indicates that the electric field around the spinneret is intensely focused at the tip of the core needle.

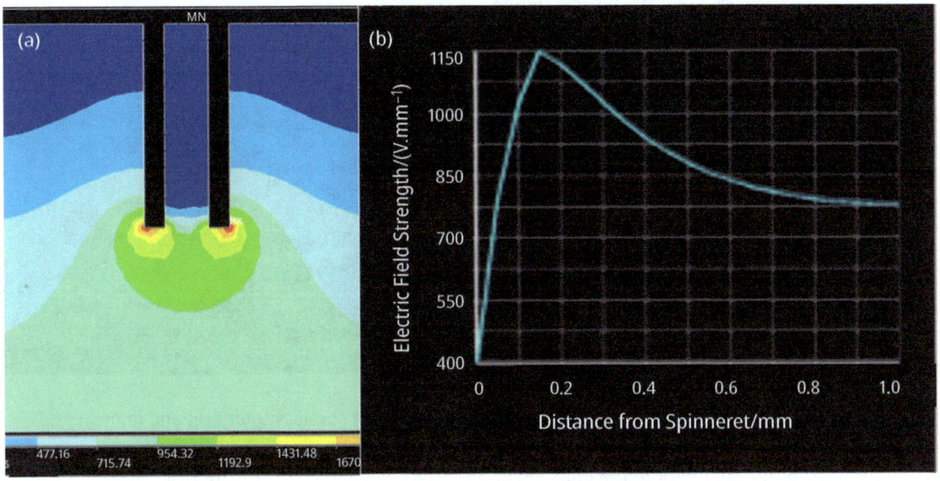

Figure 4.62: Electric field simulation for a coreless spinneret paired with a flat electrode: (a) cloud map depicting the distribution of electric field strength and (b) a curve representing the variation of electric field strength along the central axis of the needle tube, extending from the needle tip to the flat electrode. The distance between the spinneret and the collector is 1 mm, and the applied voltage is 1.0 kV is used in the simulation.

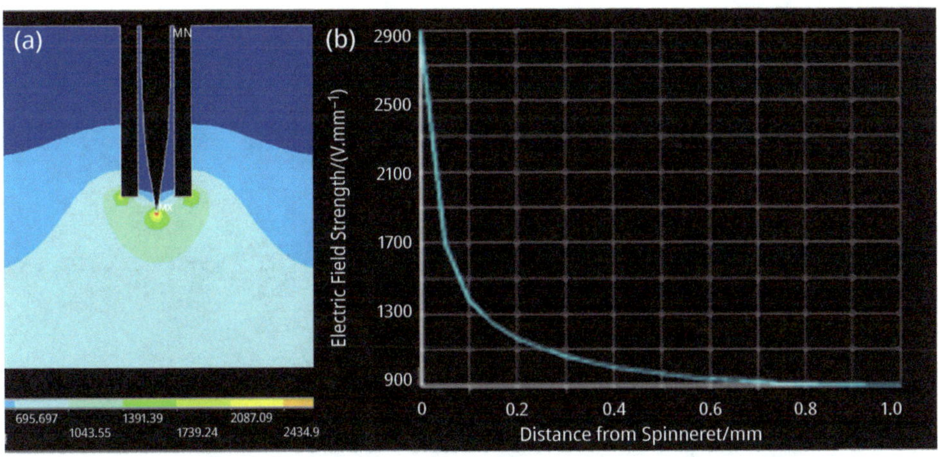

Figure 4.63: Simulation of electric field for a spinneret equipped with a core in conjunction with a planar electrode: (a) cloud map of electric field strength distribution and (b) electric field strength curve from the needle tip to the planar electrode (spinneret-to-collector distance is 1 mm and applied voltage is 1.0 kV).

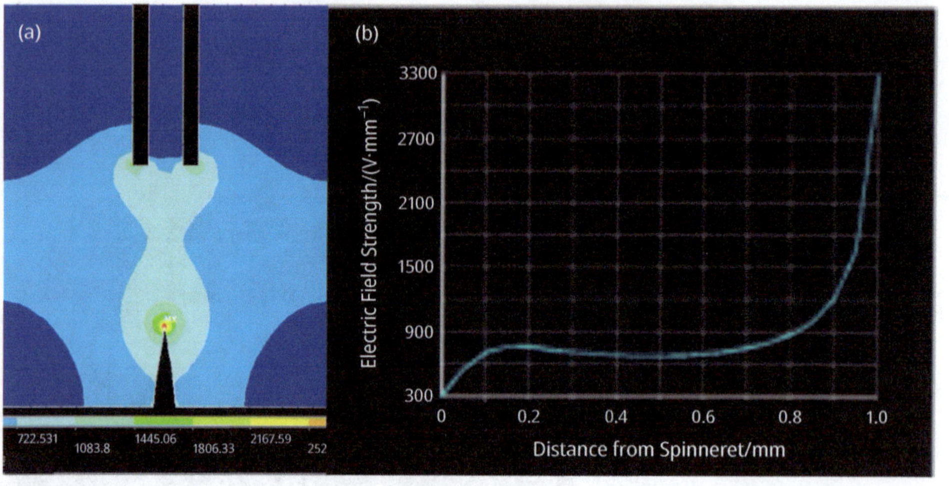

Figure 4.64: Simulation of the electric field for the combination of coreless spinneret and inducing tip electrode: (a) cloud map of electric field strength distribution and (b) electric field strength curve along the central line of the spinneret to the inducing tip electrode (spinneret-to-collector distance is 1 mm and applied voltage is 1.0 kV).

Combining the cored spinneret with the inducing tip electrode, the electric field in both the core region and the induced electrode region is enhanced significantly, as shown in Figure 4.65.

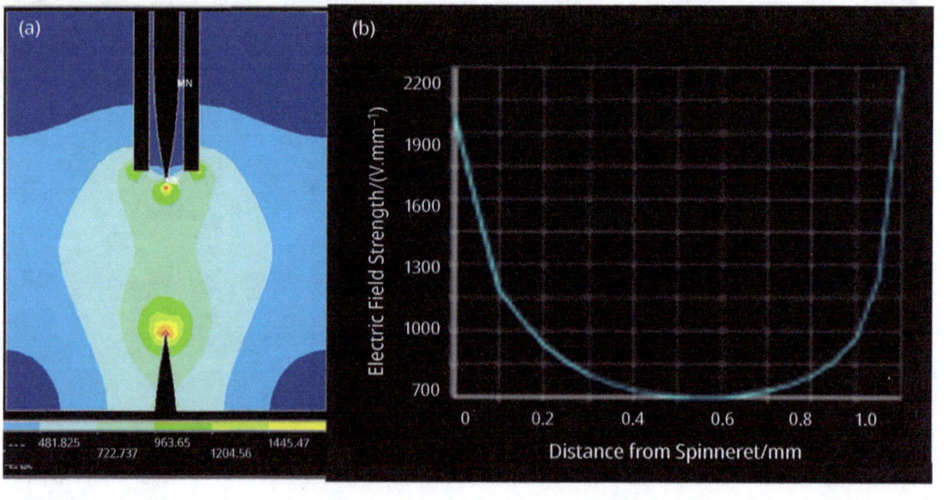

Figure 4.65: Electric field simulation for the combination of a cored spinneret and an inducing tip electrode: (a) cloud map of electric field strength distribution and (b) electric field strength curve from the core tip to the inducing tip electrode (spinneret -to-collector distance is 1 mm and applied voltage is 1.0 kV).

4.3.2.1.2 Simulation of space electric field with different tip angles

The distinctive feature of tip-induced EDW is the concentrated distribution of the electric field around the tip electrode, in contrast to the more dispersed field around a planar electrode. The geometry of the tip electrode plays a pivotal role in shaping this electric field distribution. To explore the influence of the tip electrode's shape on the surrounding electric field, we conduct simulations using various tip electrode configurations. These simulations, illustrated in Figure 4.66, examined how the electric field near the electrode tip changes with different tip angles. For this analysis, we employ ANSYS simulation software. The simulation parameters include a needle tube with an outer diameter of 0.51 mm and an inner diameter of 0.26 mm. The tip angles range from 40° to 180°, and the distance from the spinneret to the tip electrode is set at 1 mm. A voltage of 1.0 kV is applied to the needle tube. Our findings indicated that the maximum electric field strength at the tip could reach 2,234 V/mm for a tip angle of 40°. As the tip angle increases, the electric field strength diminishes, reaching 2,034 V/mm once the tip electrode's end face becomes flat.

As shown in Figure 4.67, the electric field intensity on the central axis of the needle electrode is illustrated. Figure 4.67(a) shows the electric field intensity curve for a needle electrode with a 40° tip angle. The right part of the curve exhibits an exponential rise, indicating a significant variation in the electric field strength near the needle electrode. In contrast, as seen from the curve with a 180° tip angle in Figure 4.67(b), the rise in the right half of the curve is relatively slow. This indicates that when the tip of the needle electrode becomes a plane, the amplitude of the electric field change around it is not as significant as that around a needle with a smaller tip angle.

The analysis suggests that the inducing tip electrode exhibits a more pronounced tip effect, and the electric field distribution is more focused when the tip angle is reduced.

4.3.2.2 Design and construction of tip-induced EDW setup

4.3.2.2.1 Overall design of experimental setup for tip-induced EDW

To further modulate the space electric field distribution, the experimental apparatus is set up by integrating the cored spinneret and the inducing tip electrode, as depicted in Figure 4.68. The core needle inside the spinneret facilitates control over the solution reflux and intensifies the local electric field at the spinneret tip, consequently lowering the critical voltage. The inducing electrode, positioned beneath the collector, fortifies the local electric field at the collection point on the collector, thereby guiding the ejection flow. In this setup, the inner diameter of the spinneret measures 260 μm. Both the core needle and the inducing electrode are made of stainless-steel needles of the same type, with a diameter of 170 μm and a tip angle of 40°.

The experimental setup of tip-induced EDW primarily consists of a spinneret clamping device, a bracket, and an inducing tip adjustment mechanism, as shown in

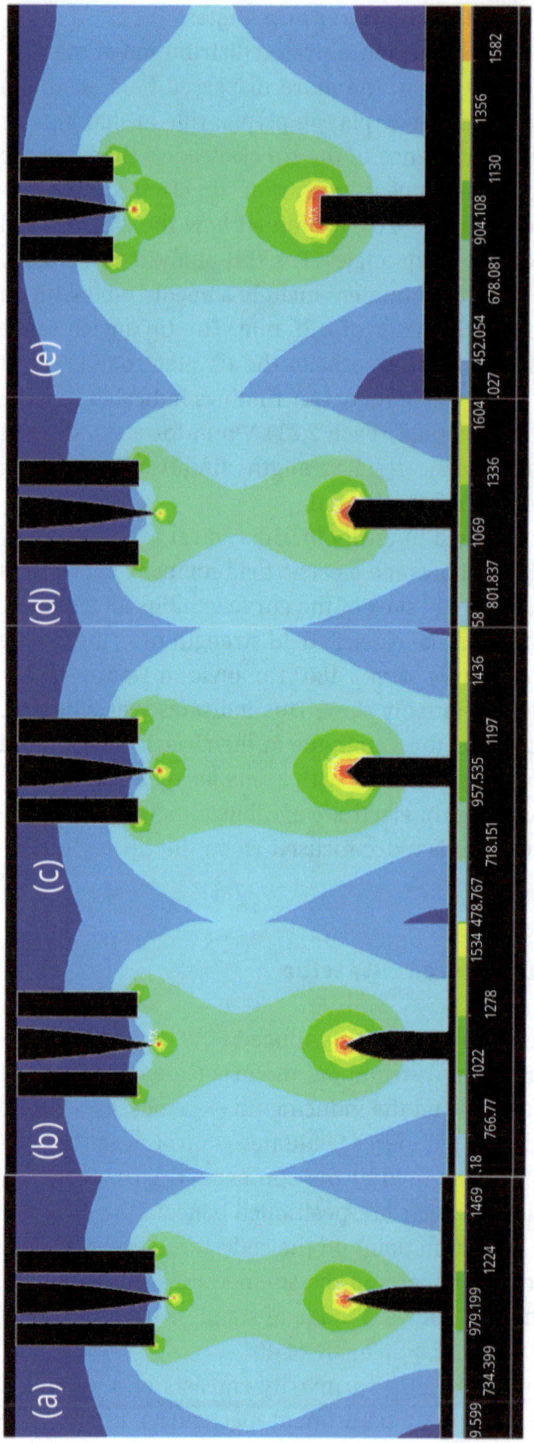

Figure 4.66: The distribution of electric field around the tip electrode with varied tip angles: (a) 40°; (b) 60°; (c) 80°; (d) 120°; and (e) 180°.

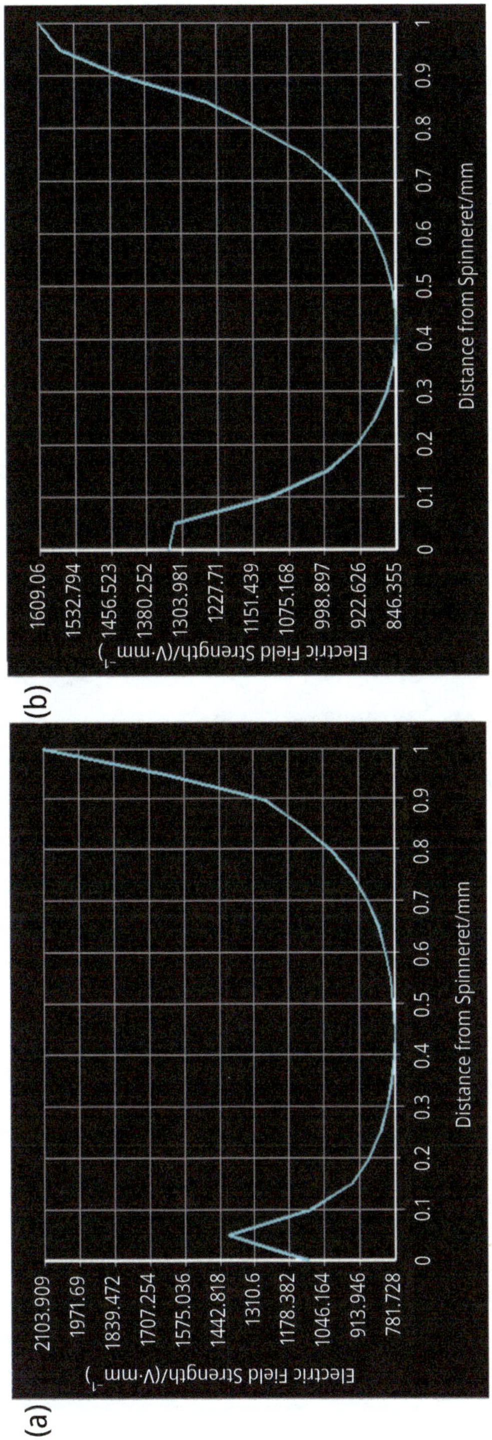

Figure 4.67: Electric field strength curves on the center line of the tip electrode with different tip angles: (a) 40° and (b) 180°.

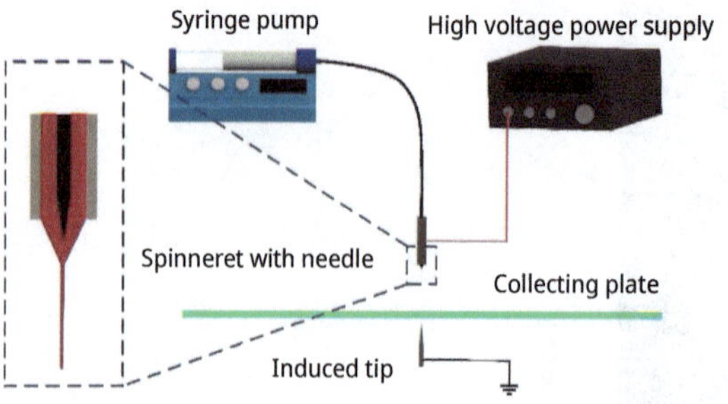

Figure 4.68: Combination of needle-core spinneret and inducing tip.

Figure 4.69. The bracket is used to secure the spinneret clamping device, with reinforcing ribs added in the middle to increase the load-bearing capacity. The bracket has cylindrical countersunk holes at its bottom for fixing to the lifting table. In terms of material selection for the device, it is necessary to consider the insulation. The bracket is made of polyethylene, which has good electrical insulation properties, excellent dimensional stability, good chemical stability, low water absorption, and resistance to weak acids and organic solvents. It can withstand short-term corrosion by organic solvents during the experiment.

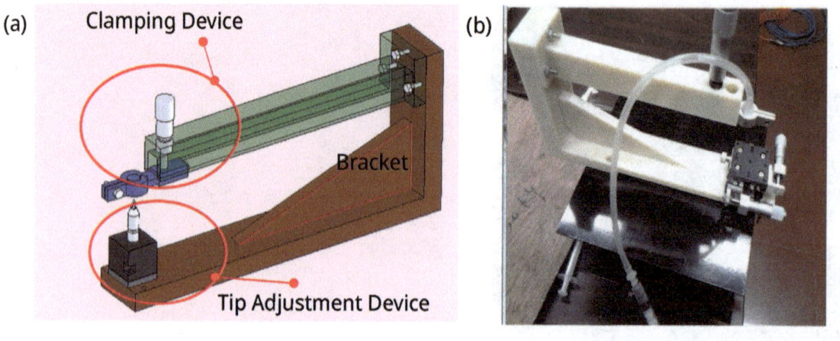

Figure 4.69: Experimental setup for tip-induced EDW: (a) diagram of the experimental setup and (b) physical image of the experimental equipment.

4.3.2.2.2 Spinneret clamping device

The spinneret clamping device is used for mounting and adjusting the vertical position of the spinneret, as shown in Figure 4.70. The small square in Figure 4.70(a) shows the spinneret and connector used, with the outer diameter of the connector being 10 mm. After inserting the spinneret into the mounting hole of the clamping device, tightening the screw bolt shown in the figure can secure the spinneret, as demonstrated in Figure 4.70(b). The adjustment of the spinneret is achieved through a through hole at the back end, which is locked with a screw at the back end. A spiral micrometer head is fixed on the support. When adjusting the spiral micrometer head, the entire clamping device moves up and down accordingly, with the minimum adjustment of the spiral micrometer head being 0.01 mm.

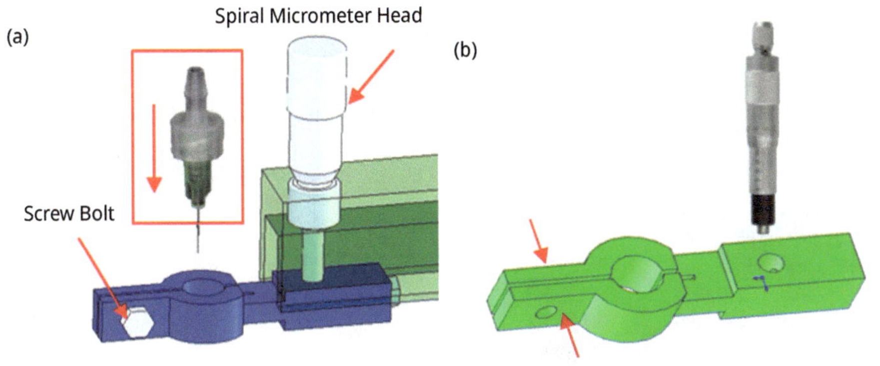

Figure 4.70: The spinneret clamping device.

4.3.2.2.3 Experimental setup for tip-induced EDW

The experimental setup of tip-induced EDW is shown in Figure 4.71. In this setup, the inner diameter of the needle tube is 60 μm, and the adjustable range of the spinneret-to-collector distance is 1–10 mm. The inducing tip is mounted on a miniature *X*–*Y* platform below, allowing for manual adjustment of its position. The collector is installed on a precision motion platform, enabling high-precision positioning and movement as the platform moves. The distance between the spinneret and the collector can also be adjusted via the adjusting knob on the spinneret. The needle tube of the spinneret is connected to the positive pole of a high-voltage DC power supply to keep it at a high potential, while the inducing tip electrode below the collector is grounded, keeping it at zero electric potential. The physical photograph of the experimental setup is shown in Figure 4.72.

4.3.2.3 Ejection behavior of tip-induced EDW jet

4.3.2.3.1 Critical conditions for tip-induced EDW jet ejection

As the jet forms at the cored spinneret, the maintaining voltage decreases due to the enhancement of local electric field. As shown in Figure 4.73, the maintaining voltage curves under different concentration solutions indicate that as the solution concentration increases, the surface tension of the solution also increases, requiring a higher maintaining voltage. Meanwhile, due to the enhanced electric field effect of the inducing tip electrode, the maintaining voltage is smaller than that required for the flat electrode.

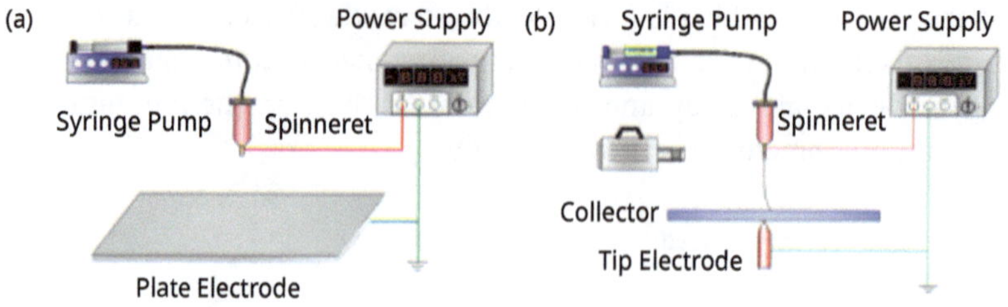

Figure 4.71: Schematic diagram of EDW experimental device: (a) flat electrode and (b) inducing tip electrode.

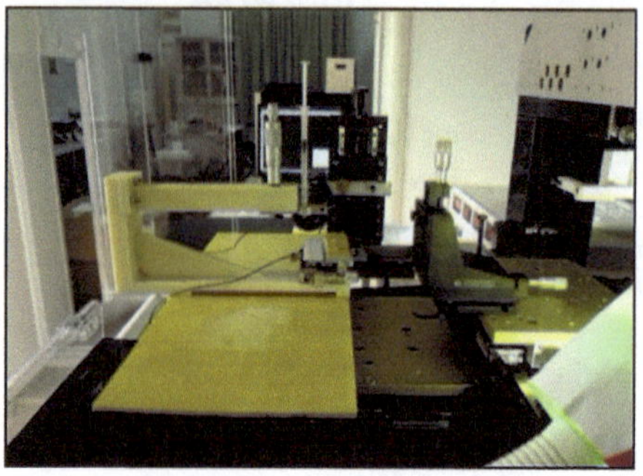

Figure 4.72: Physical diagram of tip-induced EDW experimental device: (a) side view and (b) the overall EDW system.

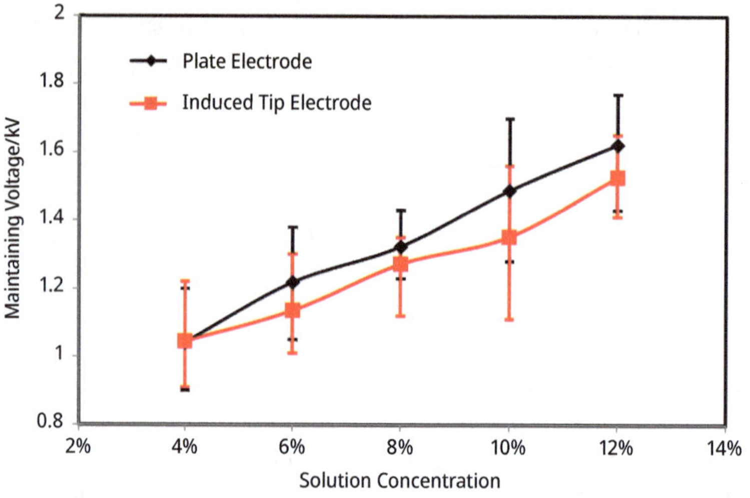

Figure 4.73: The relationship between the maintaining voltage and solution concentration (spinneret-to-collector distance is 2.5 mm and solution feeding rate is 50 μL/h).

4.3.2.3.2 Guiding effect of tip electrode on EDW jet

The direction of electric field lines is always from high potential to low potential. When a high voltage is applied to the spinneret and the inducing tip electrode is grounded, the electric field between the spinneret and the inducing tip electrode points from the spinneret to the inducing tip. Since the inducing tip electrode is the nearest zero-potential point to the spinneret, the electric field in the space will converge on the inducing tip. The main function of the inducing tip electrode is to enhance the local electric field and simultaneously guide the jet. Figure 4.74 shows the jets formation under different positions of the inducing tip electrode. To facilitate the observation, the collector is removed during filming, and the fibers are directly deposited on the inducing electrode. When the inducing electrode is located right below the spinneret, the jet moves vertically downward, and after reaching the tip electrode, the jet stacks upward layer by layer, as shown in Figure 4.74(a). When the position of the tip electrode is biased to the lower left or lower right of the spinneret, the jet is induced to move toward the tip on its formation, and deposits along this direction, as shown in Figure 4.74(b) and (c). This indicates that the deposition position of the jet is affected by the position of the tip electrode, confirming the guiding role of the inducing tip electrode on the jet.

In the middle path of the jet, introducing a glass object as a disturbance source on the left induces charges on its surface. These charges interfere with the jet, generating a force directed to the right, causing the jet to deviate, as illustrated in Figure 4.75(a). Consequently, the jet bypasses the disturbance object and deposits on the inducing tip electrode, resulting in a fiber deposition pattern as shown in Figure 4.75(b), with the arrow indicating the fiber deposition direction. Due to the presence of the disturbance object, fibers deposit away from it, forming an envelope shape.

After removing the disturbance object, the jet immediately returns to its vertical downward trajectory. As a result, the fibers deposit upward in layers, as depicted in Figure 4.75(c). This demonstrates that the inducing tip electrode possesses certain anti-interference ability and can guide the jet to overcome the electric field or external force disturbances in the surrounding environment, allowing deposition to occur on the tip.

4.3.2.4 Patterned deposition by tip-induced EDW

4.3.2.4.1 Positioning deposition of nanofibers by tip-induced EDW

Simulation studies reveal that the electric field around the inducing tip electrode converges toward the tip. Consequently, under the influence of this electric field, fibers can be effectively confined within a specific area, with the inducing tip serving to concentrate the electric field and restrict the deposition range of the fibers.

The electric field around the inducing electrode is always directed toward the tip, so the jet will deposit at the tip under the action of electric field force when it ap-

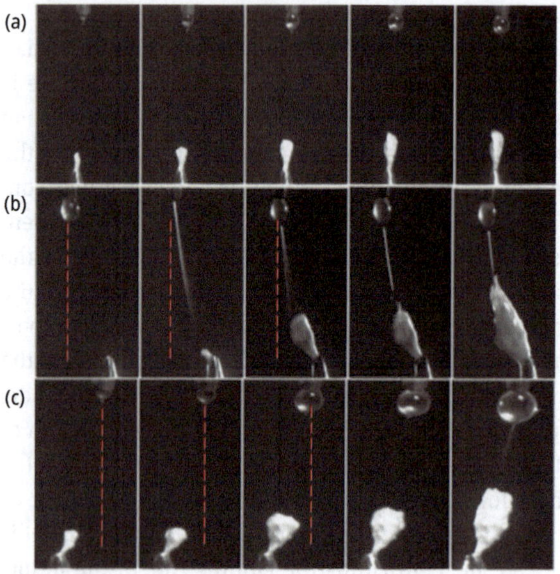

Figure 4.74: Guiding effect of tip electrode on the jet with the inducing electrode located: (a) right below the spinneret; (b) at the lower right of the spinneret; and (c) at the lower left of the spinneret (solution concentration is 10%, applied voltage is 2 kV, the spinneret-to-collector distance is 2.5 mm, and the liquid supply speed is 50 μL/h).

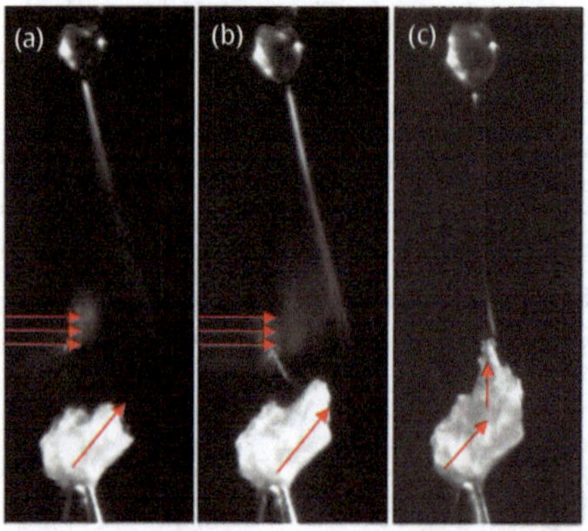

Figure 4.75: The anti-interference effect of the inducing tip electrode on the jet: (a) introducing a disturbance object to the middle of the jet; (b) the jet deviates to the right; and (c) withdrawing the disturbance object (solution concentration is 10%, applied voltage is 2 kV, spinneret-to-collector distance is 2 mm, and liquid supply rate is 50 μL/h).

proaches the collector. Therefore, when the spinneret is suspended at some point, the jet will deposit in a very small area, as shown in Figure 4.76(a). The deposition zone is circular with a diameter of about 350 μm. For flat electrode, however, the deposition area is much larger, as shown in Figure 4.76(b). By comparison, it is obvious that the fiber deposition area induced by tip electrode is much smaller than that without tip electrode.

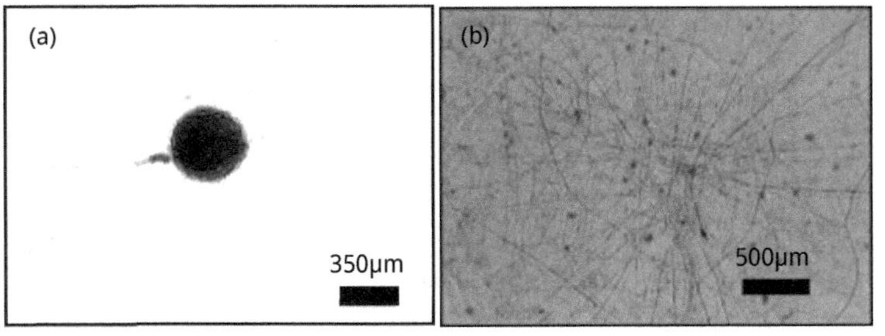

Figure 4.76: Comparison of positioning deposition area between inducing tip electrode and flat electrode: (a) inducing tip electrode and (b) flat electrode (solution concentration is 8%, spinneret-to-collector distance is 5 mm, applied voltage is 2.0 kV, silica substrate, and deposition time is 3 s).

The oscillation of the jet is greatly influenced by the electric field force; the higher the voltage, the more unstable the jet becomes. As shown in Figure 4.77, as the applied voltage increases, the fiber deposition area gradually expands, and the fiber morphology becomes increasingly disordered. From the statistical results of the diameter of the fiber deposition area in Figure 4.78, it can also be seen that when the voltage is less than 2.0 kV, the diameter of the fiber deposition area ranges from 500 to 600 μm, and the deviation of the deposition area diameter is relatively small. However, when the voltage is increased to 2.5 kV or even higher, both the diameter and deviation of the fiber deposition area increase.

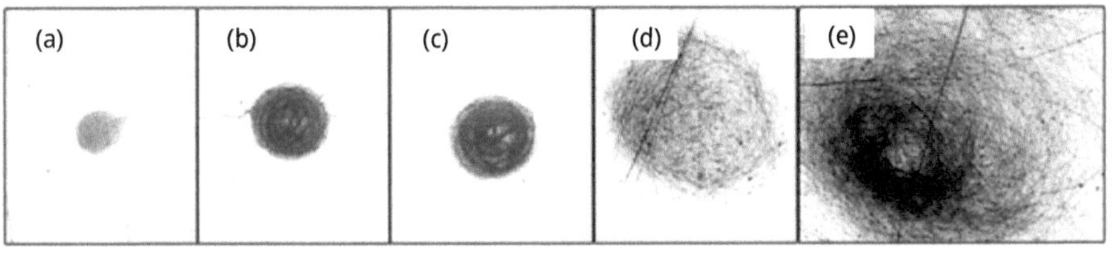

Figure 4.77: Tip-induced fiber deposition areas at different applied voltages: (a) 1.5; (b) 2.0; (c) 2.5; (d) 3.0; and (e) 3.5 kV (solution concentration is 8%, spinneret-to-collector distance is 5 mm, and deposition time is 3 s).

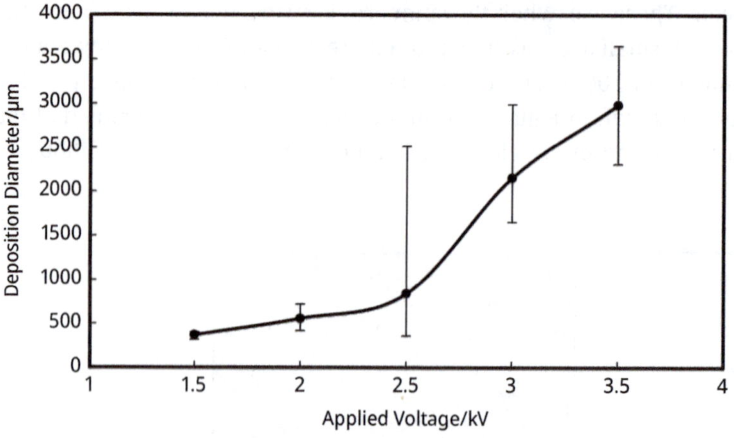

Figure 4.78: Relation between diameter of fiber deposition area and applied voltage (the solution concentration is 8%, spinneret-to-collector distance is 5 mm, and the deposition time is 3 s).

Figure 4.79 displays the fiber morphology of the tip-induced positioning deposition at a spinneret-to-collector distance of 5 mm. When the deposition duration is set to 0.5 s, the fibers are distributed within a circular area, with the diameters of the fiber deposition areas in the figure being 237 and 587 μm, respectively. Due to the relatively large spinneret-to-collector distance, the jet exhibits a noticeable whipping behavior, resulting in a disordered distribution of fibers within the deposition area on the collector.

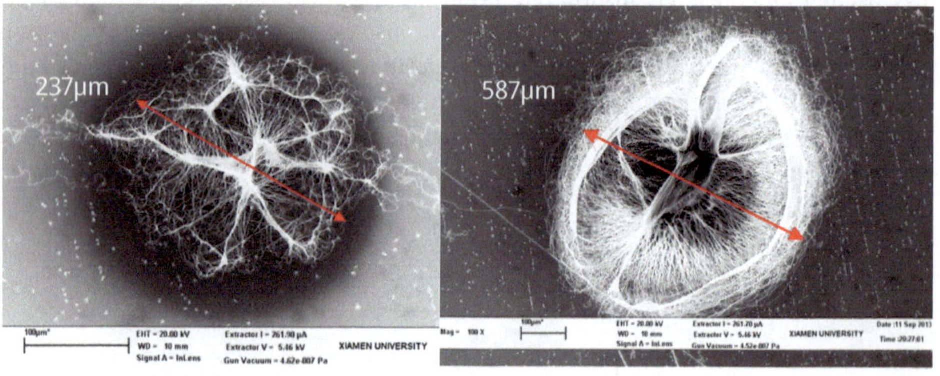

Figure 4.79: Agglomerated fibers deposited at fixed points (the solution concentration is 8%, the spinneret-to-collector distance is 5 mm, the applied voltage is 2.0 kV, and the deposition duration is 0.5 s).

The study then focuses on dot arrays generated by deposition with continuous pauses, measuring the distance between each pair of dots to evaluate the precision of fiber's positioning deposition. The resulting positioning error curves are shown in Figures 4.80 and 4.81.

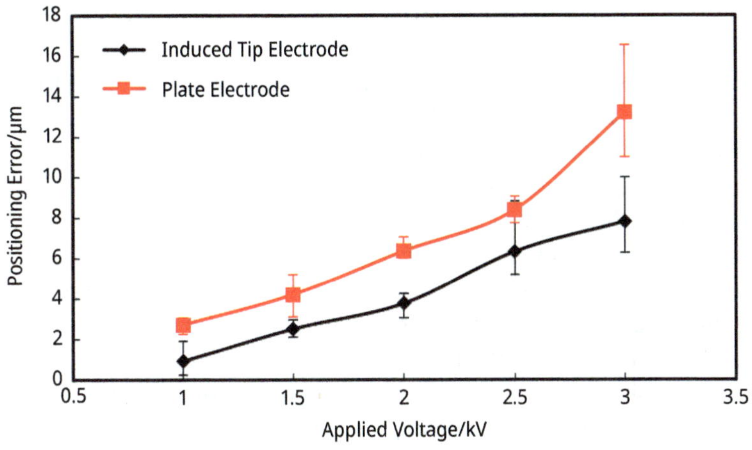

Figure 4.80: Relationship between applied voltage and positioning error (solution concentration is 8%, spinneret-to-collector distance is 3 mm, and deposition time is 1 s).

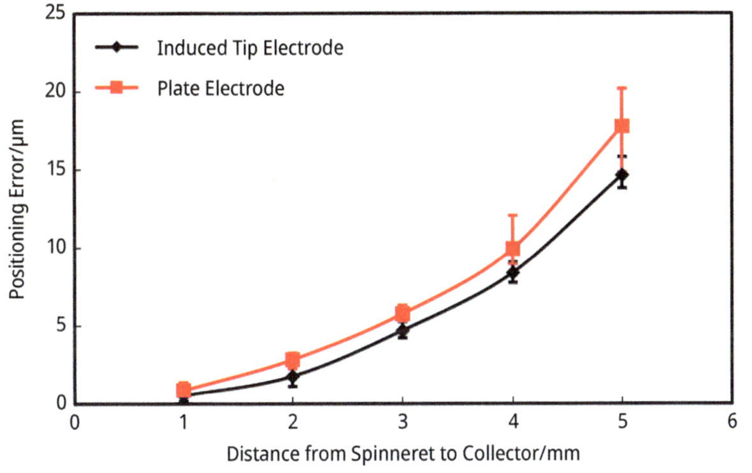

Figure 4.81: The relationship of positioning error and the spinneret-to-collector distance (solution concentration is 8%, deposition time is 1 s, and applied voltage is 2.0 kV).

Increasing the applied voltage can increase the charge density on the surface of the jet, enhance the charge repulsion force, and thus reduce the positioning accuracy of fiber deposition. Increasing the spinneret-to-collector extends the time it takes for the jet to reach the collector, resulting in a decrease in the positioning accuracy of fiber deposition. The tip electrode has an inductive effect on the jet, which can eliminate the influence of external disturbances on the positioning accuracy of jet deposition. Compared to the plate electrode, the positioning error is smaller.

4.3.2.4.2 Direct-writing of micro-nanofiber structures through tip-induced EDW

The fibers show different morphology at different collector moving speeds. The charge density on the jet surface affects the degree of jet deflection. Upon reaching the collector, the jet does not immediately release all its surface charges, particularly when the collector is composed of low-conductivity materials such as silicon dioxide. The slow migration of charges results in previously deposited fibers exerting a repulsive electrostatic force on subsequent jets, thereby affecting the positional accuracy of fiber deposition.

Figure 4.82 shows the deposition morphology of fibers electrospun at different collector moving speeds, with a designed fiber spacing of 200 μm. When the collector moves slowly, the time interval between the deposition of the first fiber and the second fiber is longer, resulting in less residual charge on the collector and less interference with the subsequent deposition of the second fiber. Therefore, when the collector moving speed is low, the fluctuation in fiber spacing is small. When the collector moving speed is below 30 mm/s, the fluctuation range of fiber spacing is less than 100 μm. However, as the collector moving speed increases, the residual charge on the collector increases, the subsequent jet oscillates under the charge interference, resulting in increased fluctuations in fiber spacing. When the collector moving speed is raised to 60 mm/s, the fluctuation range increases to 300 μm, as shown in Figure 4.83.

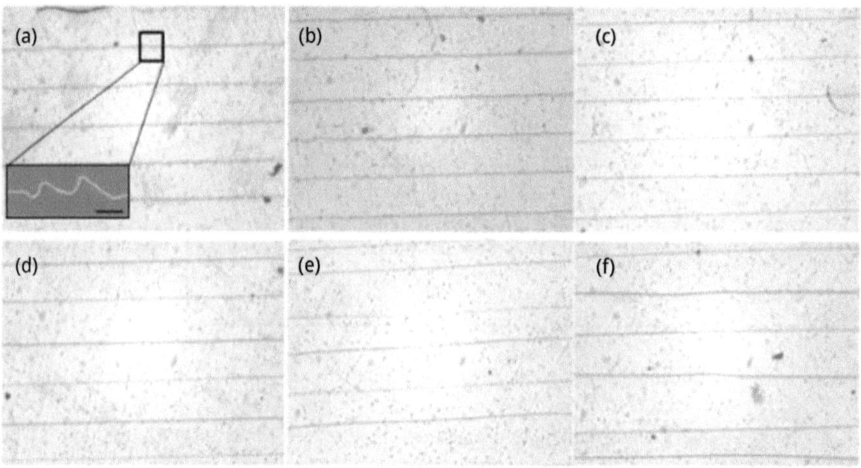

Figure 4.82: Morphology of fibers collected at different collector speeds: (a) 10 mm/s, the scale for the small box is 10 μm; (b) 20; (c) 30; (d) 40; (e) 50; and (f) 60 mm/s (applied voltage is 1.0 kV, spinneret-to-collector distance is 1 mm, and solution concentration is 10%).

The collector moving speed not only affects the fiber spacing but also affects the fluctuation of individual fibers. When the collector moves slowly, the jet deposition speed is much greater than the collector moving speed, causing the jet to whip and form wavy fibers on the collector. Figure 4.84 shows the wavy fibers obtained at a collector moving speed of 1 mm/s.

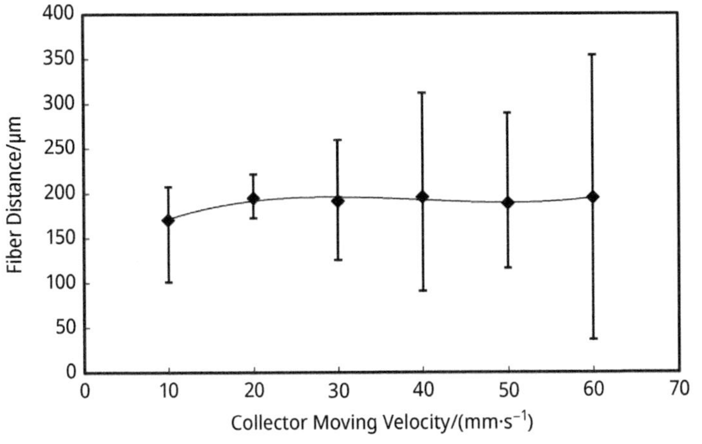

Figure 4.83: Variation of the fiber spacing at different collector speeds (applied voltage, spinneret-to-collector distance, and solution concentration are 1.0 kV, 1 mm, and 10%, respectively).

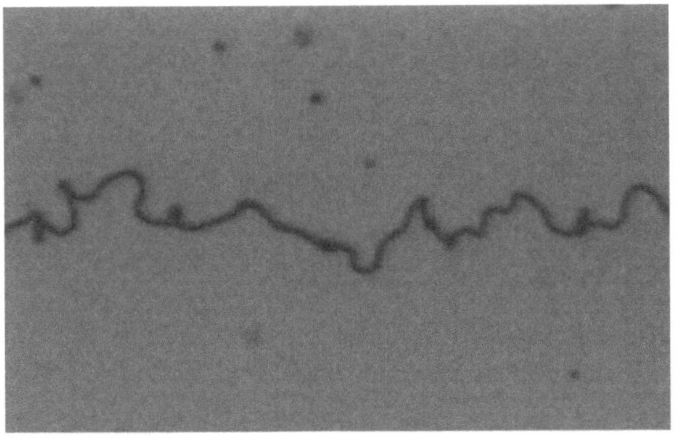

Figure 4.84: Wavy fibers obtained at a collector speed of 1 mm (applied voltage, spinneret-to-collector distance, and solution concentration are 1.0 kV, 1 mm, and 10%, respectively).

The positioning errors of fibers at different collector moving speeds are shown in Figure 4.85. When the collector moves slowly, wavy fibers are produced. With the increase of the collector moving speed, it gradually matches the jet deposition speed, the positioning error gradually decreases, resulting in straight fibers.

The morphology of a single fiber is shown in Figure 4.86. It can be seen that the diameter of the single fiber is at the nanometer range, but the surface of the fibers is not smooth. This is because the solvent continues to evaporate in a certain period after the PEO solution reaches the collector, leaving many PEO solid particles on the surface of the fiber.

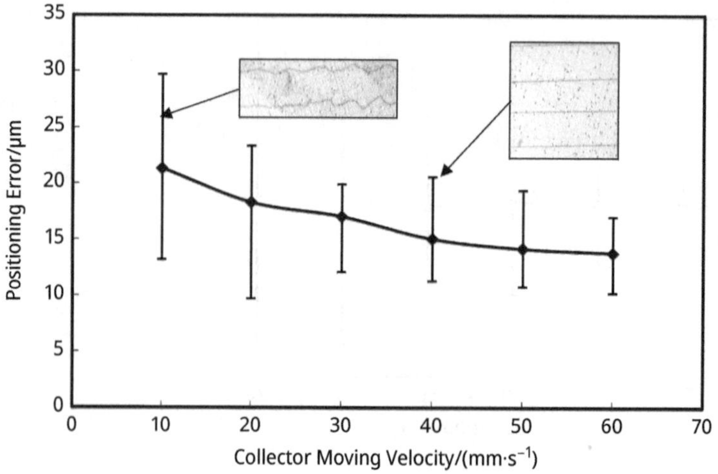

Figure 4.85: Influence of collector velocity on the fiber positioning error (applied voltage, spinneret-to-collector distance, and solution concentration are 1.0 kV, 1 mm, and 10%, respectively).

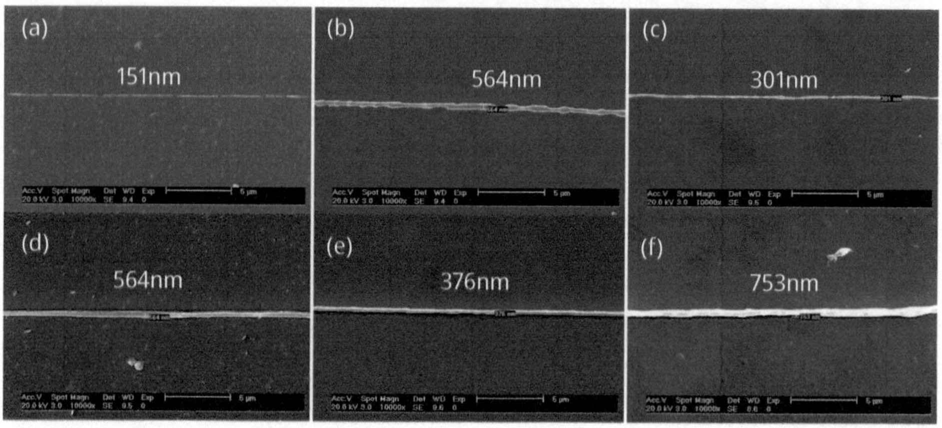

Figure 4.86: The morphology of a single fiber at different collector speeds: (a) 10; (b) 20; (c) 30; (d) 40; (e) 50; and (f) 60 mm/s (applied voltage is 1.0 kV, spinneret-to-collector distance is 1 mm, and solution concentration is 10%).

4.3.2.4.3 Patterned micro-nanostructures prepared by tip-induced EDW

The alphabet patterns "X" and "M" are designed and converted into codes, which are then imported into the ACR-View software and downloaded into the Parker control card for precise control of the motion platform. As the platform moves in the designated trajectories and the jet ejects, the fibers deposit on the collector to form corresponding graphics, as illustrated in Figures 4.87 and 4.88. The arrows in the figures denote the starting line segment of the graph and the direction of direct-writing.

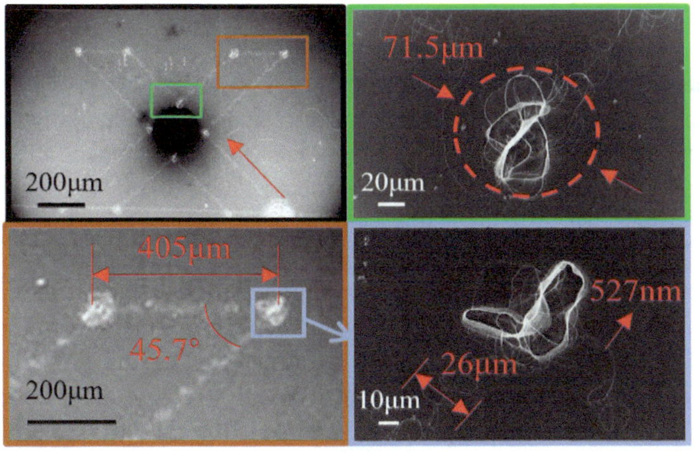

Figure 4.87: Letter "X" prepared by tip-induced EDW (solution concentration, collector speed, spinneret-to -collector distance, applied voltage are 10%, 10 mm/s, 3 mm, and 1.0 kV, respectively, and the pause time at the corner is 0.2 s).

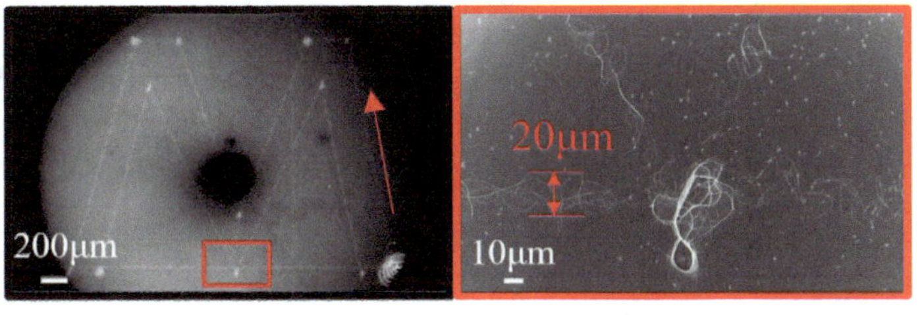

Figure 4.88: Letter "M" generated by tip-induced EDW (solution concentration, collector speed, spinneret-to-collector, and applied voltage are 10%, 10 mm/s, 3 mm, and 1.0 kV, respectively, and the pause time at the corner is 0.2 s).

As shown in Figure 4.87(a), the pattern "X," with a pause of 0.2 s at each corner, has relatively clear corner points. Measurements indicate that the fibers at these corner points are distributed within an area of approximately 71.5 μm in diameter. The diameter of individual fibers in the straight sections of the letter measures about 527 nm, while the distribution width of the fibers is roughly 26 μm.

Figure 4.88 displays the letter "M" pattern, with arrows indicating the starting line segment of the graph. The collector moves at 10 mm/s with a pause of 0.2 s at each turning point. Due to the slower collector moving speed, the line segments of the graph consist of continuously curved fibers, with a line width of about 20 μm.

4.3.3 Electrohydrodynamic direct-writing with electric field focusing

4.3.3.1 Basic principle of EDW with electric field focusing

The introduction of an auxiliary guiding tip facilitates the creation of a spatially focused electric field, enabling the confinement of charged jets. This development offers a novel approach to mitigate jet whipping and ensure stable jet deposition. The principle is shown in Figure 4.89(a). The grounded guiding tip is positioned beneath the collector, and an insulating glass sheet is used as the collector. The spinneret is connected to the positive pole of the high-voltage power supply. When subjected to the high electric field, a focused electric field emerges between the spinneret and the guiding tip with the electric field line pointing to the tip, as depicted in Figure 4.89(b). The collector, mounted on a precision motion platform, can traverse designated paths. The juncture of the auxiliary guiding tip with the insulating collector accumulates charge. This interaction results in a heterogeneous distribution of charges over the collector, demarcating areas of elevated and reduced charge densities. Charges coalesce at the nexus of the guiding tip and collector, forming a concentrated charge zone. Given the limited charge mobility within the insulating collector plate, charges are localized near the guiding tip, while other regions remain comparatively charge-deficient. Influenced by electric field forces, the jet gravitates toward the concentrated charge area, ensuring that nanofibers are preferentially deposited in these regions of the insulating collector plate. This technique enhances the precise deposition of direct-written nanofibers.

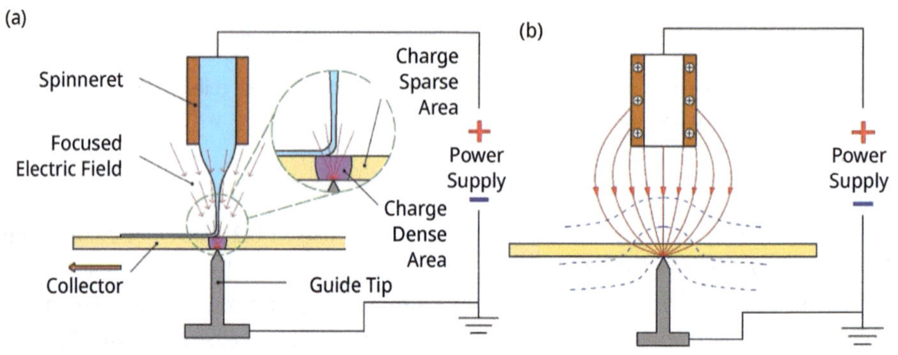

Figure 4.89: EDW system with focused electric field: (a) schematic diagram of the EDW with electric field focusing and (b) schematic diagram of focused electric field distribution.

The guiding tip facilitates the focusing of the space electric field and induces the formation of local charge-dense area on the insulating collector, thereby guiding the precise deposition of nanofibers and enhancing the positioning accuracy. Increasing the gradient of the space electric field distribution and the charge density within the charge-dense area is beneficial for strengthening the focusing constraints and the deposition inducing effects of the electric field. Analyzing the movement and deposition behavior of charged jet under the effect of electric field focusing, mastering the control strategy of the focused electric field, and identifying the mechanisms by which the shape, size, and spatial position of the guiding tip affect the electric fields have become the research focus for the precise patterned printing through electric field-focused EDW.

4.3.3.2 Simulation of space electric field for EDW with electric field focusing

Figure 4.90 illustrates the space electric field simulation model for EDW with electric field focusing. The fundamental dimensions of this simulation model include a spinneret with an inner diameter of 0.5 mm, positioned 1 mm away from the collector, which is an insulating glass slide measuring 0.2 mm in thickness.

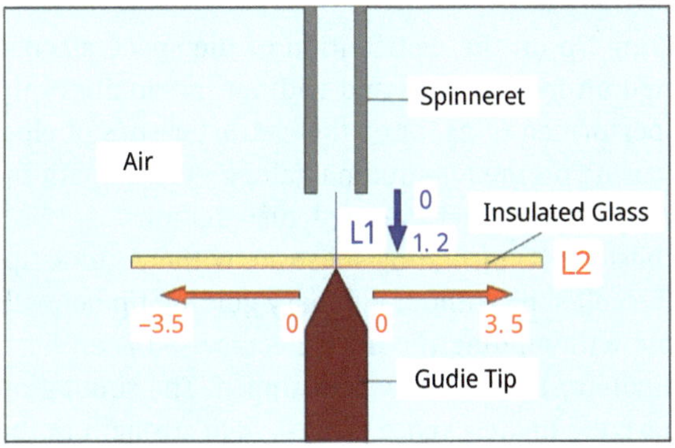

Figure 4.90: Schematic diagram of the space electric field simulation model for electric field-focused EDW.

The simulation conditions are set as follows:

(1) Simulation type: electric field simulation;
(2) Material properties: two materials are set, namely air and insulating glass sheet, with relative permittivity of 1.0 and 2.2, respectively; and
(3) Boundary conditions: the voltage potentials applied on the boundary lines of the spinneret and the guiding tip are 1.0 kV and 0 V, respectively.

In EDW experiments, the focused space electric field helps to improve the stability of jet motion and the accuracy of jet deposition. Therefore, the longitudinal electric field distribution from the spinneret to the collector and the transverse electric field distribution across the collector surface are investigated. The analysis is conducted along two directions: the vertical path L_1 and the horizontal path L_2, as shown in Figure 4.90.

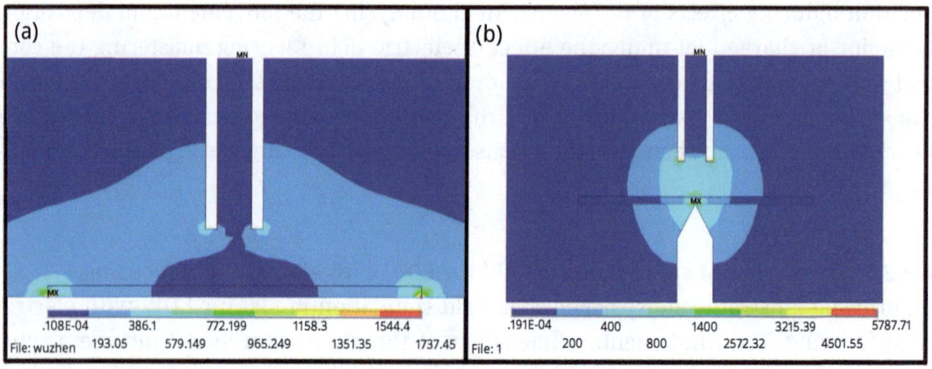

Figure 4.91: Electric field distribution of EDW: (a) insulating substrate without guiding tip and (b) insulating substrate with guiding tip.

To investigate the effect of the guiding tip on the distribution of the space electric field, simulation studies are conducted on models with and without the auxiliary tip electrodes. Comparative analysis is performed to examine the characteristics of electric field distribution, specifically focusing on the longitudinal (along vertical path L_1) and transverse (along horizontal path L_2) electric field strength distributions.

Two simulation models are established: (1) insulating substrate without guide tip: an insulating glass sheet is used as the collector without auxiliary guiding tip beneath it and (2) induced insulating substrate with guiding tip: the collector used is an insulating glass sheet, under which an auxiliary tip electrode is equipped. The simulation results are shown in Figure 4.91, where the highest space electric field strength in the insulating substrate without a tip electrode is on both sides of the collector, at 1,737.4 V/mm, while the electric field strength directly below the spinneret is very small. The electric field mainly concentrates on the side walls of the collector, making it difficult to form a downward focusing electric field to ensure the steady jet ejection. Figure 4.91(b) shows that when an auxiliary tip electrode is placed beneath the insulating collector, a strong electric field in an arc shape is formed between the spinneret and the electrode tip. The closer to the tip, the greater the electric field strength, and the maximum electric field strength occurs at the top of the tip (where it contacts the collector), reaching 5,787.7 V/mm. The focused electric field lines point toward the electrode tip, inducing the charged jet to stably deposit toward the charge-dense area.

Comparative analysis of electric field simulation results between silicon substrates without and with a tip electrode is shown in Figure 4.92. For the former, the electric field strength is the highest at the edge of the spinneret; and the closer it is to the collector, the smaller the electric field strength. This can form a relatively concentrated and stable electric field, guiding the charged jet toward the collector. However, as the concentrated electric field strength is small (maximum value of 1,724.4 V/mm), compared to the insulating substrate with an inducing tip, its ability to constrain the jet is somewhat reduced. The distribution of electric field on the silicon substrate with auxiliary guiding tip is shown in Figure 4.92(b). It can be seen that its electric field distribution is basically consistent with that of the silicon substrate without a guiding tip. Since the silicon chip is a good semiconductor and the needle tip is also a conductive electrode, the combination of the two can be considered as a single conductor, with its upper surface being the silicon wafer collector; thus the electric field distribution is similar to that without guiding tips.

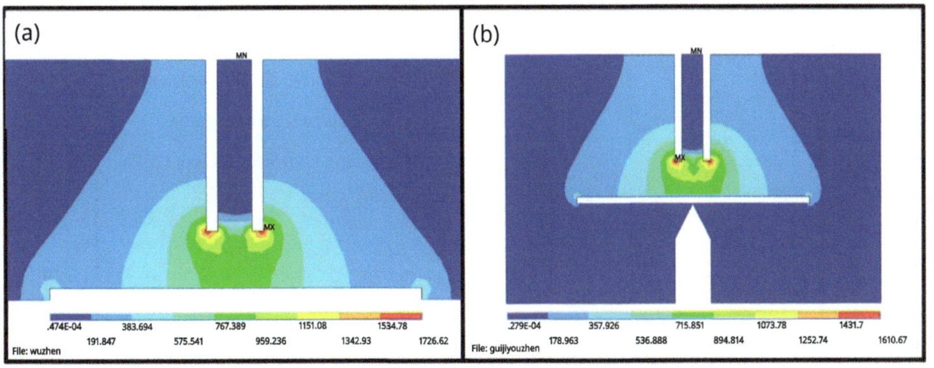

Figure 4.92: Space electric field distribution on silicon collector: (a) silicon substrate with needle tip and (b) silicon substrate with needle tip.

The variation curves of the electric field strength along the horizontal path L_2 and the vertical path L_1 help to elucidate the role of the guiding tip in the formation of a focused electric field as shown in Figures 4.93 and 4.94. The electric field distribution on the collector surface for the silicone substrate without a guiding tip is consistent with that on the silicon substrate with an inducing tip. Along the horizontal path L_2, the change in electric field strength on the collector surface is relatively gentle. The maximum value is 853.4 V/mm, within a horizontal position range of ±1 mm, and the electric field strength decreases by 302.61 V/mm. On the vertical path L_1, the electric field strength is low, and the variation gradient is very small, indicating that the focusing effect of the electric field between the spinneret and the collector is not pronounced. For the horizontal path L_2 on an insulating substrate without guiding tips, the electric field is concentrated on the outermost side of the collector, with a maximum value of 363.58 V/mm, and the electric field strength on the collector directly below the spinneret is 55.07 V/mm, resulting in a more dispersed electric field distribution. Conversely, for the insulating substrate with a guiding tip, along the horizontal path L_2, as the horizontal position varies from −2.0 to 2.0 mm, the electric field strength rises from 633.18 V/mm to 3,913.1 V/mm before decreasing to 603.56 V/mm. The electric field distribution on both sides is essentially the same, and a focused electric field forms above the guiding tip. For the vertical path L_1, the electric field strength between the spinneret and the collector changes very rapidly, with a variation of 2,800 V/mm within the vertical-position range of 0.875–1.0 mm. These results show that the guiding tip has a significant electric field focusing effect. The jet carries positive charges under the influence of the high-voltage electric field. A more concentrated electric field exerts a stronger constraining effect on the jet. The focusing effect of the guiding tip is substantial, playing a crucial role in suppressing the unstable whipping motions during jet ejection process and improving the deposition accuracy of micro-nanofibers.

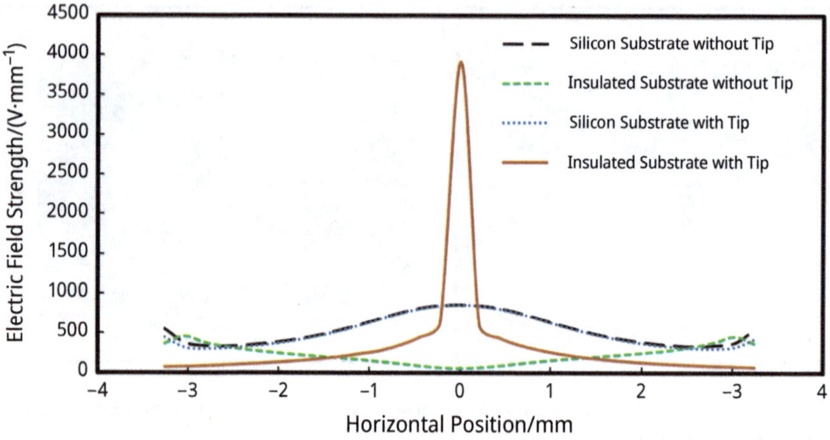

Figure 4.93: Variation curves of the lateral electric field strength (horizontal path L_2).

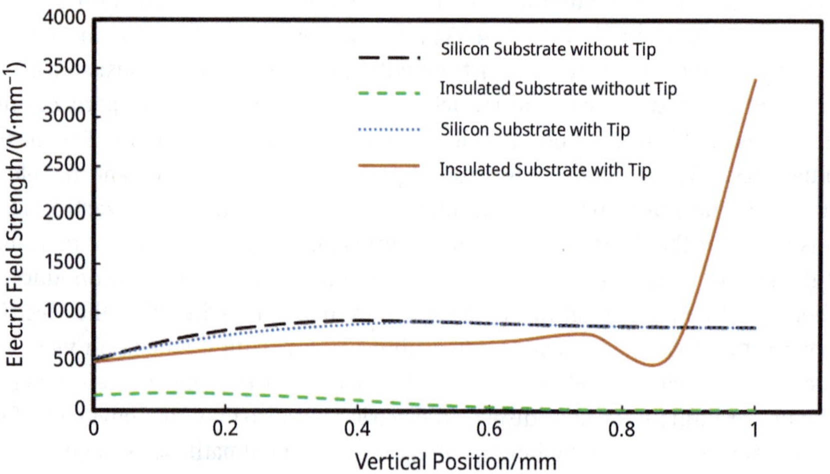

Figure 4.94: Variation curves of longitudinal electric field strength (vertical path L_1).

4.3.3.2.1 The focusing effect of the tip geometry on the space electric field

Auxiliary guiding tip can be of various tip geometries such as flat-headed, rounded-headed, or pointed-headed. The tip geometry affects the intensity of the tip effect, thereby altering the distribution of the electric field, as shown in Figure 4.95. When using a flat-headed tip, the maximum electric field strength is located at the vertices on both sides of the tip, amounting to 2,372.18 V/mm. The electric field in the middle section is symmetrically distributed, and there is a significant decrease in electric field strength near the tip, as shown in Figure 4.95(a). The maximum electric field strength along the horizontal path L_2 of the collector is 1,175.1 V/mm, while the electric field at the very center of the tip is relatively weak at 751.19 V/mm. Figure 4.96 shows the variation curve of the transverse component of the focused electric field on the collector. For flat-headed tip, since the electric field is concentrated at the vertex positions on both sides of the tip, the jet is induced to move toward these two vertices, leading to unstable whipping motions of the jet. When the needle tip is round-headed, the electric field distribution is as shown in Figure 4.95(b). The maximum electric field strength is at the top of the round head, and the electric field strength gradually diminishes on both sides. Consequently, the jet moves vertically downward under the influence of the electric field. In contrast, for a point-headed tip, the electric field simulation results are presented in Figure 4.95(c). The maximum electric field strength along the horizontal path L_2 is 3,913.1 V/mm, significantly higher than 1,355.3 V/mm for the round-headed tip. The electric field strength on both sides of the pointed tip drops sharply, resulting in a more evident electric field focusing effect. Therefore, choosing a point-headed tip in the direct-writing process of electrospinning can enhance the focusing capability of the electric field and improve the positioning accuracy of the jet deposition.

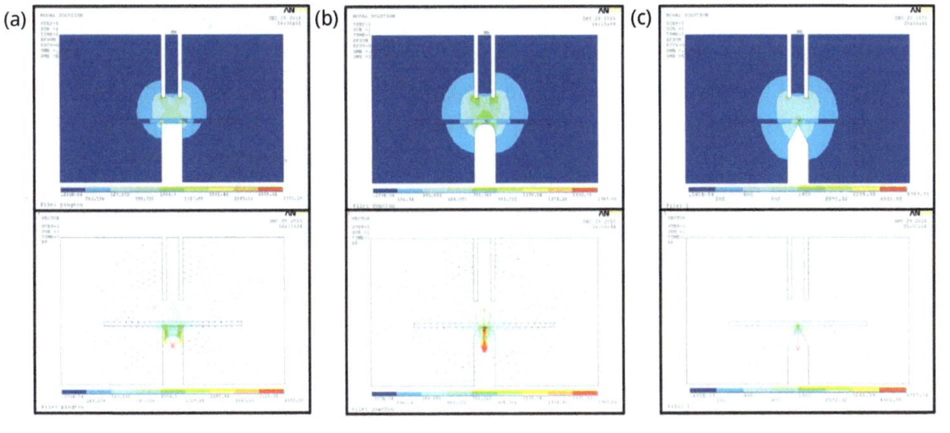

Figure 4.95: Cloud maps and vector maps of electric field distribution for electric field-focused EDW with different tip geometries: (a) flat-headed tip; (b) round-headed tip; and (c) pointed tip.

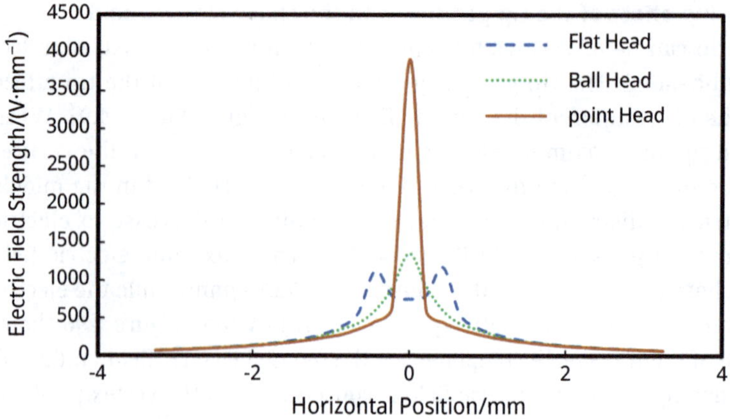

Figure 4.96: Change curves of the transverse electric field strength (horizontal path L_2) for different tip geometries.

4.3.3.2.2 The effect of the tip diameter on the focusing of space electric field

Investigating the impact of the tip diameter on the distribution of the space electric field is crucial for the selection of the tip electrode during the experiments. Research on the distribution of space electric field is conducted using electrodes with tip diameters of 0.2, 0.5, 0.8, 1.0, and 1.5 mm. The simulation results are shown in Figure 4.97. From the electric field distribution contour plot, it can be observed that as the tip diameter increases from 0.2 to 1.5 mm, the maximum value of the electric field strength decreases from 4,189.5 to 3,828.5 V/mm, a change of only 8.6%. The variation curve of the maximum values of electric field strength is shown in Figure 4.98. Comparing the simulation results for tip diameters of 0.2, 0.8, and 1.5 mm, the variation curve of transverse electric field strength on the collector is shown in Figure 4.99. It shows that the tip diameter does not affect the range of electric field focusing; the range of rapid change in the electric field is consistently −0.02–0.02 mm, indicating that the focusing effect of the electric field is basically consistent, thereby also resulting in a uniform constraining effect on the jet. Taking stainless steel needle tips as an example, considering that needle tips with diameters less than 0.5 mm are prone to deformation, which affects experimental operations, needle tips with a diameter of 0.5 mm are adopted.

4.3.3.2.3 The effect of the needle tip angle on the focusing of space electric field

According to the simulation results of the electric field distribution characteristics with different tip geometries, the point-headed tip shows the best focusing effect. Moreover, the tip angle also affects the spatial distribution of the electric field. Therefore, exploring the impact of tip angle on the electric field focusing effect is of great significance. Therefore, electric field distribution simulation for tip angles ranging from 20° to 120° are conducted. The electric field distribution contour plot and the curve of maximum electric

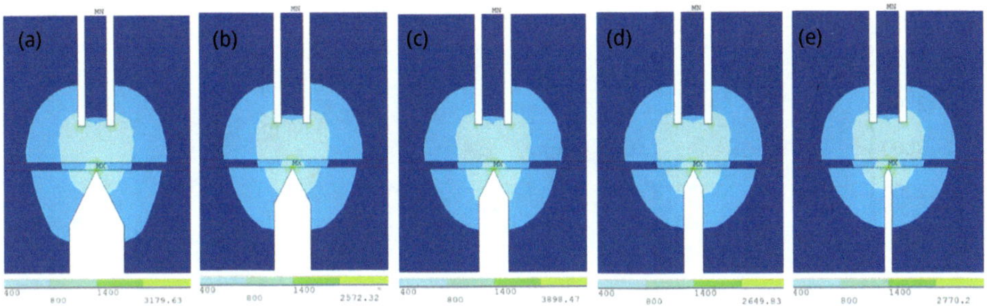

Figure 4.97: Contour plot of electric field distribution for electric field-focused EWD with different tip diameters: (a) 0.2; (b) 0.5; (c) 0.8; (d) 1.0; and (e) 1.5 mm.

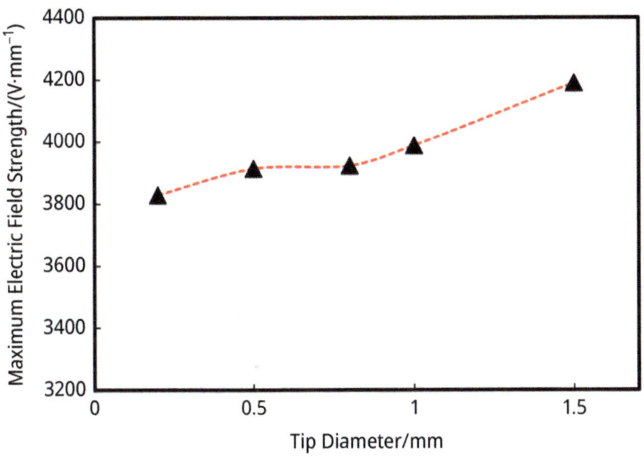

Figure 4.98: Relationship between the maximum electric field strength and the tip diameter.

field strength are shown in Figures 4.100 and 4.101, respectively. With a tip angle of 20°, the maximum value of electric field strength is the highest at 4,639.6 V/mm, and the electric field is focused around the tip. As the tip angle increases, the maximum value of electric field strength gradually decreases. Figure 4.101 shows that the maximum electric field strength decreases roughly linearly with the tip angle; that is, the smaller the tip angle, the more concentrated the electric field. Moreover, when the tip angle is greater than 60°, the space electric field distribution near the tip gradually diverges, and the electric field focusing effect weakens. At a tip angle of 120°, the maximum electric field strength drops to 2,321.6 V/mm, the transverse electric field strength changes slowly, the electric field distribution near the tip diverges, the focusing effect of the electric field becomes worse, and the stability of jet ejection decreases. For a tip angle of 180°, which is the flat-headed tip in the tip geometry simulation, the maximum electric field strength

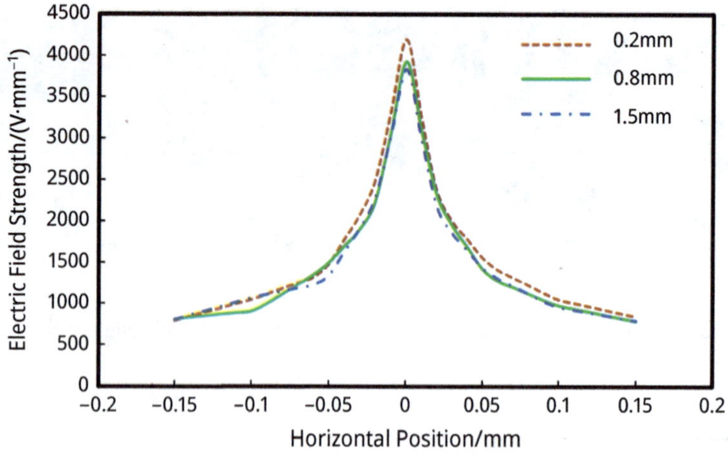

Figure 4.99: Variation curves of the electric field strength along horizontal path L_2 for different tip diameters.

decreases to 1,175.1 V/mm, and it is distributed on the vertices of both sides of the tip, indicating the most divergent electric field distribution.

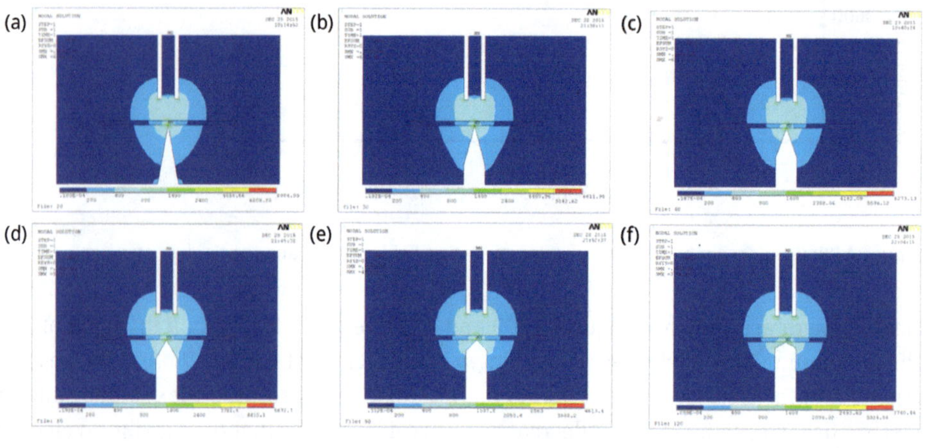

Figure 4.100: Contour plot of electric field distribution for electric field-focused EWD with different tip angles.

To further analyze the impact of tip angle on the focusing effect of electric field, the variation of the transverse electric field strength on the collector with different tip angles is investigated as shown in Figure 4.102. The horizontal position at the tip is marked as 0 mm. When the tip angle is 20°, the maximum electric field strength reaches 4,639.6 V/mm at the tip, and the electric field strength near the tip changes rapidly, with a lateral range of significant change from −0.05 to 0.05 mm, indicating that the spatial electric field near the needle tip is concentrated and the focusing effect is significant.

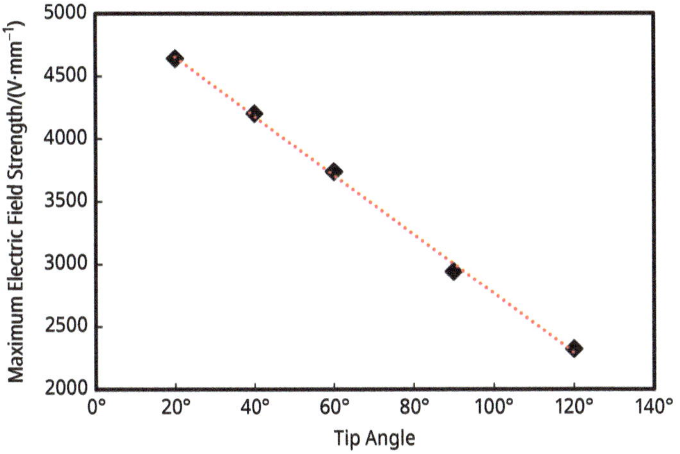

Figure 4.101: Relationship between the maximum electric field strength and the tip angle.

With the increase of the tip angle, the maximum value of the electric field gradually decreases. As the tip angle increases to 120°, the maximum value of the electric field drops to 2,321.6 V/mm. For this case, The curve exhibits the smallest slope, indicating that the electric field near the tip is highly divergent, resulting in the weakest focusing effect.

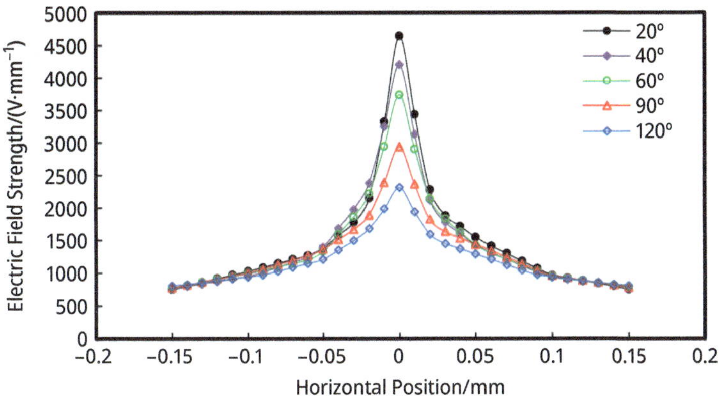

Figure 4.102: Variation curve of transverse electric field strength (horizontal path L_2) at different tip angles.

4.3.3.2.4 The effect of the guiding tip position on the focusing of the space electric field

In addition, the relative position of the guiding tip to the spinneret also has a significant effect on the distribution of the space electric field. When the guiding tip is placed right below the spinneret, the electric field around the tip is stronger, and the distribution of the electric field on both sides is basically consistent, as shown in Figure 4.103(a). When the tip electrode is shifted 0.5 mm to the right, the maximum electric field at the tip decreases slightly from 6,611.95 to 6,496.3 V/mm, a small reduction, as shown in Figure 4.103(b). However, the maximum electric field on the horizontal path L_2 decreases significantly, from 4,428.1 to 1,310.2 V/mm, as shown in Figure 4.104, and the variation in electric field intensity is slow, indicating a decrease in the focusing effect of the electric field. At the same time, the electric field distribution vector map shows that the electric field lines no longer point vertically downward, but are inclined toward the guiding tip in general, leading to deposition of the jet toward the tip along the electric field lines. When the guiding tip is further shifted to 1.0 mm, as shown in Figure 4.103 (c), the maximum electric field strength at the tip drops to 5,334.12 V/mm, and that on the horizontal path L_2 drops to 882.93 V/mm. The change in electric field intensity is even slower, and the tilt angle of the electric field lines is larger, but they still point toward the guiding tip in general. Because of the change in the focusing position of the electric field and the weakened focusing effect, the jet will experience unstable deviation, which is unfavorable to the accurate and stable deposition of the jet.

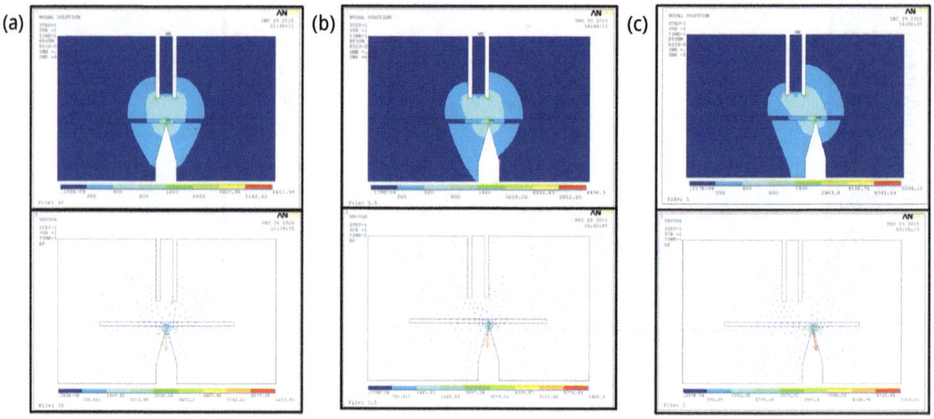

Figure 4.103: Cloud and vector maps of electric field distribution in electric field-focused EDW with different guiding tip positions: (a) the tip electrode placed right below the spinneret; (b) the tip electrode shifts 0.5 mm to the right; and (c) the tip electrode shifts 1.0 mm to the right.

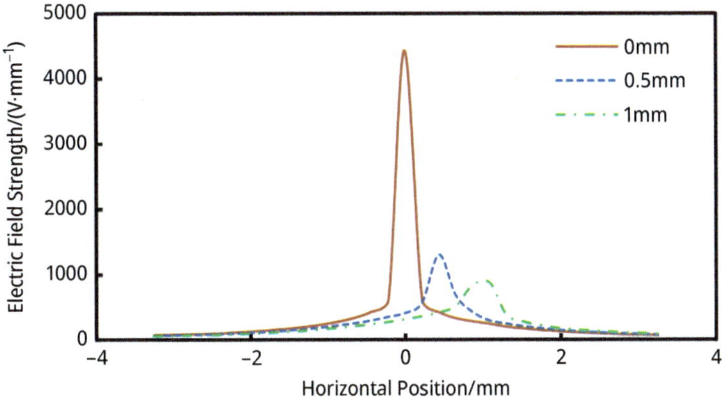

Figure 4.104: The transverse electric field strength on the horizontal path L_2 with different probe positions.

4.3.3.3 Experimental setup for electric field-focused EDW

The experimental setup for electric field-focused EDW primarily comprises high-voltage DC power supply, real-time observation equipment, a host computer, spinneret, guiding tip electrode, precision syringe pump, collector, and precision motion platform, as illustrated in Figure 4.105. During the experiment, a collector is placed on the precision motion platform, with an auxiliary tip electrode positioned below it, ensuring close contact between the collector and the tip electrode. The spinneret is located above the collector, aligned with the tip electrode along the same axis. The distance between the spinneret and the collector is adjustable from 1 to 10 mm. The precision syringe pump supplies the solution to the spinneret at a feeding rate of 2–200 μL/h. The positive output terminal of the high voltage DC power supply is connected to the spinneret, and the negative output terminal is linked to the auxiliary tip electrode and grounded. The host computer controls the motion of the precision motion platform and manage the real-time video observation and recording of the CCD. In the experiment, the solution ejects from the spinneret under the influence of high-voltage electric field to form a jet, and its motion and deposition process can be observed and recorded by the CCD.

4.3.3.4 Jet ejection behavior of EDW with electric field focusing

4.3.3.4.1 Jet formation process of EDW with electric field focusing

The formation of EDW jet with electric field focusing is illustrated in Figure 4.106. At the outset, the solution at the spinneret tip undergoes deformation due to the electric field, resulting in the formation of a Taylor cone. Following this, a jet emerges from the apex of the Taylor cone and is propelled toward the collector. Initially, the jet has a

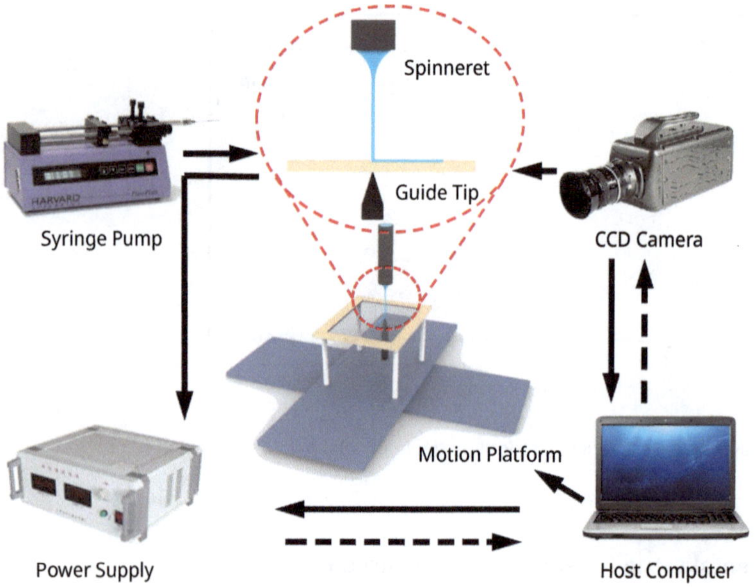

Figure 4.105: Schematic diagram of the electric field focused EDW system.

larger diameter because a significant amount of charge has accumulated during the Taylor cone's formation, leading to a greater volume of solution being carried away by the jet. As the charge on the Taylor cone diminishes, the jet diameter progressively decreases under the influence of the electric field force. Once the electric charge supplied by the electric field achieves a stable equilibrium with the charge being transported by the jet, the dimensions of the Taylor cone remain constant and the jet diameter stabilizes, ensuring a sustained stable ejection. The transition from Figure 4.106(b)–(e) reveals that it takes 65 ms to overcome the droplet's surface tension and viscoelastic forces to initiate the jet and an additional 244 ms to reach a steady state.

To better understand the confinement effect of focused electric field on the ejection of the charged jet and achieve the stable ejection state, we compare the ejection states on silicon and insulating substrates with and without auxiliary tip electrodes, respectively, as illustrated in Figure 4.107.

As shown in Figure 4.107(a), (c), and (d), stable jet ejections are observed from silicon substrates with and without a guiding tip electrode as well as from the insulating substrate equipped with a guiding tip electrode. However, the insulating substrate without guiding tip electrode exhibits poor ejection stability due to the divergence of the electric field distribution. This instability leads to significant oscillations. In addition, due to the lower intensity of the spatial electric field, the jet ejection speed is slower, resulting in the formation of larger Taylor cones, as shown in Figure 4.107(b).

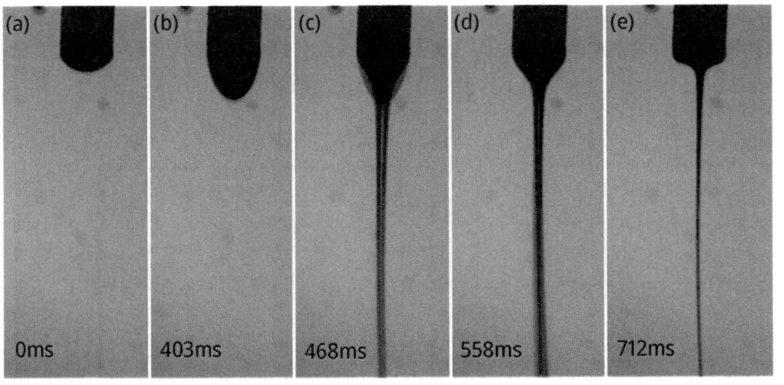

Figure 4.106: Jet ejection of EDW with electric field focusing (solution concentration is 10%, applied voltage is 2.0 kV, inner diameter of spinneret is 0.21 mm, and spinneret-to-collector distance is 4 mm).

Figure 4.107: EDW jet ejection state with different substrates: (a) silicon substrate without guiding tip electrode; (b) insulating substrate without guiding tip electrode; (c) silicon substrate with guiding tip electrode; and (d) insulating substrate with guiding tip electrode (solution concentration is 10%, applied voltage is 2.0 kV, inner diameter of the spinneret is 0.21 mm, and distance from spinneret to collector is 4 mm).

During the EDW process, the current directly reflects the charge migration process and serves as a key indicator of the rheological and motion behaviors of the jet. To further elucidate the difference between jets without tip electrode constraint and electric field-focused jets, the current characteristics during the jet deposition in both modes are tested, as shown in Figure 4.108. Observation of the current fluctuation shows that the current in EDW with electric field focusing is more stable than that without with electric field focusing. The average current value for EDW without electric field focusing is 0.596 μA, with significant current fluctuation, and the maximum variation is ±53% of the average value, whereas the average value for EDW with electric field focusing is 0.318 μA, with significantly reduced current fluctuation amplitude, as shown in Figure 4.108(b), and the maximum variation value of current is ±2.5% of the average value, indicating a significant improvement in current stability.

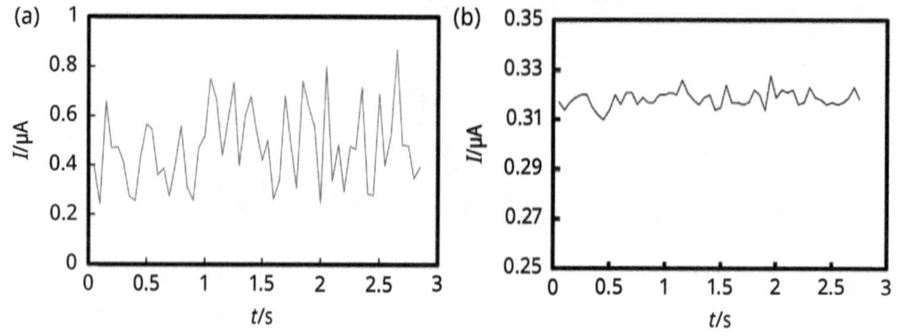

Figure 4.108: Current changes during the EDW process: (a) EDW without constraint of guiding tip and (b) EDW with electric field focusing (solution concentration is 10%, applied voltage is 2.0 kV, inner diameter of the spinneret is 0.21 mm, and distance from spinneret to collector is 4 mm).

The analysis of the electrical characteristics of the jet finds that the electric field-focused EDW jet has smaller diameter and higher stability, and its focused electric field can induce and restrain the jet. According to Figure 4.109, when the tip electrode is placed right below the spinneret, the jet shoots vertically downward toward the tip electrode; When the tip electrode is offset by 0.2 mm, the jet moves with the tip electrode, with a deflection angle of approximately 6°. As the tip offset distance increases, the jet's deflection angle also increases. And when the tip is offset by 0.5 mm, the deflection angle increases to 16°. When the guiding tip position is offset (not directly below the nozzle), the jet deflects accordingly, and its direction aligns with the tip. At

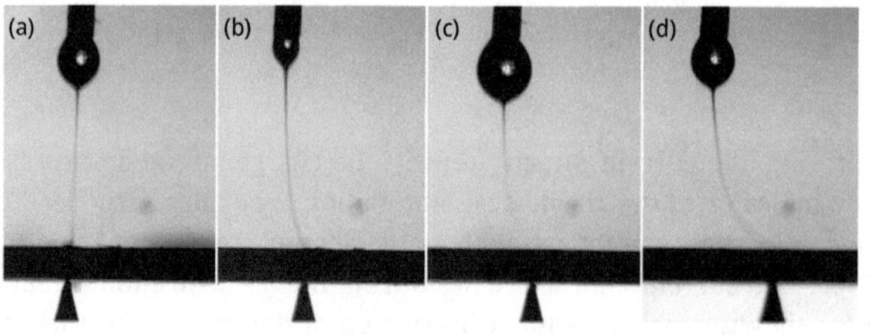

Figure 4.109: Morphology of tip-induced EDW jets with (a) tip electrode right below the spinneret; (b) tip electrode offset by 0.2 mm; (c) tip electrode offset by 0.3 mm; and (d) tip electrode offset by 0.5 mm (solution concentration is 10%, applied voltage is 2.0 kV, spinneret inner diameter is 0.21 mm, and spinneret-to-collector distance is 4 mm).

this point, the jet's deposition point is close to the tip position, indicating that the focused electric field can effectively induce and constrain the jet.

4.3.3.4.2 Critical condition for jet ejection in EDW with electric field focusing

The ejection of EDW jet is a process driven by the combined action of electric field force, solution surface tension, and viscoelastic force. The critical voltage refers to the minimum voltage required for the solution to transform from the Taylor cone shape to a jet, while the maintaining voltage refers to the voltage needed to maintain stable jet ejection. Generally, the critical voltage is higher than the maintaining voltage. Lower voltage implies less charge in the solution, resulting in more stable jets.

The critical voltage exerts an electric field force on the Taylor cone that exceeds the forces of solution surface tension and viscoelasticity. The properties of the solution can directly affect the critical voltage in EDW. Figure 4.110 illustrates the relationship between the critical voltage and the concentration of the PEO solution. As shown in the figure, the critical voltage increases with the solution concentration and is lower for the EDW with electric field focusing than that for EDW without guiding tip constraints. The guiding tip electrode enhances the focused electric field on the collector, resulting in a greater electric field tensile force on the jet in electric field-focused EDW than in guiding-tip-free EDW under the same voltage, making jet ejection easier to achieve. In electric field-focused EDW, when the PEO solution concentration increases from 6% to 10%, the critical voltage increases from 1.3 to 1.7 kV. However, when the solution concentration is further increased to 12%, the required critical voltage increases significantly from 1.7 to 2.8 kV.

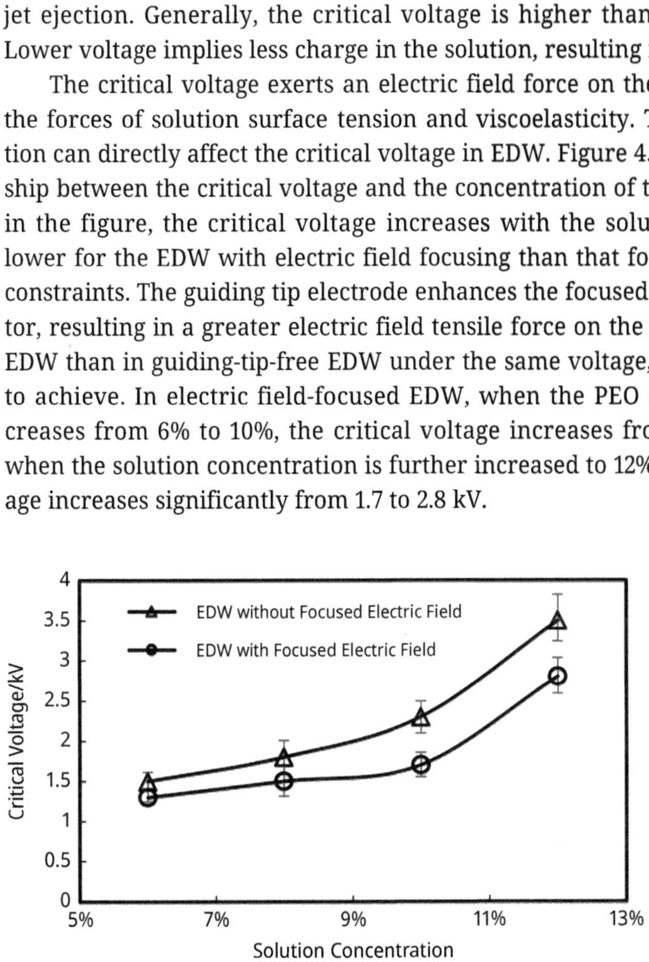

Figure 4.110: Relationship between the critical voltage and solution concentration.

The maintaining voltage is a crucial condition for sustaining stable jet ejection. Therefore, the maintaining voltages for both EDW without guiding tip and electric field-focused

EDW are investigated. The results are shown in Figure 4.111. Regardless of changes in the solution concentration, the maintaining voltage required for electric field-focused EDW is always lower than that for EDW without guiding tip. As the solution concentration increases, the viscosity also increases, leading to a higher maintaining voltage. Furthermore, regardless of the solution concentration, the maintaining voltage is always smaller than the critical voltage.

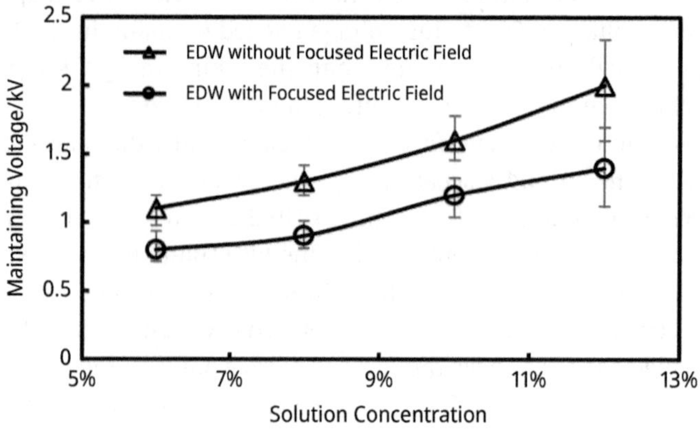

Figure 4.111: Relationship between maintaining voltage and solution concentration of PEO (solution supply rate is 100 μL/h and spinneret-to-collector distance is 2 mm).

The viscosity of the solution increases exponentially with its concentration. As the solution concentration increases, forces such as surface tension and viscoelasticity also increase. Consequently, the electric field force required for EDW increases as well. Electric field-focused EDW enhances the strength of the focused electric field and improves the electric field constraining force, thereby reducing both the critical voltage and maintaining voltage for jet ejection.

4.3.3.4.3 Jet ejection behavior in electric field-focused EDW

In the process of electric field-focused EDW, the jet ejection behavior is influenced by numerous factors including solution properties and process parameters, primarily involving applied voltage, spinneret-to-collector distance, spinneret inner diameter, and solution viscosity.

The applied voltage directly determines the amount of charge carried by the jet and the strength of the space electric field. By adjusting the applied voltage, the jet morphology shown in Figure 4.112 can be obtained. When the applied voltage is low (1.0–3.0 kV), as the voltage increases, the amount of charge carried by the jet and the strength of the space electric field gradually increase. Consequently, the tensile force exerted by the electric field on the jet is enhanced, leading to an increase in the volume of stretched solution and a larger jet diameter.

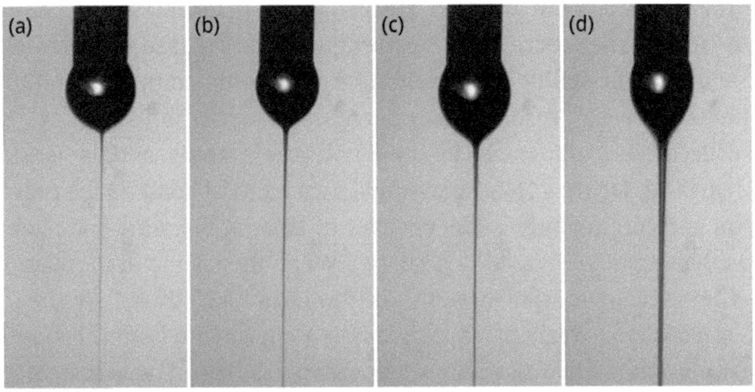

Figure 4.112: EDW jet with electric field focusing at different applied voltages: (a) 1.0; (b) 1.5; (c) 2.0; and (d) 3.0 kV (solution concentration is 10%, diameter of inner spinneret is 0.21 mm, and spinneret-to-collector distance is 4 mm).

When the applied voltage continues to increase (>3.0 kV), the charged jet under the focusing effect of electric field is shown in Figure 4.113. If the solution supply rate remains constant, the larger electric field force causes more solution to be ejected than supplied, resulting in a smaller jet diameter. When the voltage increases to 4.0 kV, the jet transitions from a stable ejection mode to an unstable oscillation mode, as shown in Figure 4.113(c) and (d). The mismatch between the solution feeding speed and jet ejection speed is not conductive to maintain a stable Tylor cone, causing the jet to oscillate unstably back and forth along the edge of the spinneret. In general, this unstable oscillation leads to unstable jet deposition, resulting in significant positioning errors, which is unfavorable to the precise fabrication of micro-nanostructures through EDW. Therefore, it is necessary to avoid the occurrence of this mode.

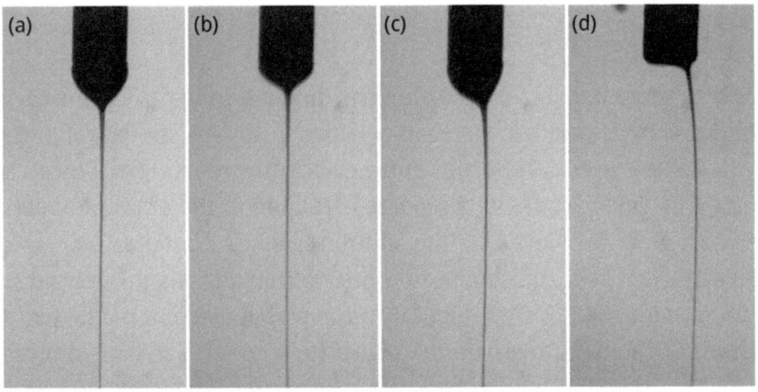

Figure 4.113: Electric field-focused EDW jet under different voltages: (a) 3.5; (b) 3.8; (c) 4.0; and (d) 4.3 kV (solution concentration is 10%, spinneret inner diameter is 0.21 mm, and spinneret-to-collector is 4 mm).

Figure 4.114 shows the electric field focused EDW jets at different spinneret-to-collector distances. As the distance increases, the strength of the focused electric field weakens, reducing the induction ability of the guiding tip electrode on the jet, and the motion distance of the jet increases, resulting in a decrease in deposition accuracy.

At a spinneret-to-collector of 2 mm, the focused electric field is strong, and the jet stably ejects vertically downward. Figure 4.114(a) shows that, when the jet contacts the collector, it deposits at a point and continuously stacks up layer by layer at the same point, with the jet being stable and having high deposition accuracy. When the spinneret-to-collector distance increases to 4 mm, the jet morphology is shown in Figure 4.114(b), where the intensity of the focused electric field weakens, causing the jet to deviate to both sides, with a deviation angle of approximately 2.13°, and a deposition radius of approximately 0.149 mm. As the spinneret-to-collector increases to 6 mm, the deviation angle continues to increase, reaching about 3.11°, and the deposition radius becomes 0.326 mm.

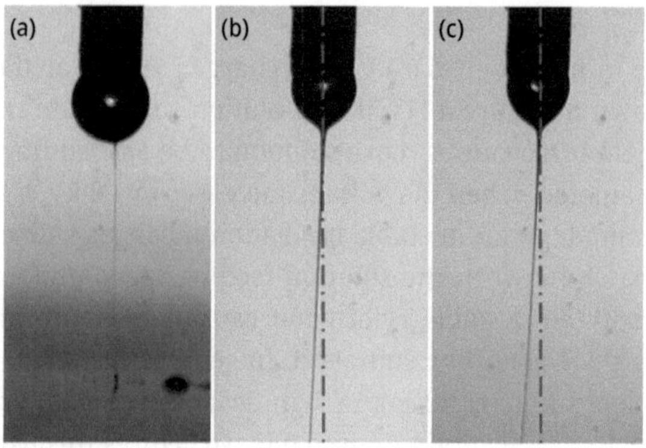

Figure 4.114: EDW jet with electric field focusing at different spinneret-to-collector distances: (a) 2; (b) 4; and (c) 6 mm (solution concentration is 10%, inner diameter of the spinneret is 0.21 mm, and applied voltage is 2.0 kV).

The morphology of EDW jet under different spinneret inner diameters is shown in Figure 4.115. At the same voltage and spinneret-to-collector distance, the jet diameter remains essentially consistent across different spinneret inner diameters. Since the jet is ejected from the tip of Taylor cone, its diameter is related to the amount of solution drawn by the electric field. However, as the spinneret inner diameter increases, the Taylor cone gradually enlarges, and the jet is prone to fluctuations influenced by the shape of the Taylor cone, which is detrimental to the stable ejection of the jet, as shown in Figure 4.115(d). When the inner diameter of the spinneret is 0.41 mm, the overall diameter of the Taylor cone is about 1.00 mm, and the jet flow is likely to fluctuate and deviate, or even bifurcate, as shown in Figure 4.115(e).

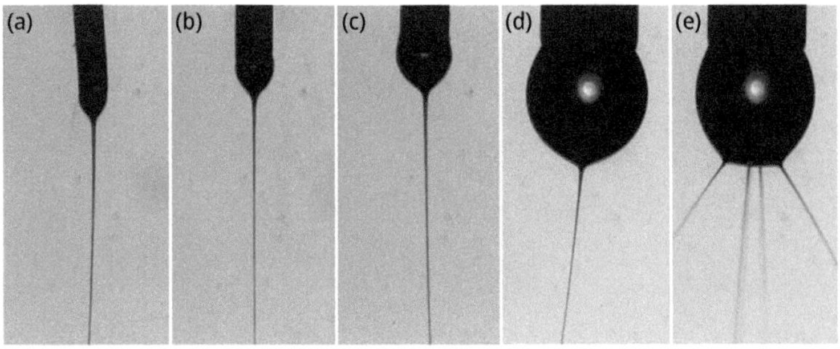

Figure 4.115: Electric field-focused EDW jet at different spinneret inner diameters: (a) 0.06 mm; (b) 0.11 mm; (c) 0.21 mm; (d) 0.34 mm; and (e) 0.41 mm (solution concentration is 10%, applied voltage is 2.0 kV, and spinneret-to-collector distance is 4 mm).

Due to the exponential increase in solution viscosity with concentration, the solution viscosity influences the morphology of the EDW jet and the fiber deposition outcome. Figure 4.116 illustrates the morphology of the electric field-focused EDW jet at different PEO solution concentrations. At lower solution concentrations, such as 6%, 8%, and 10%, the jet remains in a stable ejection state. For solution concentration of 6% and 8%, due to the low solution viscosity and slow solvent evaporation, the jet is prone to form large droplets and bead-on-string structures; whereas at a solution concentration of 10%, the volatility of the solvent is somewhat enhanced, allowing for stable straight jet deposition. At a concentration of 12%, the viscosity increases from 8,595 to 14,641 mPa s, while at 14% it increases from 8,595 to 30,752 mPa s respectively, making it difficult to initiate a jet and resulting in larger jet diameters, as shown in Figure 4.116(d) and (e). At a solution concentration of 12%, the jet stretching becomes challenging, and the larger jet diameter significantly reduces deposition accuracy. When the solution concentration reaches 14%, the excessively high viscosity causes the surface tension and viscoelastic forces of the Taylor cone to far exceed the tensile force exerted by the electric field on the solution, thereby suppressing jet formation.

4.3.3.5 Patterned deposition of EDW with electric field focusing

4.3.3.5.1 Electrohydrodynamic direct-writing fiber structures with electric field focusing

To achieve patterned printing of micro-nanostructures, it is particularly crucial to develop techniques for precisely controlling the diameter of direct-written fibers and the accuracy of fiber deposition. In the EDW process, there are many factors that can affect the morphology and deposition accuracy of the fibers such as solution properties, applied voltage, and collector movement speed. The movement of the collector provides a lateral tensile force to the jet and affects the deposition morphology of the

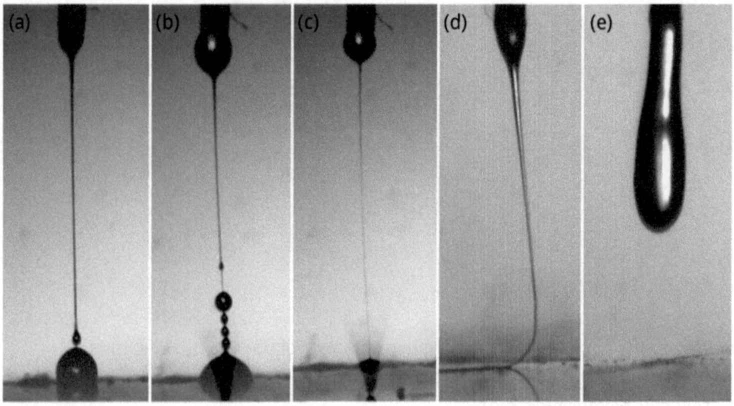

Figure 4.116: EDW jet with electric field focusing at different solution concentrations: (a) 6%; (b) 8%; (c) 10%; (d) 12%; and (e) 14% (applied voltage is 2.0 kV, spinneret inner diameter is 0.21 mm, and spinneret-to-collector distance is 4 mm).

jet. Clarifying the effects of collector motion on fiber deposition morphology helps in the precise patterned printing through EDW with electric field focusing.

During the electrospinning process, when the charged jet is deposited onto the collector, subsequent charged jets are prone to whipping motion influenced by charge repulsion during the ejection process, thereby affecting the morphology of the deposited fibers. In EDW, the spinneret-to-collector distance is reduced, making the residual charge carried by the already deposited fibers significantly impact the deposition behavior and morphology of the charged jet. Therefore, it is important to study the influence of collector moving speed on jet deposition and fiber morphology during EDW jet's patterned deposition process.

By controlling the motion trajectory and collector moving speed through computer programming software, the designed patterns can be fabricated. Pre-written trajectory codes for the designed patterns control the motion path of the precision motion platform as shown in Figure 4.117, with a parallel line spacing of 0.2 mm and a single straight-line distance of 20 mm.

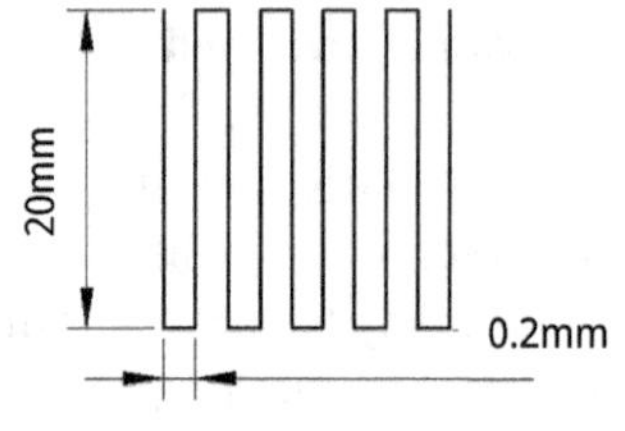

Figure 4.117: Diagram of parallel line motion trajectory for the precision motion platform.

Figure 4.118 shows the morphology of fibers deposited by EDW with electric field focusing at different collector moving speeds. At a collector moving speed of 2 mm/s, the jet whips significantly due to the existence of the charge repulsion force. As the collector moves, the fiber deposition exhibits a spiral wavy morphology, resulting in lower deposition accuracy. With an increase in the collector moving speed, the tensile force exerted by the collector on the jet increases, reducing the spiral waviness in jet deposition and resulting in undulating fibers. At a collector moving speed of 10 mm/s, the fiber waviness is mostly eliminated, and the deposition morphology appears as nearly straight fibers with a small portion of spiral waviness. When the collector moving speed is raised to 20 mm/s, the spiral waviness of the deposited fibers completely disappears, resulting in continuous straight fibers. With further increases in the collector moving speed, the tensile force exerted by the collector surpasses the influence of the jet's charge repulsion forces, dominating the process. The jet will be stretched to deposit as straight fibers, with the average fiber diameter decreasing from 5.64 μm at a collector moving speed of 2 mm/s to 1.85 μm at 60 mm/s.

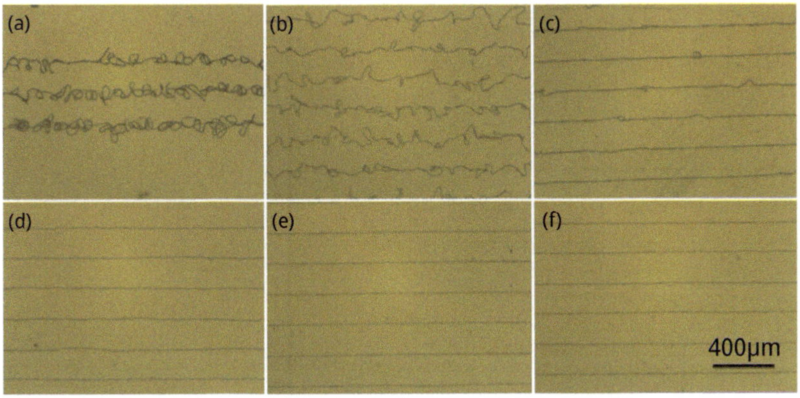

Figure 4.118: Fiber morphology deposited by EDW with electric field focusing at different collector speeds: (a) 2; (b) 5; (c) 10; (d) 20; (e) 40; and (f) 60 mm/s (solution concentration is 10%, spinneret inner diameter is 0.11 mm, spinneret to collector distance is 2 mm, and applied voltage is 1.5 kV).

Simultaneously, experiments on fiber deposition via EDW without guiding tip under the same conditions are conducted at different collector moving speeds. The corresponding fiber deposition morphologies are shown in Figure 4.119. A comparison reveals that, in electric field-focused EDW, both the jet fluctuation and the fiber diameter are smaller. At a collector moving speed of 10 mm/s, fiber fluctuations are essentially eliminated, resulting in straight fibers with uniform diameters. In contrast, for guiding-tip-free EDW, a collector moving speed of 20 mm/s is required to eliminate fiber fluctuations, and the uniformity of fiber diameter is poor. A collector moving speed of approximately 40 mm/s is needed to form uniformly straight fibers. As the collector moving speed increases from 2 to 60 mm/s, the average fiber diameter decreases from 7.51 μm to 3.23 μm. The variation curve of fiber diameter with collector moving speed is shown in Figure 4.120. Compared to tip-free EDW, electric field-focused EDW demonstrates significant reductions in both fiber diameter and jet oscillation amplitude.

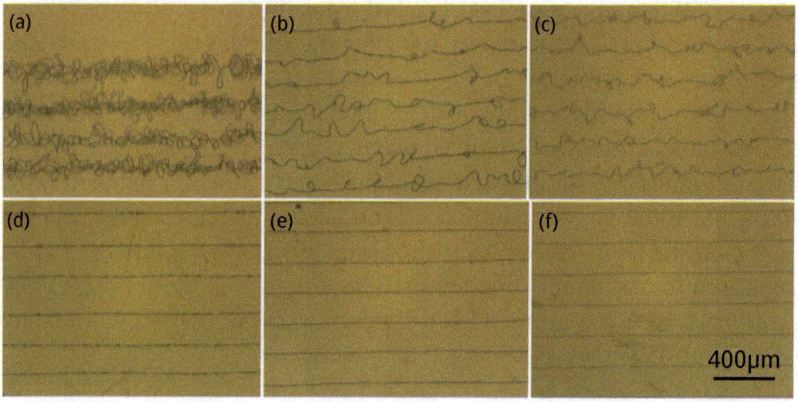

Figure 4.119: Fiber morphology deposited by EDW without guiding tip constraint at different collector speeds: (a) 2; (b) 5; (c) 10; (d) 20; (e) 40; and (f) 60 mm/s (solution concentration is 10%, inner diameter of spinneret is 0.11 mm, spinneret-to-collector distance is 2 mm, and applied voltage is 1.5 kV).

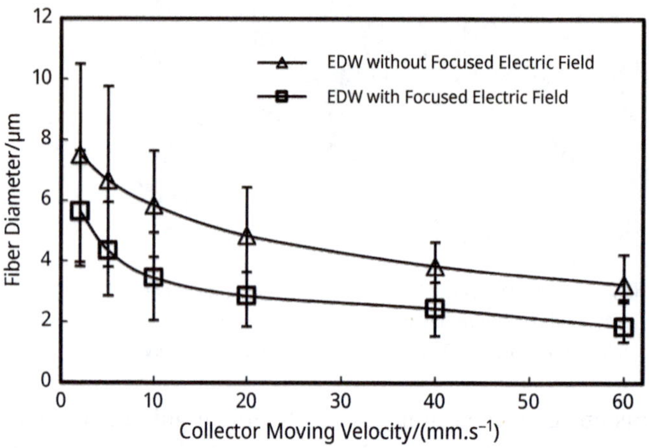

Figure 4.120: Relationship between fiber diameter and collector moving velocity (solution concentration is 10%, spinneret inner diameter is 0.11 mm, spinneret-to-collector distance is 2 mm, and applied voltage is 1.5 kV).

To achieve precise patterned deposition of electrospun fibers, it is crucial to control the distribution width of the fibers, which is defined by the deposition range of a single fiber, as shown in Figure 4.121.

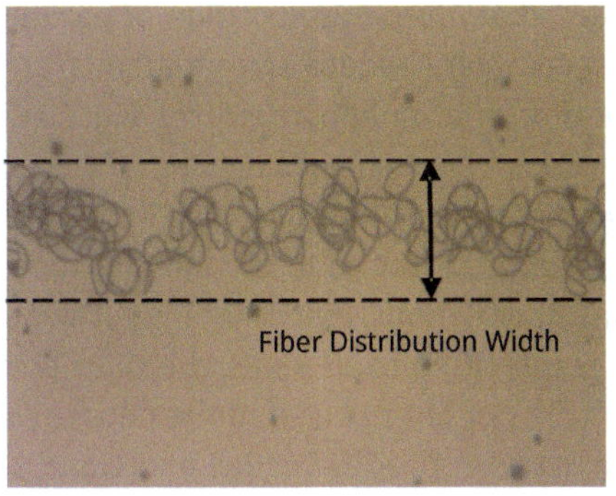

Figure 4.121: Diagram of fiber distribution width.

The relationship between the fiber distribution width and the collector moving speed is shown in Figure 4.122. The fiber distribution width decreases as the collector moving speed increases. In EDW with field focusing, when the collector moving speed is less than 10 mm/s, fibers are obviously affected by the helical whiplash of the jet and forms a spiral wave-like fiber, with a wide fiber distribution width of about 34.36 μm. As the collector moving speed increases, the fibers are gradually drawn into straight line shape, with smaller fiber fluctuations and a decreased distribution width. Ultimately, the fiber distribution width becomes close to the fiber diameter. On the other hand, the distribution width of the fibers formed by EDW without tip guiding is larger than that prepared by the electric field-focused EDW. The fiber distribution width is about 42.54 μm when the collector moving speed is 2 mm/s, and only when the collector moving speed reaches 20 mm/s can uniform linear fibers be obtained.

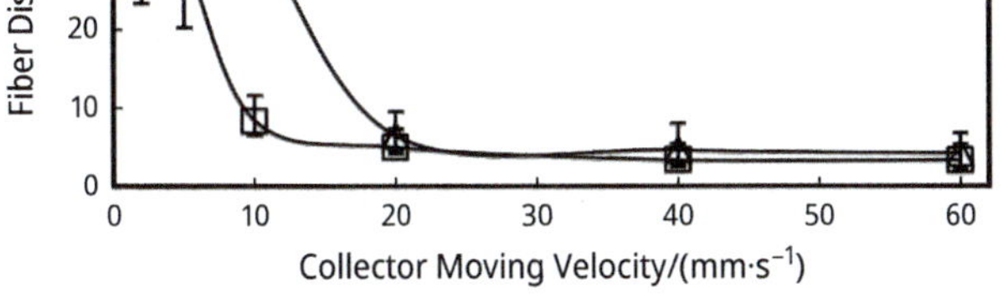

Figure 4.122: The relationship between the fiber distribution width and the collector moving velocity (solution concentration is 10%, the inner diameter of the spinneret is 0.11 mm, the spinneret-to-collector distance is 2 mm, and the applied voltage is 1.5 kV).

From the above experiments, it can be concluded that, in the preparation of parallel line patterns, increasing the collector moving speed helps reduce fiber diameter and distribution width. However, in the fabrication of micro-nanostructures via EDW, complex patterned structures often include corners, where excessively high collector moving speeds can have detrimental effects. Figure 4.123 shows the fiber deposition morphology at the corner of a parallel-line matrix. At a low collector moving speed (5 mm/s), the deposited fibers exhibit large spiral fluctuations, resulting in poor deposition accuracy. As the collector moving speed is increased to 20 mm/s, the fiber fluctuations at the corners is small, the corner patterns remain intact, and the deposition accuracy is high. However, when the collector moving speed is further increased to 40 and 60 mm/s, accurate corner patterns cannot be obtained. The faster the collector moving speed, the greater the deformation of the corner patterns.

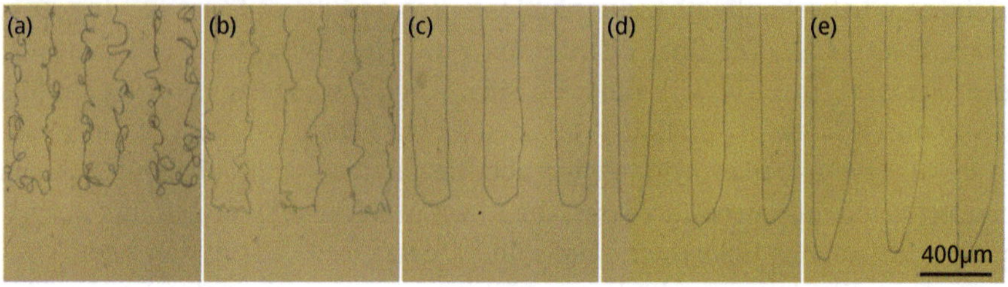

Figure 4.123: Corner patterns of the parallel lines direct-written by electric field-focused EDW at different collector speeds: (a) 5; (b) 10; (c) 20; (d) 40; and (e) 60 mm/s (solution concentration is 10%, spinneret inner diameter is 0.11 mm, spinneret-to-collector distance is 2 mm, and applied voltage is 1.5 kV).

4.3.3.5.2 Printing of patterned structure by EDW with electric field focusing

In micro-nano bio-tissue engineering, cross-grid scaffolds are the most commonly used structures. The fabrication of patterned grid structures based on electric field-focused EDW enables precise control over grid size, promoting cell adhesion and differentiation, thereby promoting the application of EDW technology in the biological field. By designing the platform motion trajectory for horizontal and vertical intersections in the host computer, the corresponding deposited cross-grid patterns can be deposited on the collector, as shown in Figure 4.124. It can be observed that when at a collector moving speed of 20 mm/s, straight fiber structures are achieved without any bending or fluctuations. Additionally, the diameter of each fiber is uniform, and the fibers in both horizontal and vertical directions are well-aligned at the intersections.

Next, we investigate the precision of patterned deposition under different gird sizes. The designed grid pattern size is 200 μm × 200 μm. As shown in Figure 4.125(a), the actual measured transverse width of the grid is 204.022 μm and the longitudinal width is 194.062 μm, with an average error of 2.49%. The fiber deposition trajectory aligns well with the motion trajectory of the platform. At the intersections, the longitudinal fibers overlap the transverse fibers, and both the horizontal and vertical fibers

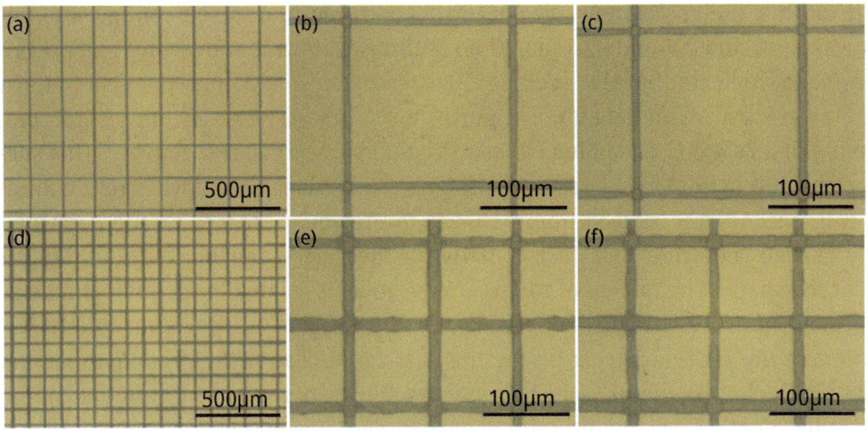

Figure 4.124: Cross-grid patterns prepared by electric field-focusing EDW: (a)–(c) 200 μm × 200 μm; (d) ~; and (f) 100 μm × 100 μm (solution concentration is 10%, inner diameter of spinneret is 0.11 mm, spinneret-to-collector distance is 2 mm, applied voltage is 1.5 kV, and the collector speed is 20 mm/s).

maintain their original morphology at the overlap, as shown in Figure 4.125(b) and (c). The diameter of the fibers is about 3.007 μm, as shown in Figure 4.125(d).

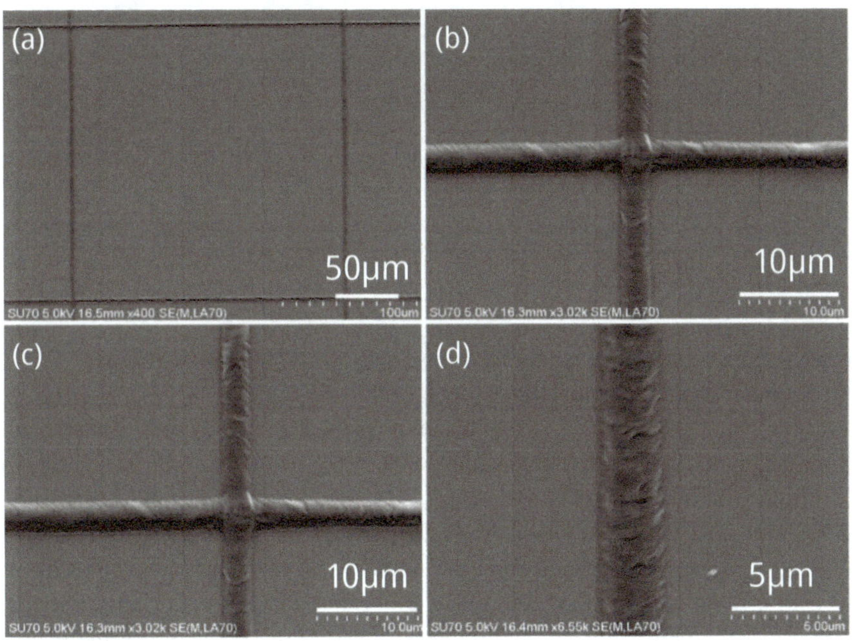

Figure 4.125: 200 μm × 200 μm cross-grid pattern: (a) overall cross-grid pattern; (b) and (c) intersecting fibers; and (d) longitudinal fiber (solution concentration is 10%, inner diameter of spinneret is 0.11 mm, spinneret-to-collector distance is 2 mm, applied voltage is 1.5 kV, and collector speed is 20 mm/s).

Building on this, the grid pattern size is further reduced to a mixed configuration of 200 μm × 120 μm and 200 μm × 80 μm. The resulting pattern is shown in Figure 4.126. Measurements indicate that the average fiber diameter is 3.124 μm, the grid pattern dimensions are approximately 207.245 μm × 113.650 μm and 204.124 μm × 87.160 μm, with errors of 5.29% and 8.95% for 120 and 80 μm grid widths, respectively. This suggests a reduction in deposition accuracy. The obtained fibers show two states: continuous straight fibers and curved fibers. Due to the residual charges on the deposited fibers, a charge repulsion force arises between the fibers, causing them to deposit into a curved shape. The degree of the fiber curvature is related to the total charge of the deposited fibers. The deposition trajectory of the fibers is from right to left. Initially, most of the charge carried by the fibers is conducted to the grounded guiding tip via the collector. As the number of deposited fibers increases, the residual charge gradually increases. Consequently, the fibers to be deposited experience repulsion from the already deposited fibers, resulting in the formation of curved fibers, as shown in Figure 4.126(b).

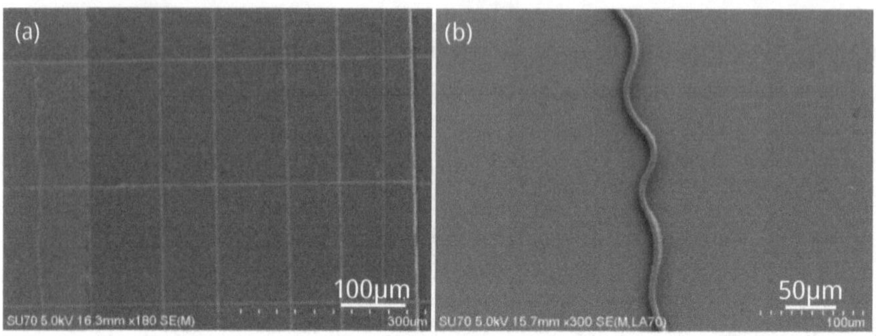

Figure 4.126: Irregular cross-grid pattern prepared by EDW with electric field focusing: (a) irregularly cross-grid and (b) curved fibers (solution concentration is 10%, spinneret inner diameter is 0.11 mm, spinneret-to-collector distance is 2 mm, applied voltage is 1.5 kV, and collector speed is 20 mm/s).

Reducing the collector moving speed to 15 mm/s, cross-grid structures of various sizes as shown in Figure 4.127 are fabricated by electric field-focused EDW. Since the jet deposition speed is faster than the collector moving speed, the jet cannot be stretched into smooth, straight fiber structures, which will result in the formation of uneven fibers. The edge lengths of the grid are measured, and the deviation and positioning errors of the deposited fibers are shown in Figure 4.128. With the increase of designed grid edge length, the average deviation of the deposited fiber increases from 3.262 to 4.786 μm, with a fluctuation range of about 1.5 μm. However, at smaller edge lengths, the deviation has a greater impact on the overall positioning error of the pattern. A deviation of 3.262 μm can lead to an error of 4.66%. As the grid size increases, the overall positioning error of the pattern reduces although the deviation increases. It is reduced to 2.39% when the designed edge length reaches 200 μm.

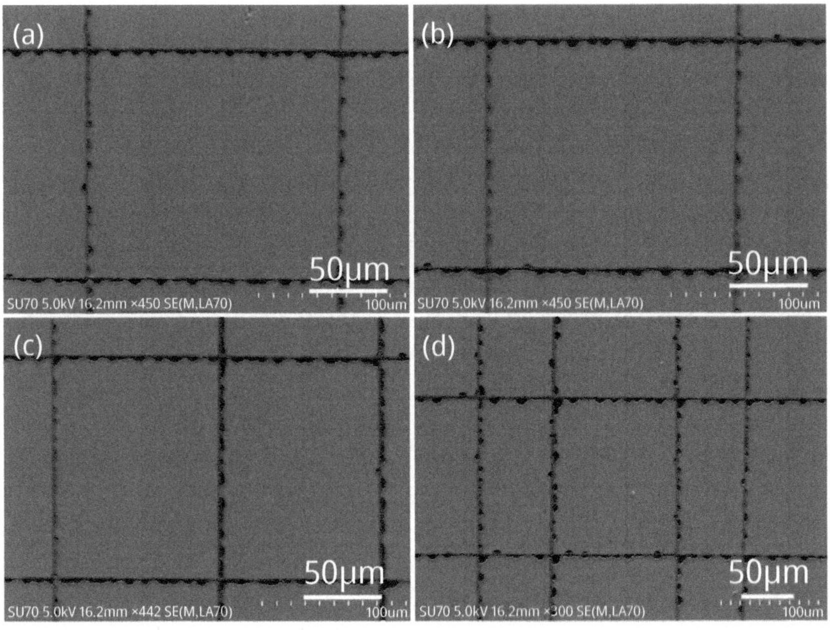

Figure 4.127: Irregular cross-grid pattern at a collector speed of 15 mm/s: (a) and (b) 160 μm × 160 μm; (c) 160 μm × 120 μm; (d) 160 μm × 70 μm and 160 μm × 130 μm (solution concentration is 10%, inner diameter of spinneret is 0.11 mm, spinneret-to-collector distance is 2 mm, and applied voltage is 1.5 kV).

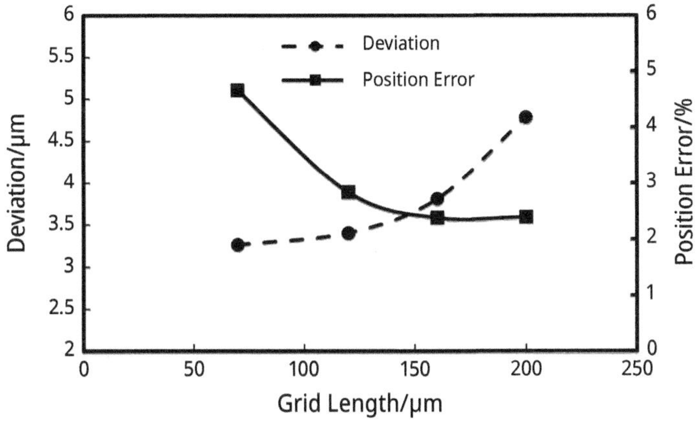

Figure 4.128: Deviation and positioning error of the grid pattern written by EDW with electric field focusing (solution concentration is 10%, inner diameter of spinneret is 0.11 mm, spinneret-to-collector distance is 2 mm, and applied voltage is 1.5 kV).

In addition to the grid patterns, the motion path of the collector can also be designed into any desired patterns. Figure 4.129(a) shows a pentagram motion trajectory, with arrows indicating the direction of motion and each side measuring 1 mm. Figure 4.129 (b)–(d) illustrates the fiber deposition morphology at collector moving speeds of 10, 15, and 20 mm/s, respectively. At a collector moving speed of 10 mm/s, the diameter of a single fiber is about 4.84 μm, the fiber exhibits fluctuations, resulting in poor deposition accuracy. The deposition error is significant at the corners. For instance, in corner 1, two fibers intersect instead of being continuous, with an error of approximately 1.24%. In corners labeled 2–5, the fibers form continuous arcs with errors of 4.87%, 6.74%, 8.21%, and 12.67%, respectively. However, when the collector moving speed is increased to 15 mm/s, the fiber diameter decreases to about 3.67 μm, and the overall error decreases to 6.97%. Further increasing the collector moving speed to 20 mm/s results in the most accurate pentagram pattern, with side lengths of 1.034, 1.079, 0.984, 1.027, and 0.977 mm, and an error of approximately 3.58%. In terms of positioning error, the deposition points of the fibers align well with the five vertices of the pentagram. However, as shown in Figure 4.129(d), all five sides exhibit slight fluctuations, which affect the integrity of the pattern. This is because during the electrospinning process, the motion platform needs to undergo deceleration and acceleration at the corners of the polyline, resulting in variable speeds that are not conducive to the patterned deposition with EDW.

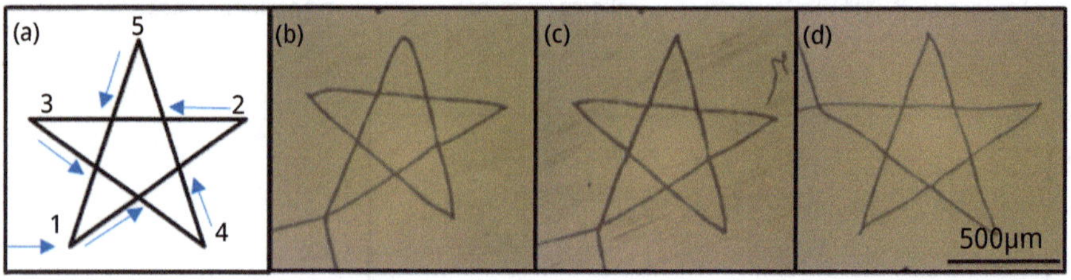

Figure 4.129: Fiber deposition morphology of pentagram pattern: (a) pentagram trajectory; (b) at collector speed of 10 mm/s; (c) at collector speed of 15 mm/s; and (d) at collector speed of 20 mm/s (solution concentration is 10%, spinneret inner diameter is 0.11 mm, spinneret-to-collector distance is 2 mm, and applied voltage is 1.5 kV).

In order to reduce the sudden change in the collector moving speed at the corner, a 10 μm fillet is set at the corner of the pentagram. With the transition of the fillet, the collector does not need to decelerate significantly when passing through the corner. The designed trajectory of the rounded-corner pentagram is shown in Figure 4.130(a). An accurate pentagram pattern can be obtained at a collector moving speed of 20 mm/s. The diameter of a single fiber is 3.48 μm, and the lengths of the five sides are 1.012, 1.008, 1.011, 0.998, and 0.997 mm, respectively, with an error of about 0.72%. The pattern deposition accuracy is high. A stable collector moving speed ensures that the deposited fibers at the corner are consistent with the designed pattern, as shown in Figure 4.130(b) and (c).

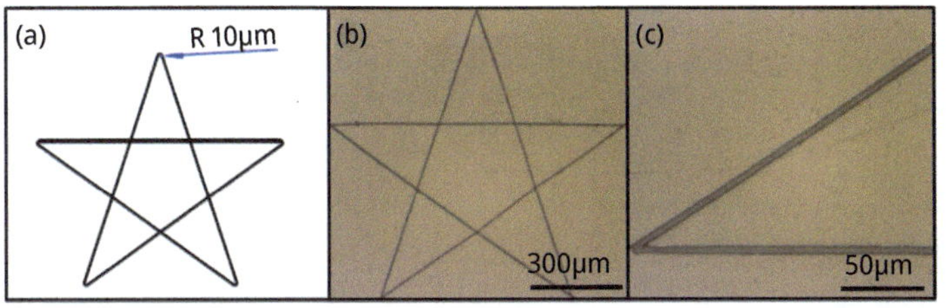

Figure 4.130: Morphology of the deposited pentagram pattern with fillet transition at the corners: (a) designed pentagram trajectory; (b) electrospun pentagram pattern; and (c) morphology of the rounded-corner fibers (solution concentration is 10%, spinneret inner diameter is 0.11 mm, spinneret-to-collector distance is 2 mm, and applied voltage is 1.5 kV).

The electric field-focused EDW can not only produce simple patterns like a pentagram but also more complex patterns. Figure 4.131(a) shows a designed badge pattern, with a total diameter of 16 mm. To complete the entire badge pattern in one step, auxiliary connecting lines are set as shown in Figure 4.131(b) and (c) and small radius corners are added at the figure's sharp turns to reduce sudden change in the collector moving speed.

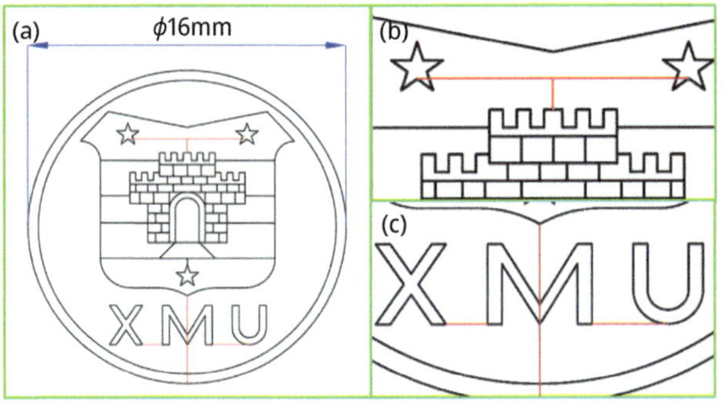

Figure 4.131: Preset trajectory of the badge pattern: (a) the overall badge pattern; (b) the upper part of the badge pattern; and (c) the lower part of the badge pattern.

The badge pattern trajectory is compiled into a control program recognized by the precision motion platform. Through electric field-focused EDW, the complete complex badge pattern is fabricated, as shown in Figure 4.132(a). The rounded corners at the bends reduce abrupt changes in the collector moving speed, thereby enabling precise micro-patterns, as shown in Figure 4.132(b) and (c).

Figure 4.133 shows a phoenix flower pattern produced by EDW with electric field focusing. As shown in Figure 4.133(b), the pattern of the phoenix flower is compiled into a trajectory code for the movement of the collector, according to which a flower is completely deposited on the collector, as shown in Figure 4.133(a). The petals and

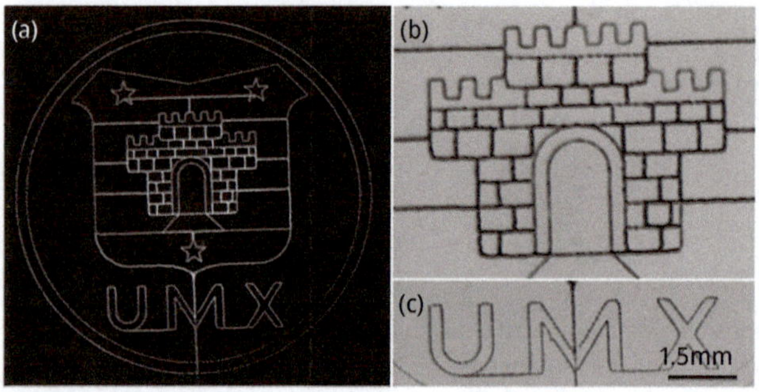

Figure 4.132: Deposition morphology of the badge pattern directly written by EDW with electric field focusing (a) overall pattern; (b) the gate pattern in the middle part; and (c) the letter pattern at the lower part (solution concentration is 10%, spinneret inner diameter is 0.11 mm, spinneret-to-collector distance is 2 mm, applied voltage is 1.5 kV, and collector speed is 20 mm/s).

the center of the phoenix flower pattern are composed of arcs. During the electrospinning process, the collector moves stably, so the deposited fiber pattern is smooth and continuous, as shown in Figure 4.133(c).

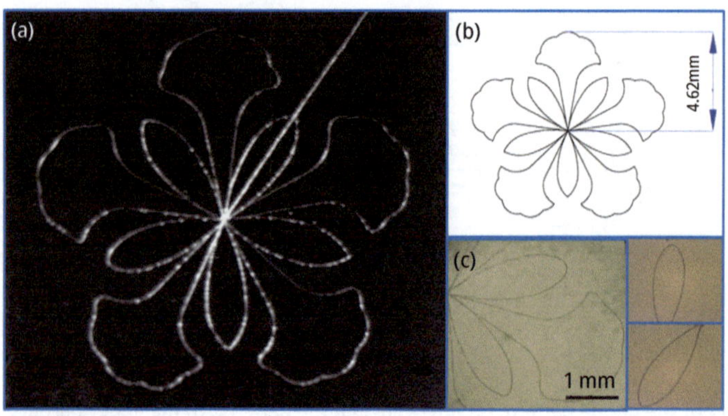

Figure 4.133: Deposition morphology of the phoenix flower produced by EDW with electric field focusing: (a) the overall phoenix flower fiber pattern; (b) the preset trajectory of phoenix flower pattern; and (c) flower petal and flower center patterns (10% solution concentration, spinneret inner diameter is 0.11 mm, spinneret to collector distance is 2 mm, applied voltage is 1.5 kV, and collector speed is 20 mm/s).

4.3.4 Sheath-gas-confined electrohydrodynamic direct-writing

4.3.4.1 Basic principles of sheath-gas-confined EDW

Sheath-gas-confined EDW is a novel technique that introduces sheath gas into the EDW process, as shown in Figure 4.134(a), thereby suppressing jet whipping instability and enabling precise deposition of micro-nanostructures [3–5]. The schematic diagram is shown in Figure 4.134(b). A gas delivery system (including air compressor and pressure regulating valve) is integrated to the spinneret. The gas flow enters into the sheath-gas-constrained EDW spinneret, passes through a buffer channel, and then ejects into the air in a laminar flow through the gas outlet.

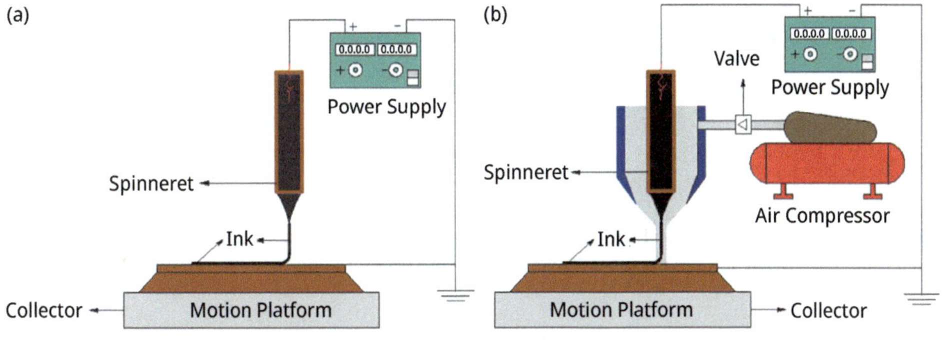

Figure 4.134: Schematic diagram of EDW experimental device: (a) EDW without sheath gas constraint and (b) sheath-gas-confined EDW.

4.3.4.2 Simulation of gas flow field in coaxial spinneret for sheath-gas-confined EDW

To design a sheath-gas-confined EDW spinneret, the gas flow field in coaxial spinneret is simulated using the fluid simulation software ANSYS CFX. The effects of different gas channel structures on gas flow characteristics are compared, the evolution of steady laminar flow during the sheath gas flow process is analyzed, and the stable ejection distance of laminar gas is also investigated. Based on the characteristics of the flow field, the spinneret structure is further optimized a to enhance the confinement capability of the sheath gas.

The simulation model for gas flow field in the spinneret, as shown in Figure 4.135, is constructed. Figure 4.136 presents the global velocity distribution cloud image of *X*–*Z* cross section and the basic dimensions of the model (unit: mm). During the simulation, the inlet pressure is set to 25 kPa (matching the sheath gas pressure in the experiment), and the distance between the spinneret and the collector is 2 mm (with the hollow spinneret protruding 1 mm from the gas outlet to facilitate experimental observation and imaging). The figure shows a uniform flow field, with the sheath gas evenly ejected into the air, thereby enhancing the stability of the jet movement.

The simulation conditions are set as follows:

(1) Domain type: Select the fluid domain option for the domain type. Set the fluid material to AirIdeal gas and keep all other settings as default.
(2) Inlet boundary setting: Name the boundary condition “inlet.” Set the relative pressure to 12.5 kPa (25, 37.5, 50, and 70 kPa) and keep all other settings as default. Click the “OK” button to generate the inlet boundary.

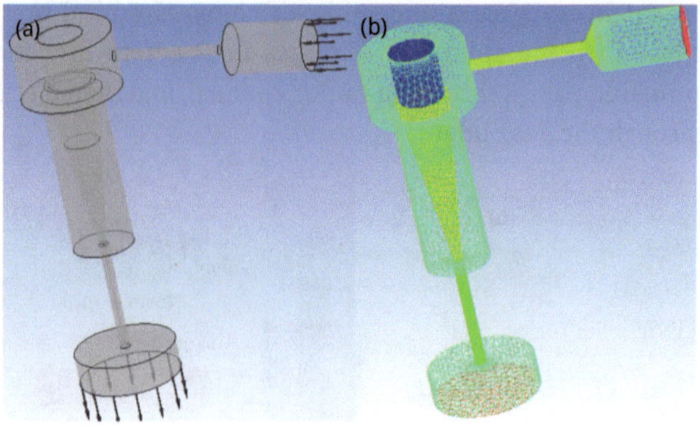

Figure 4.135: 3D simulation model for gas flow region with the spinneret: (a) schematic diagram of the fluid region in the sheath-gas-confined EDW spinneret and (b) grid-generated discrete flow region.

(3) Outlet boundary setting: Enter the boundary condition name "Outlet," set the average static pressure to 0 kPa, and keep other parameters as the default. Click the "OK" button to generate the outlet boundary.
(4) Wall boundary setting: Set the boundary type as a wall, which is a fixed interface that the fluid cannot penetrate. The impact of the wall on the fluid is set as "No Slip," meaning the velocity of the fluid near the wall is assumed to be zero, with other settings remaining at their default values.
(5) Initial condition setting: Select automatic setting, the system will use the initial value automatically set by the solver to begin calculation.
(6) Solution control setting: Choose high resolution for solution control, set the maximum number of iterations (max iterations) to 100, set the residual type as RMS for convergence criteria, and set the residual target to 1.E-4. Keep other settings as default.
(7) Solver: Click "Start Solver Manager" to open "Solver Manager" dialog box. After clicking "confirm," the system will start the automatic solving process. Following post-processing, the simulation results shown in Figure 4.136(a) can be obtained.

The velocity distribution of the sheath gas at the outlet of the hollow nozzle is illustrated in Figure 4.137. At the nozzle outlet, the sheath gas interacts with the jet, forming a stable laminar structure around the jet. This reduces the relative velocity between the jet and the surrounding medium, which can suppress the onset of unstable motion of the jet. The sheath gas exerts a stretching effect on the jet, which lowers the critical voltage and maintaining voltage for jet ejection. This reduction in voltage minimizes interference caused by high voltage and high charge density, improving the stability of the jet ejection. In addition, the stretching effect of the sheath gas can accelerate the jet thinning and improve the ejection efficiency.

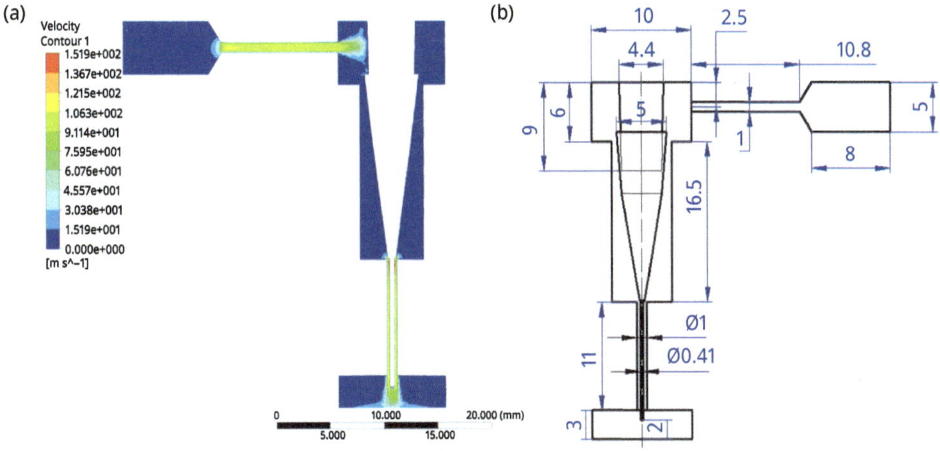

Figure 4.136: Simulation results of the sheath gas flow field: (a) global velocity distribution contour of the *X–Z* cross section in the simulation model and (b) basic dimensions of the simulation model (unit: mm).

Figures 4.137 and 4.138 illustrate the velocity distribution of the sheath gas at the spinneret exit. At an intake air pressure of 25 kPa, the gas flows into the EDW spinneret to form a stable laminar flow after passing through the double-layer stepped buffer airflow region and is uniformly sprayed into the air via the gas outlet.

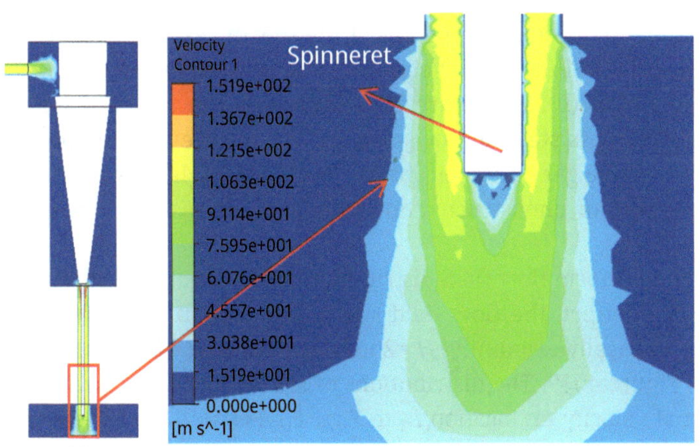

Figure 4.137: Sheath gas velocity contour at the *X–Z* section of nozzle exit.

To evaluate the confinement effect of the sheath gas, a simulation study is conducted to analyze several factors influencing the uniformity of the sheath gas flow field. A series of simulation experiments are designed to access the performance of the sheath gas flow field. The selected experimental variables include the height of the air hole

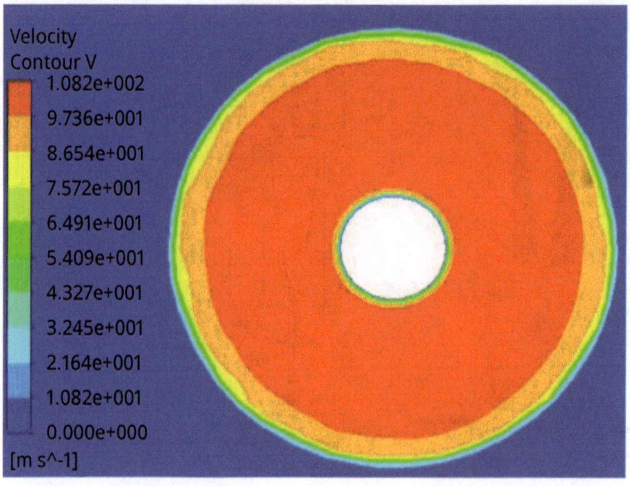

Figure 4.138: Sheath gas velocity contour at the *X–Y* section of the nozzle exit.

H, the inner diameter of the air hole *D*, and the inlet gas pressure *P*. Stable and uniform distribution of the sheath gas flow field is the evaluation indicator.

4.3.4.2.1 The influence of air hole height on the spatial distribution of sheath gas flow field

Figures 4.137 and 4.138 show the simulation results of the sheath gas flow field at the maximum air hole height of $H = 11$ mm. This height is derived from the model dimensions illustrated in Figure 4.136(b), which are constrained by the attached nozzle length of 13 mm for the stainless steel hollow spinneret. To ensure that the hollow nozzle extends 1 mm beyond the gas outlet for observation and imaging purposes and to maintain a 1 mm clearance for unobstructed flow, the resulting gas outlet height is set to 11 mm.

To obtain a stable and uniform flow field distribution in the sheath gas layer and strengthen its confinement effect, the flow field simulations for different air hole heights are conducted, specifically examining $H = 2$ mm, $H = 5$ mm, $H = 8$ mm, and $H = 11$ mm, as illustrated in Figure 4.139. The distributions of axial velocity (at location L_1 in Figure 4.139) and radial velocity (at location L_2 in Figure 4.139) at different air hole heights *H* are investigated, as shown in Figures 4.140 and 4.141.

As the air hole height *H* increases, both the axial and radial velocities of the gas flow decrease, causing the sheath gas to predominantly remain in a laminar state. Even at $H = 11$ mm, the maximum gas flow velocity remains at 90 m/s, ensuring effective confinement performance of the sheath gas. At $H = 11$ mm, the gas flow field exhibits high uniformity, with a small and stable velocity difference in the central region. This minimizes the impact on the jet, thereby facilitating precise control of the jet.

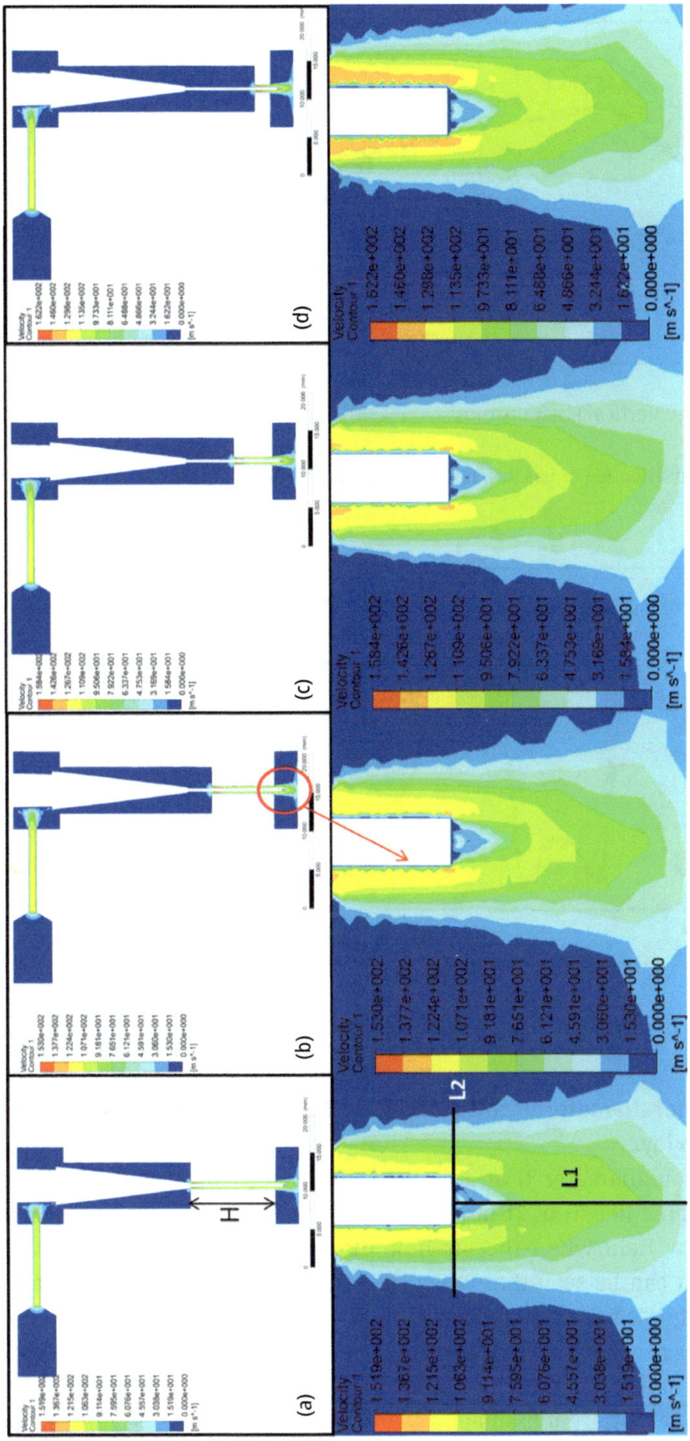

Figure 4.139: Velocity distribution of gas flow field at different air hole heights *H*: (a) *H* = 11 mm; (b) *H* = 8 mm; (c) *H* = 5 mm; and (d) *H* = 2 mm.

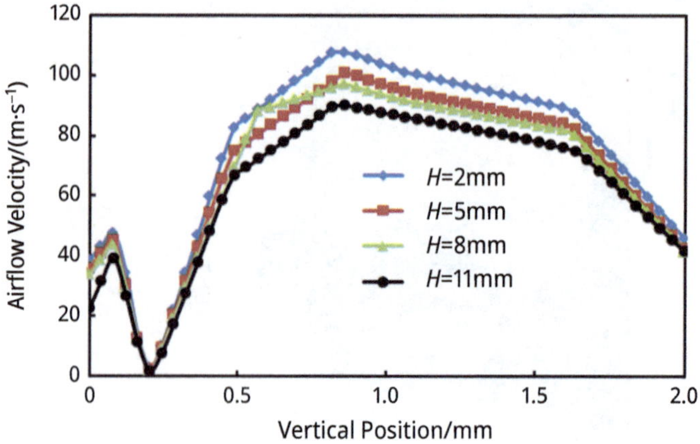

Figure 4.140: Axial velocity distribution of gas flow field at different air hole heights *H*.

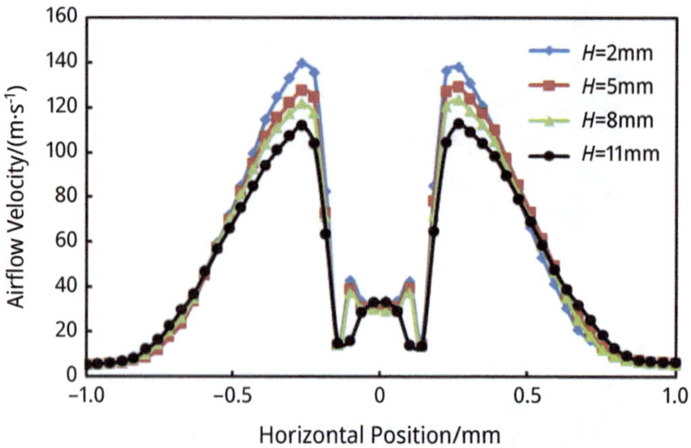

Figure 4.141: Radial velocity distribution of gas flow field at different air hole heights *H*.

4.3.4.2.2 Impact of inner diameter of the air hole on distribution of gas flow field in the sheath layer

The influence of air hole diameter D on the distribution of sheath gas flow field is investigated at an air hole height of 11 mm. Figure 4.142 shows the velocity distribution of sheath gas flow field when the air hole diameters D are 1.0, 1.5, 2.0, and 3.0 mm, respectively. It can be seen that when the inner diameter $D = 1.0$ or 1.5 mm, the flow field distribution is relatively uniform, the flow velocity in the middle part is stable, and the sheath gas can effectively constrain the jet. The axial velocity (at position L_1 in Figure 4.142) and radial velocity (at position L_2 in Figure 4.142) distributions of the gas flow field under different diameters D are shown in Figures 4.143 and 4.144.

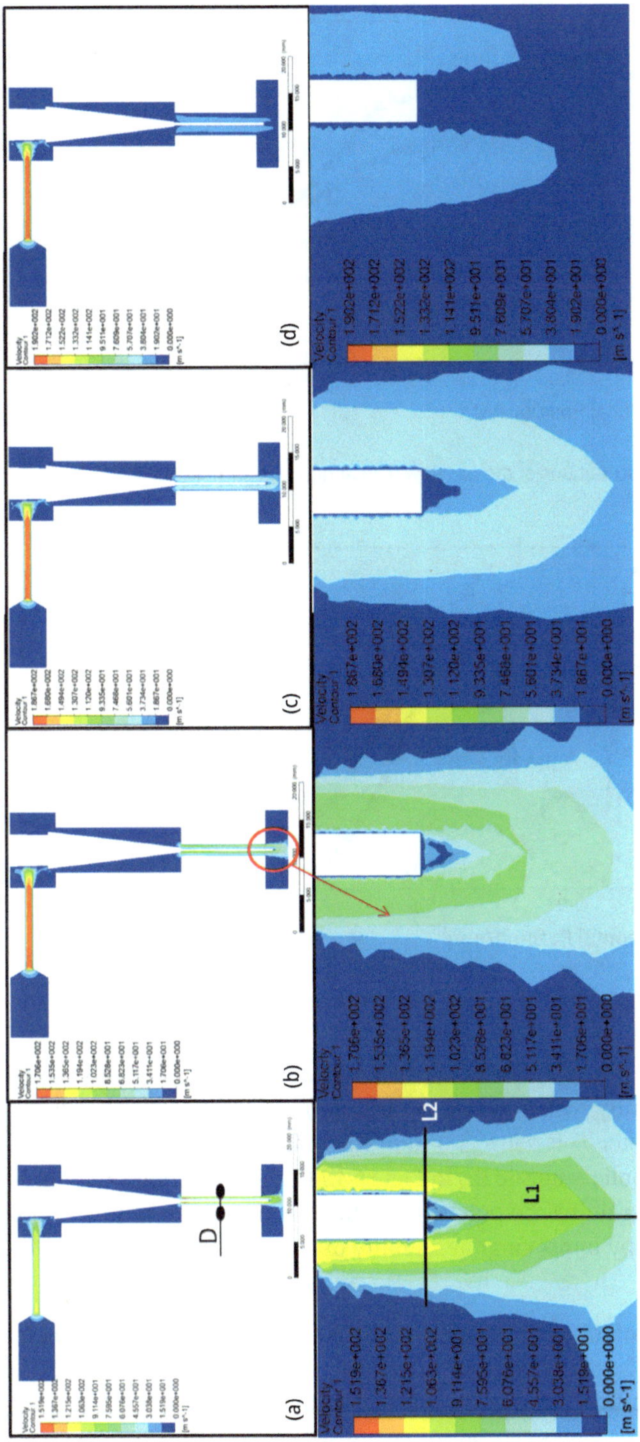

Figure 4.142: Velocity distribution of flow filed at different air hole inner diameters D: (a) $D = 1.0$ mm; (b) $D = 1.5$ mm; (c) $D = 2.0$ mm; and (d) $D = 3.0$ mm.

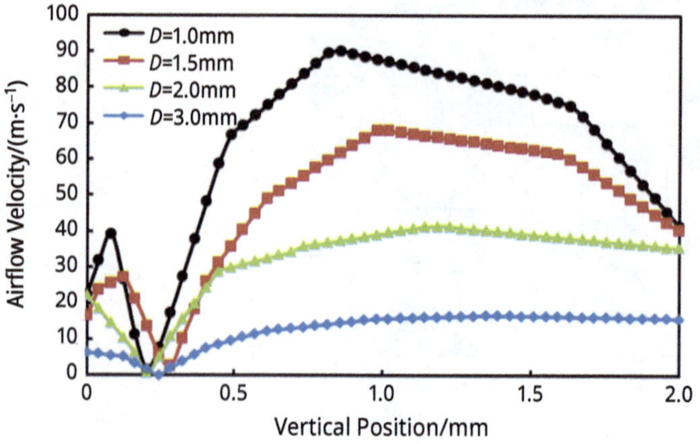

Figure 4.143: The axial velocity distributions of the gas flow field at different air hole diameter.

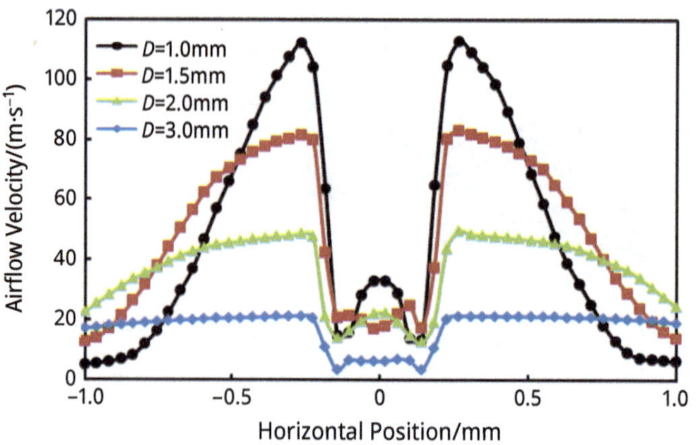

Figure 4.144: The radial velocity distribution of gas flow field at different air hole diameter.

With the increase of the air hole diameter D, both the axial and radial velocities of the gas flow decrease significantly, which weakens the confinement effect of the sheath gas. When the inner diameter D is set to 1.0 mm, the maximum velocity difference occurs near the exit of the hollow nozzle (L_2 in Figure 4.142). This facilitates the formation of Taylor cone by constraining the polymer solution droplet at the spinneret during the direct-writing, thereby improving the stability of the jet ejection.

4.3.4.2.3 The influence of inlet gas pressure on the distribution of gas flow field in the sheath layer

After optimizing the structure parameters of the spinneret ($H = 11$ mm and $D = 1.0$ mm), the inlet pressure becomes crucial influencing parameter for the sheath gas to be mainly in a laminar flow state and be stably and uniformly ejected into the air. The effects of different inlet pressure (P = 12.5, 25.0, 37.5, and 50.0 kPa) on the distribution of sheath gas flow field are investigated. The maximum velocities of sheath gas are 106, 151, 186.8, and 216.4 mm/s, respectively, as shown in Figures 4.145. When the inlet pressure is 12.5 or 25.0 kPa, the airflow velocity distribution is uniform with small velocity differences, allowing stable and uniform injection of the sheath gas into the air. As the inlet pressure increases, the maximum velocity of the sheath gas significantly increases, which is unfavorable for stable jet ejection. To enable the sheath gas to form a steady laminar flow around the jet and constrain the jet, the sheath gas pressure can be set to 25.0 kPa during the experiment, with some fluctuations based on actual experimental conditions).

4.3.4.2.4 The impact of coaxiality on the flow field distribution of the sheath gas

The introduction of sheath gas confinement can resist the unstable ejection of the jet, providing a new technical approach for precise jet control. However, during actual experiments, achieving strict coaxial alignment of the sheath-gas-constrained EDW spinneret is challenging. Although the coaxiality can be adjusted by rotating the adjustment screw, this may affect the uniformity of the flow field and the repeatability of experimental data.

To address this, simulation analysis of the sheath gas flow field distribution under different coaxiality is conducted. Four simulation models are established with eccentricity distances (ΔL) of 0, d/4, d/2, and d, where d represents the outer diameter of the hollow nozzle (set to 0.23 mm in the simulation). As shown in Figures 4.147 and 4.148, when the eccentricity distance ΔL is 0 mm (the ideal case of strict alignment), the sheath gas flow field is very uniform and stable. At an eccentricity distance of d/4, the radial velocity distribution contour at the hollow nozzle exit (Figure 4.148(b)) indicates that the sheath gas flow field remains relatively uniform on both sides of the nozzle exit. However, as the eccentricity distance increases to d/2 or d, a significant velocity difference arises between the left and right sides of the hollow nozzle exit, causing chaotic sheath gas flow and impairing its ability to constrain the jet.

4.3.4.3 Spinneret design and experimental setup for sheath-gas-confined EDW

4.3.4.3.1 Model of sheath-gas-confined EDW spinneret

The sheath-gas-confined EDW spinneret features a dual-layer structure, consisting of an inner liquid spinneret and an outer gas spinneret. The inner diameter of the liquid spinneret ranges from 60 to 300 μm, while that of the gas spinneret is 1 mm, as shown in Figure 4.149. A high-voltage electric field is applied between the liquid spinneret and the collector. Under the combined action of the electric field and the sheath gas, the solution ejected from the inner nozzle deforms to form a Taylor cone. The sheath gas facilitates the formation of the Taylor cone and the subsequent generation of the jet.

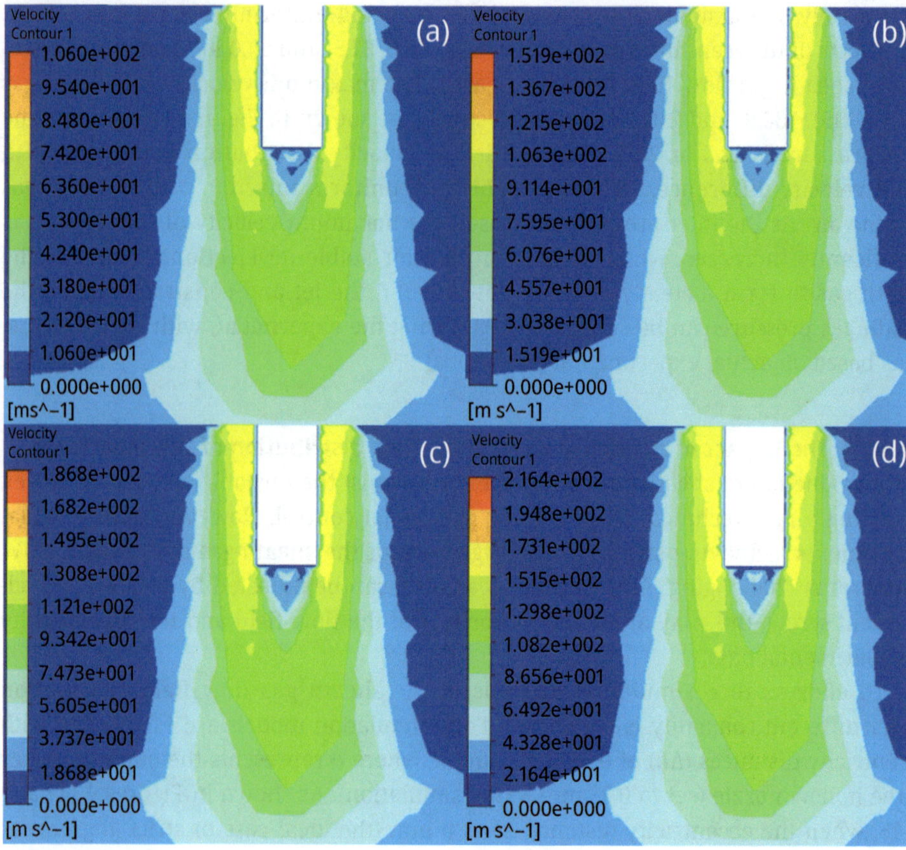

Figure 4.145: Axial velocity distribution of gas flow field under different inlet pressure P: (a) 12.5; (b) 25.0; (c) 37.5; and (d) 50.0 kPa.

4.3.4.3.2 Structure of the sheath-gas-confined EDW spinneret

Based on the basic principle of sheath-gas-confined EDW, we designed a sheath-gas-confined EDW spinneret, as shown in Figure 4.150. The spinneret comprises (1) bracket, (2) active block, (3) adjusting screw, (4) connecting screw, (5) air chamber, (6) spinneret holder, (7) flat washer, (8) wire connector, (9) sealing ring, (10) through hole, (11) solution inlet, (12) solution supply interface, (13) sealing ring groove, and (14) air inlet and nozzle (stainless steel hollow spinneret, connected to the liquid supply interface).

The wire connector is connected to the high-voltage power supply. The liquid inlet is connected to an external liquid supply device, with the liquid supply interface connected to the liquid spinneret. The air inlet is linked to an external gas supply device. The collector is connected to the negative terminal of the power supply and grounded, positioned directly beneath the spinneret. The air chamber is equipped

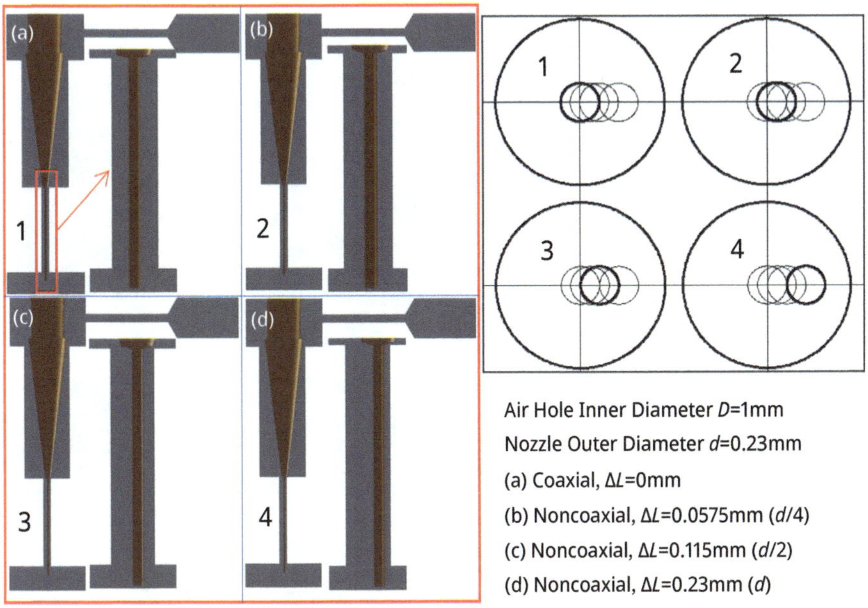

Figure 4.146: Coaxiality simulation model of EDW spinneret with sheath gas constraint.

with a double-layer stepped buffer structure that cushions the airflow and make it form a stable laminar flow. The sheath gas jetting from the air hole in the laminar flow state maintains a good alignment with the jet. The bracket is evenly fitted with three adjustment screws in a circular arrangement. Adjusting these screws allows for the alignment of the spinneret and the air hole, ensuring that they are as coaxial as possible during experiments. This alignment facilitates a more uniform gas flow field, enhancing the stability of the jet and the precision of deposition positioning.

To prevent the device from interfering with the electric field and affecting the stability of the jet, apart from using metals with good electrical conductivity for the terminals and spinnerets, other structures should be fabricated from transparent polymethyl methacrylate (organic glass) with excellent insulation properties. Organic glass offers outstanding light transmittance, allowing more than 92% of sunlight to pass through. This facilitates the observation of experimental phenomena and the collection of experimental data. It also boasts good insulation, high voltage resistance, considerable mechanical strength, and dimensional stability, making it suitable for repeated experiments without deformation.

To facilitate sheath-gas-confined EDW experiments, address the issue of repeated assembly and disassembly of equipment, and streamline the operation process, we designed and manufactured an integrated sheath-gas-confined spinneret. This spinneret combines a liquid supply interface, gas supply interface, electrode, and coaxiality adjustment function into a single unit. As illustrated in Figure 4.151, the spinneret

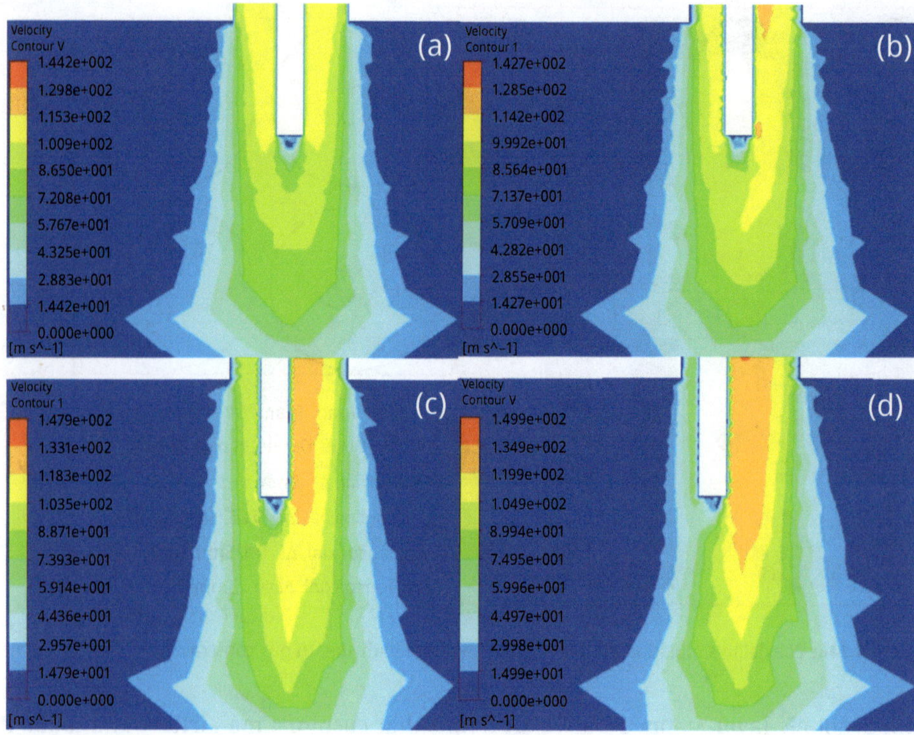

Figure 4.147: Axial velocity distribution of gas flow field at the nozzle exit under different coaxiality: (a) ΔL = 0 mm; (b) ΔL = 0.0575 mm; (c) ΔL = 0.1150 mm; and (d) ΔL = 0.2300 mm.

primarily consists of a liquid spinneret holder, an air chamber, an active block, and a bracket. The machining drawings and physical images of each component are provided in Figures 4.152–4.155, respectively.

4.3.4.3.2.1 Liquid spinneret holder and its assembly

Figure 4.152 illustrates the liquid spinneret holder. The solution inlet is connected to an external liquid supply device (a precision syringe pump), which is used to regulate the liquid supply rate within a range of 0–5,000 µL/h. The solution supply interface features a 1°–2° taper, which is designed to externally connect and secure the spinneret. The spinneret is a stainless-steel hollow needle. At the top of the spinneret holder, there is a wire hole (Figure 4.152(b)), through which the wire electrode from the terminal block is embedded into the liquid supply channel. This wire is externally connected to a high-voltage power supply for adjusting the electrospinning voltage.

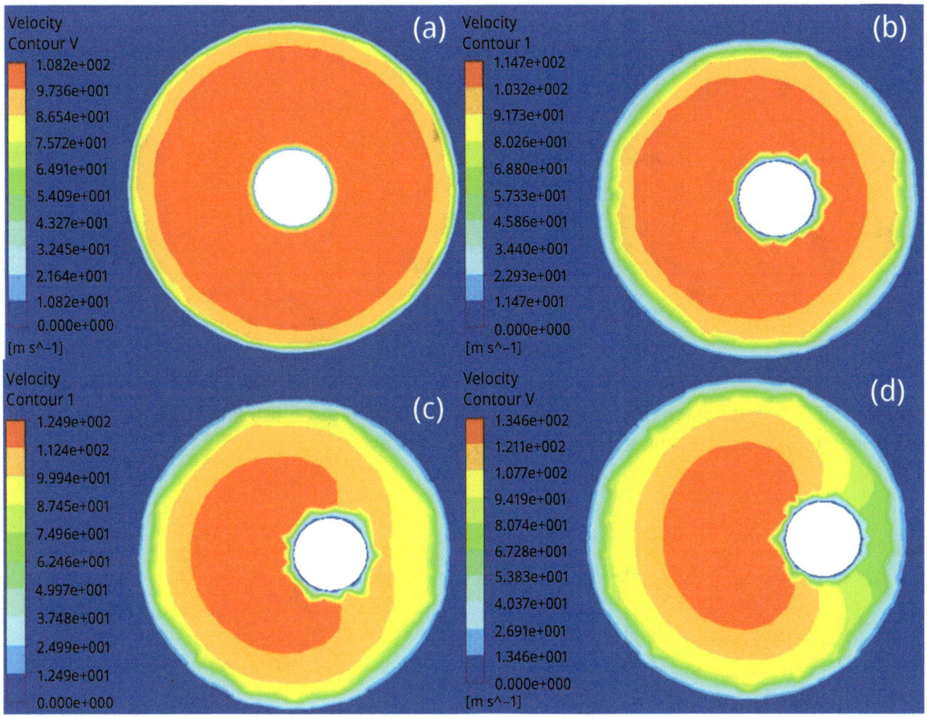

Figure 4.148: Radial velocity distribution of gas flow field at nozzle exit under different coaxialities: (a) ΔL = 0 mm; (b) ΔL = 0.0575 mm; (c) ΔL = 0.1150 mm; and (d) ΔL = 0.2300 mm.

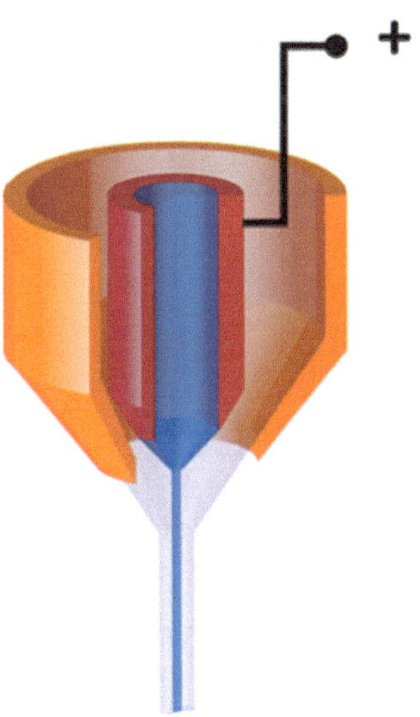

Figure 4.149: Schematic diagram of the core-shell spinneret assembly.

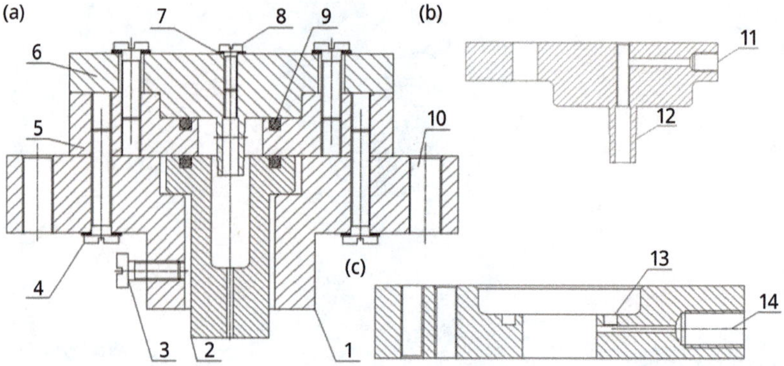

Figure 4.150: Schematic diagram of spinneret structure for sheath-gas-constrained EDW: (a) diagram of the spinneret structure; (b) spinneret holder; and (c) air chamber.

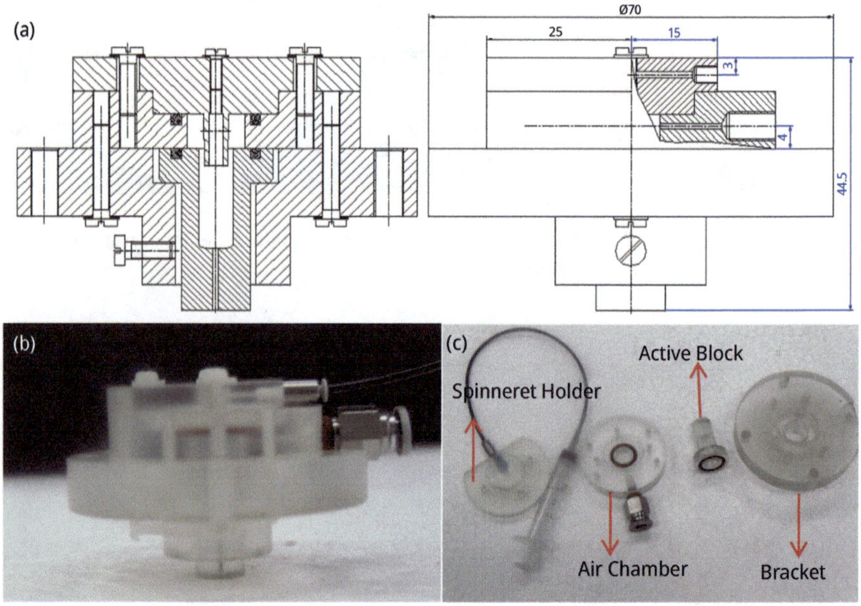

Figure 4.151: Spinneret of sheath-gas-confined EDW: (a) CAD drawing, (b) physical image, and (c) various components.

4.3.4.3.2.2 The air chamber and its assembly

Figure 4.153 shows the air chamber. The gas inlet hole is connected to an external gas supply device as shown in Figure 4.153(b), which is used to regulate the inlet gas pressure during experiments, with a pressure range of 0–0.1 MPa. The liquid spinneret holder is evenly equipped with three through-holes, while the corresponding positions

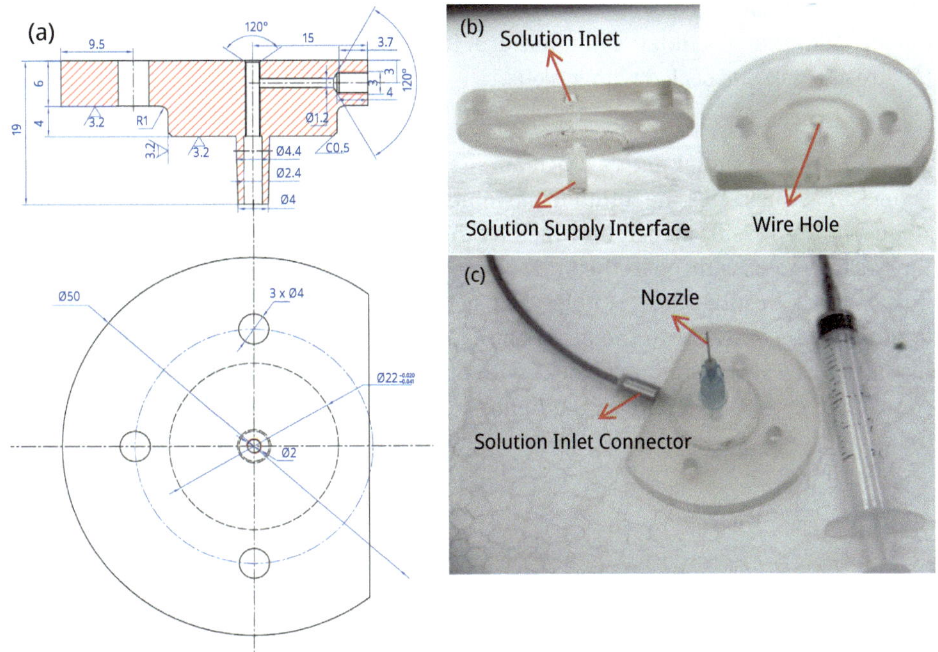

Figure 4.152: Schematic diagram of liquid spinneret holder and its assembly: (a) CAD drawings; (b) physical drawing; and (c) assembly diagram.

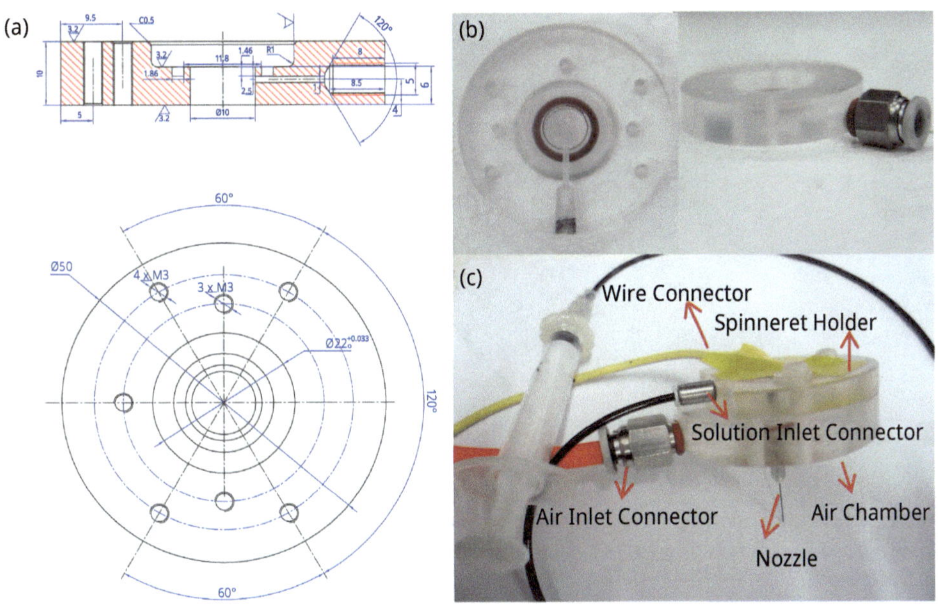

Figure 4.153: Schematic diagram of gas inlet chamber and its assembly: (a) CAD drawings; (b) physical drawing; and (c) assembly diagram of the spinneret holder and air chamber.

on the air chamber are evenly arranged with three threaded holes. The liquid spinneret holder and the gas inlet chamber are fastened together by connecting screws (Figure 4.153(c)) and sealed with gaskets.

4.3.4.3.2.3 The annular active block and its assembly

Figure 4.154 illustrates the annular movable block and its assembly. The double-layer stepped buffer gas flow region formed by the active block and the air chamber allows the gas flow to pass through the buffer zone and spray into the air in a stable laminar flow manner through the gas holes. The active block is equipped with a sealing ring to ensure a tight seal with the air chamber (Figure 4.154(c)). We designed and manufactured three sets of annular blocks with different heights (27.5, 28.5, and 29.5 mm) to control the extension length of the hollow nozzle from the active block, which are −1 (nozzle inside the active block), 0, and 1 mm, respectively, to meet various experimental requirements.

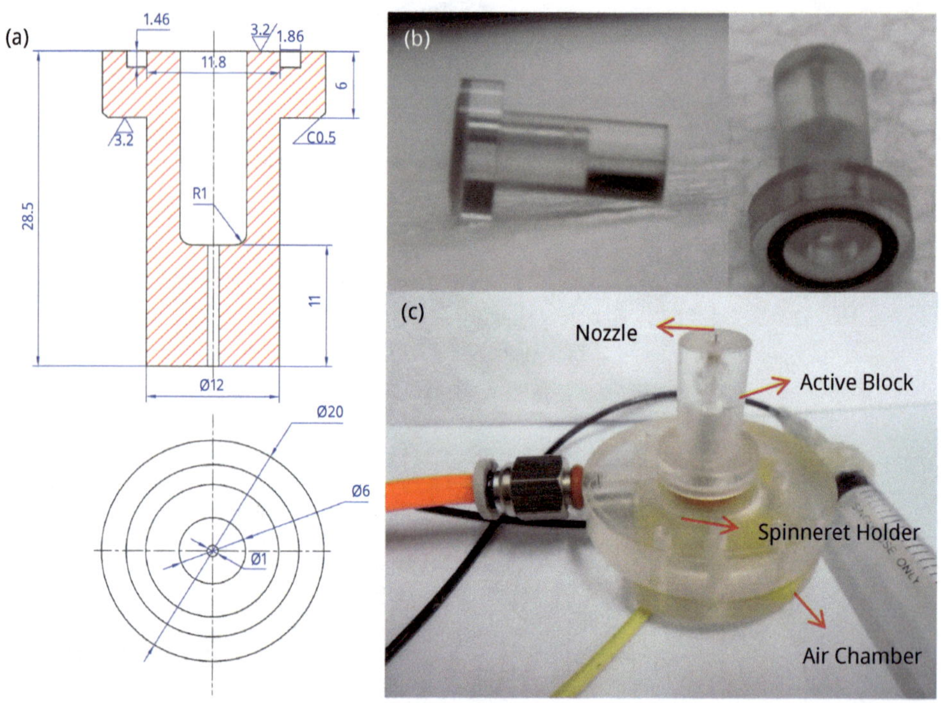

Figure 4.154: Schematic diagram of the active block and its assembly: (a) CAD drawings; (b) physical drawing; and (c) assembly diagram of the liquid spinneret holder, gas inlet chamber, and annular movable block.

4.3.4.3.2.4 The bracket and its assembly

As shown in Figure 4.155, the bracket is equipped with four evenly placed through holes, while four threaded holes are uniformly positioned at the corresponding positions on the air chamber. The bracket and the air chamber are connected via screws, securing the active block in place. This active block is sealed to the air chamber by a sealing ring. An adjusting screw is fixed to the bracket, and rotating this screw allows for the adjustment of the coaxiality between the spinneret and the gas hole. This ensures that the gas hole and the spinneret are as aligned as possible during experiments, resulting in a uniform gas flow field that guarantees the stability of the jet and the precision of deposition positioning. Four through-holes are evenly distributed along the edge of the bracket to secure the entire spinneret device. The bracket, active block, air chamber, and liquid inlet spinneret holder all feature center through-holes and are coaxially arranged. The active block is situated within the central through-hole of the bracket. The air chamber is affixed to the upper surface of the bracket, while the liquid intake spinneret holder is fixed to the upper surface of the air chamber. The lower portion of the liquid intake spinneret holder extends into the center through-holes of both the air chamber and the active block, as illustrated in Figure 4.155(c).

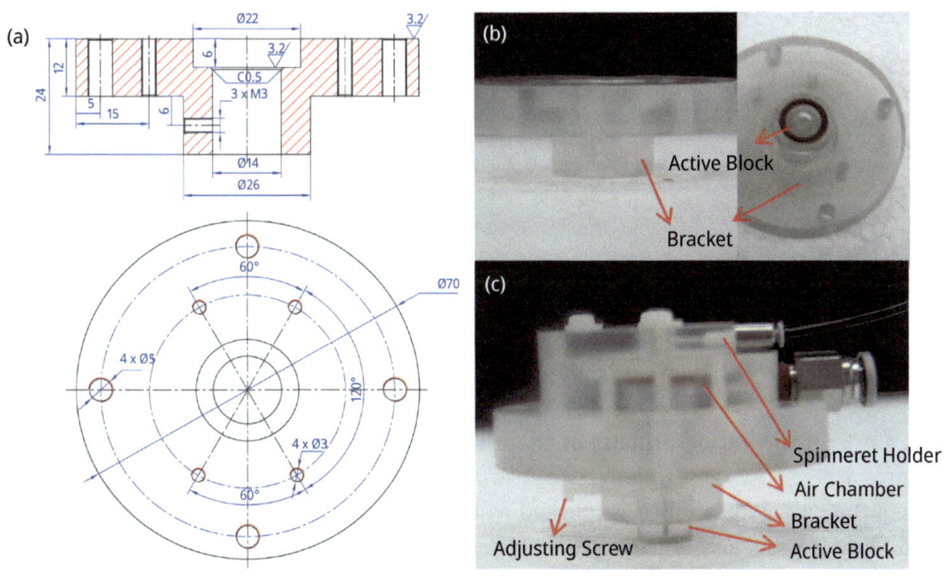

Figure 4.155: Diagram of the bracket and assembly: (a) CAD drawings; (b) physical drawings of bracket and annular movable block; and (c) assembly diagram of the spinneret components.

4.3.4.3.3 Experimental setup for the sheath-gas-confined EDW

According to the basic principle of sheath-gas-confined EDW technique, the experimental platform, as shown in Figure 4.156, is set up. It mainly includes a high-voltage DC power supply, a collector, an air compressor, a precision syringe pump, and a sheath-gas-confined EDW spinneret. The distance between the spinneret and the collector is adjustable from 0 to 10 mm, and the air compressor is used to adjust the inlet gas pressure (adjustable from 0 to 50 kPa), enabling laminar airflow ejection from the outlet. The collector is fixed on a precision motion platform, with an adjustable movement speed of 0–2 m/s. A CCD camera is used for observation and recording. An industrial host computer controls the motion trajectory and speed of the precision motion platform as well as the real-time video observation and recording with the CCD camera.

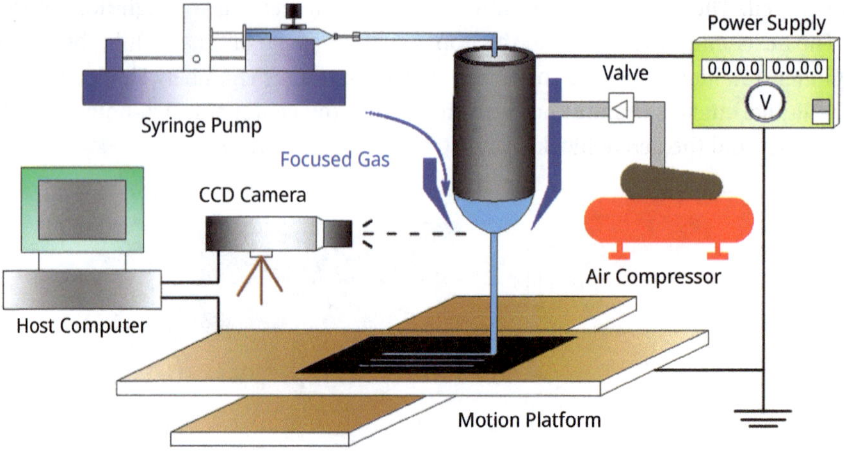

Figure 4.156: Experimental setup for the sheath-gas-confined EDW.

4.3.4.4 Ejection behaviors of sheath-gas-confined EDW jet

4.3.4.4.1 Jet formation process of sheath-gas-confined EDW

Figure 4.157 shows the Taylor cone formation and jet ejection process. The moment when the voltage is applied is marked as 0 s. The droplet on the spinneret, under the influence of surface tension, contracts into a hemispherical shape, as shown in Figure 4.157(a). When the electric field force increases sufficiently to overcome the surface tension of the solution, a jet ejects from the tip of the Taylor cone, as shown in Figure 4.157(c). The jet continues to move forward under the stretching effect of the electric field force, and its diameter gradually decreases. The higher the voltage, the stronger the electric field force, and thus the time for the jet to generate from the tip of the Taylor cone decreases with increasing applied voltage. The solution concentration affects the surface tension of the droplets. The higher the solution concentration, the greater the surface tension, and the shorter the time for the jet to be ejected from the tip of the Taylor cone.

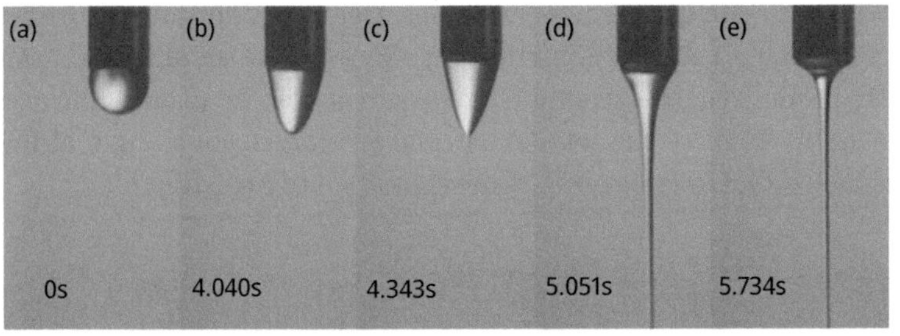

Figure 4.157: Taylor cone formation and jet ejection process: (a) voltage application begins; (b) droplet deformation; (c) critical ejection; (d) jet ejection; and (e) stable Taylor cone jet.

During the sheath-gas-confined EDW process, experimental parameters such as applied voltage, sheath gas pressure, solution supply rate, and solution properties may vary. Consequently, the repulsive forces between free charges in the polymer solution and the surface tension will change, leading to corresponding variations in the ejection behavior and morphology of the droplet. Apart from the stable Taylor cone jet mode, other ejection modes include oscillating jet, whipping jet, bifurcating jet, multi-jet, and entangled multi-jet. These unstable ejection modes can significantly affect the precise positioning deposition of sheath-gas-confined EDW and its application.

Experimental observation finds that the fluctuation of the liquid surface in these unstable ejection modes is quite unstable. This is primarily due to mismatched or unstable experimental parameters, uneven collectors or the external interferences such as airflow and external electric field. During experiments, disturbances should be avoided as much as possible to achieve stable Taylor cone ejection, thereby enabling the precise deposition of micro-nanostructures.

During the electrospinning process, the interplay between the repulsive forces among free charges and the surface tension within the polymer solution causes the droplet to assume a conical shape (Taylor cone). When the surface charge density of the Taylor cone formed by the polymer solution exceeds a critical threshold, the Taylor cone becomes unstable, leading to the ejection of a charged jet from its tip. Researchers have utilized tracer particles to unveil the presence of axisymmetric toroidal flows within the Taylor cone, indicating that the stable jet is formed by shearing the fluid on the surface of the cone [6].

The formation of a stable Taylor cone jet in the EDW process is conducive to the stable, continuous, and controllable jet flow, enabling the precise positioning and deposition of micro-nanostructures. Figure 4.158(a) shows the jet ejection without sheath gas confinement, where relatively large pendant droplet forms at the spinneret. Since the maximum electric field strength on the droplet surface does not always occur at the droplet tip, environmental factors such as the uneven surface of the collector can cause the root of the jet to oscillate across the droplet surface, thereby affecting positioning accuracy. Figure 4.158(b) depicts the jetting process with sheath gas confinement. The introduction of sheath gas applies additional stretching and shearing forces to the polymer droplet surface, reducing the initial jet diameter and enhancing jet stability. This effectively minimizes the oscillation of the jet root on the droplet surface, resulting in a more stable Taylor cone jet. Furthermore, a larger cone angle of the Taylor cone reduces the jet diameter, leading to the formation of finer fibers.

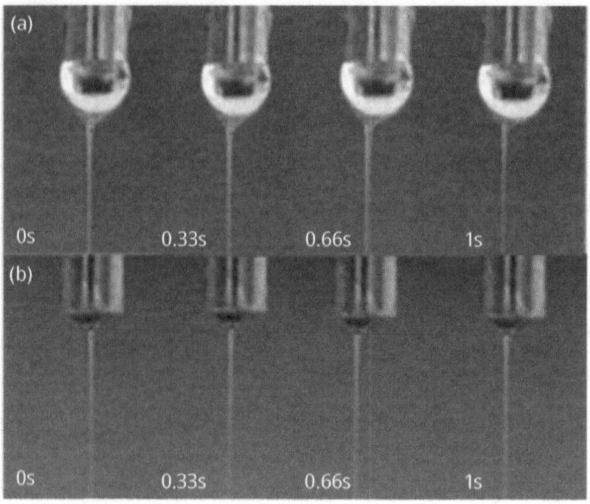

Figure 4.158: Taylor cone ejection: (a) EDW without sheath gas confinement; (b) EDW with sheath gas confinement (solution concentration, nozzle inner diameter, spinneret-to-collector distance, liquid supply speed, applied voltage, and sheath gas pressure are 0.11 mm, 2 mm, 100 μL/h, 2.3 kV, 10%, and 25 kPa, respectively).

The additional tensile and shear forces exerted by the sheath gas deform the pendant droplet at the spinneret, forming a stable Taylor cone ejection mode with a larger cone angle. We examine the Taylor cone morphology under different sheath gas pressures, as shown in Figure 4.159. Without sheath gas (0 kPa), a large pendant droplet forms at the spinneret. When a gas pressure of 5 kPa is applied, the droplet is confined by the sheath gas, resulting in a stable Taylor cone jet. The initial jet diameter decreases as the sheath gas pressure increases. The constraining effect of the sheath gas enhances jet stability by creating a stable laminar flow around the jet, reducing the relative velocity between the jet and the surrounding medium, and thereby suppressing the onset of jet instability.

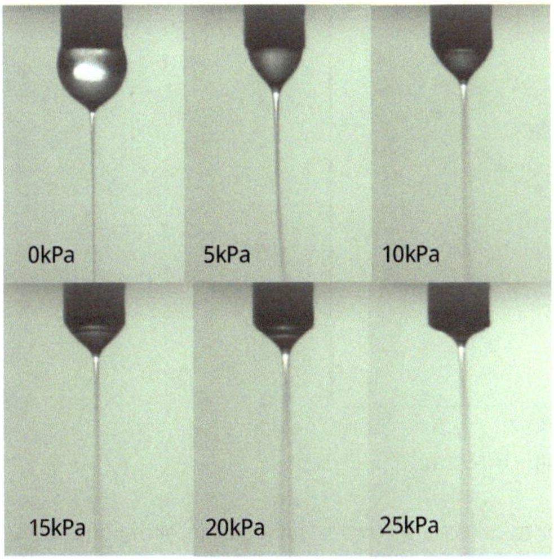

Figure 4.159: Taylor cone morphology under different sheath gas pressures (solution concentration is 10%, inner diameter of the nozzle is 0.21 mm, distance from the spinneret to the collector is 2 mm, liquid supply rate is 100 μL/h, and applied voltage is 2.0 kV).

4.3.4.4.2 Critical conditions for jet ejection in sheath-gas-confined EDW

In the electrospinning process, the applied voltage primarily affects jet ejection through two factors: the electric field strength at the spinneret and the charge density concentrated on the solution surface. When the applied voltage increases to a critical value (where the charge density on the Taylor cone surface exceeds the critical charge density), the combined force of the electric field and gravity acting on the pendant droplet exceeds the surface tension. As a result, the tip of the Taylor cone formed by the polymer droplet begins to form a jet, which then moves along the direction of the electric field. This critical voltage is referred to as the onset voltage. When the applied voltage exceeds the onset voltage, the electric field force acting on the droplet becomes sufficient to overcome the surface tension; thus a jet is ejected from the tip of the Taylor cone.

Since increasing the spinneret-to-collector distance reduces the space electric field strength, the critical voltage required for both EDW without and with sheath gas confinement will increase with the spinneret-to-collector distance, as shown in Figure 4.160. With the introduction of sheath gas, under the same conditions, the critical voltage of sheath-gas-confined EDW will decrease significantly, and it further decreases as the sheath gas pressure increases, as shown in Figure 4.161. Experimental results demonstrate that sheath gas provides additional force on the suspended droplets at the spinneret, making it easier for the droplet to overcome the surface tension and form a jet. The introduction of sheath gas can reduce the voltage required for jetting, thereby lowering the electric field intensity at the spinneret and promoting stable jetting and precise fiber deposition.

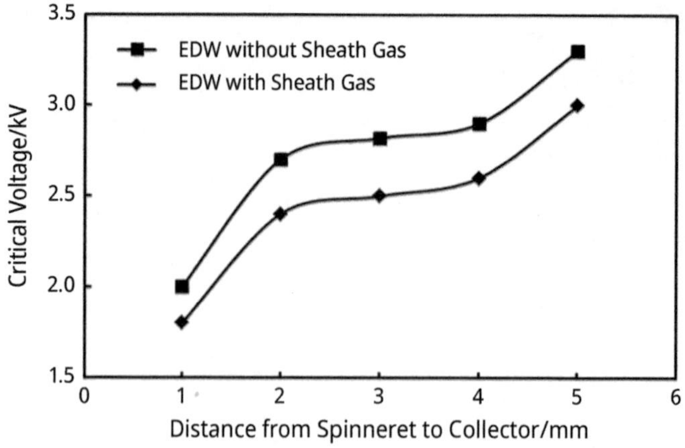

Figure 4.160: Critical voltage for different spinneret-to-collector distances (solution concentration is 12%, needle inner diameter is 0.21 mm, liquid supply rate is 100 μL/h, and sheath gas pressure is 25 kPa).

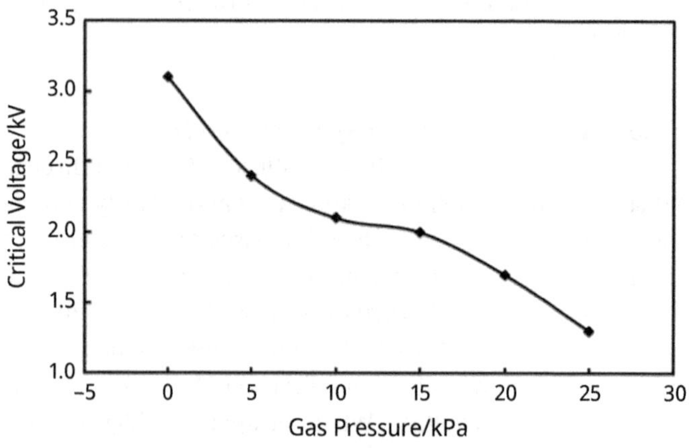

Figure 4.161: Critical voltage for sheath-gas-constrained EDW under different sheath gas pressures (solution concentration 12%, needle inner diameter 0.21 mm, and liquid supply speed 100 μL/h).

4.3.4.4.3 Induction effect of sheath gas confinement on fiber deposition

During electrohydrodynamic direct-writing, although the distance from the spinneret to the collector is reduced, it is still susceptible to environmental factors such as uneven collector surface and deposited fibers. If the collector surface is uneven, the distance between the spinneret and the collector will reduce at the protruding regions, resulting in increased electric field strength between these regions and the spinneret. This induces the jet to deposit preferentially onto the protruding areas. The distribution of the electric field near the collector surface directly influences the deposition behavior of the charged jet and nanofibers.

In the experiment, a substrate with a protruding structure is used as the collector to investigate the deposition behavior of nanofibers in sheath-gas-constrained EDW and to explore the positioning deposition control technology. The schematic diagram of the principle is shown in Figure 4.162. When a substrate with a protruding structure is used as the collector, the 3D protruding structure alters the space electric field distribution. The local non-uniformity of the electric field induces the nanofibers to deflect and deposit onto the surface of the protruding structures. However, with the introduction of sheath gas, the influence of the protruding structures is shielded, external interference is reduced, and the stability of jet ejection is improved. The sheath gas confines the jet, enhancing the anti-interference capability of the charged jet.

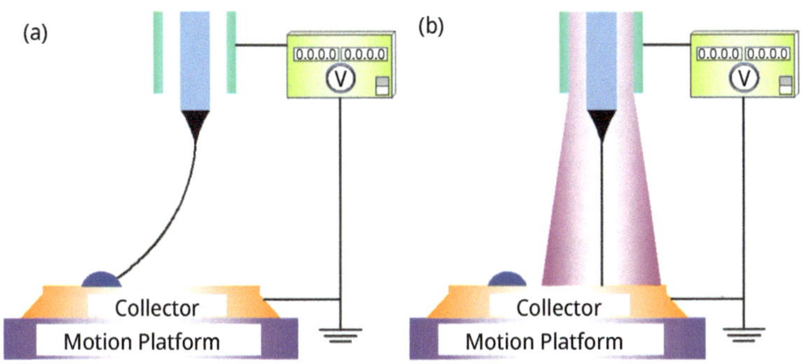

Figure 4.162: Schematic diagram of sheath-gas-constrained EDW on substrates with raised structures: (a) without sheath gas constraint and (b) with sheath gas constraint.

As shown in Figure 4.163, space electric field simulation analysis reveals that the electric field strength on the surface of the 3D protruding structure is significantly higher than that in other regions of the collector. The electric field lines are directed toward the edges of the protruding structure, altering the direction of the force acting on the charged jet and inducing the charged jet to deposit onto the surface of the protruding structure.

On the flat collector without protruding structures, the electric field distribution is uniform and the electric field lines is perpendicularly directed to the collector surface. Therefore, the jet deposits at the intersection of the spinneret's center axis extension line and the collector surface under the electric field, as shown in Figure 4.164(a). When subjected to a locally non-uniform electric field, the charged jet bends, causing its deposition trajectory to deviate, as shown in Figure 4.164(b)–(d). The protruding structures on the collector alter the space electric field distribution, exerting an inductive effect on the charged jet. This causes the jet to deflect toward the protruding structures (e.g., glass, iron, or polymer droplets). Figure 4.164 demonstrates that the protruding structures modify the distribution of space electric field, thereby changing the direction of the force acting on the charged jet and inducing its deposition onto

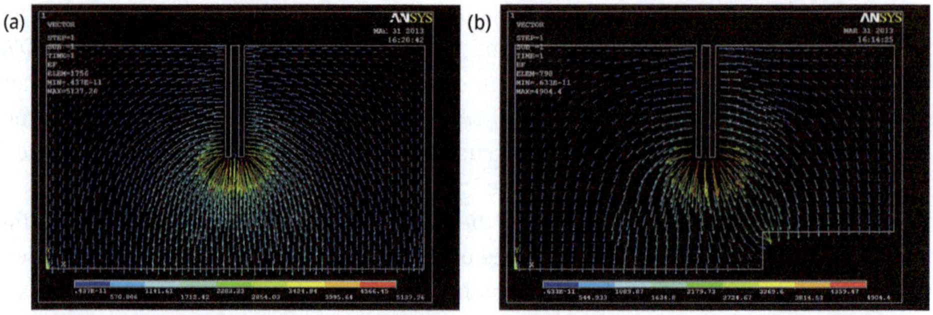

Figure 4.163: Simulation of space electric field distribution on the surface of the collector with different structures: (a) flat structure and (b) raised structure.

the raised surfaces. This highlights the inductive role of protruding structures in the deposition of charged jets during electrohydrodynamic direct-writing.

To investigate the deposition behavior of nanofibers electrospun by sheath-gas-confined EDW on a collector with protruding structures, we define the *critical distance* as the horizontal distance between the protruding structure and the spinneret's central axis when jet deflection occurs. For example, as shown in Figure 4.164(b), the critical distance is 493 μm for a glass sheet as the protruding structure. We further examine the effects of applied voltage and spinneret-to-collector distance on the critical distance.

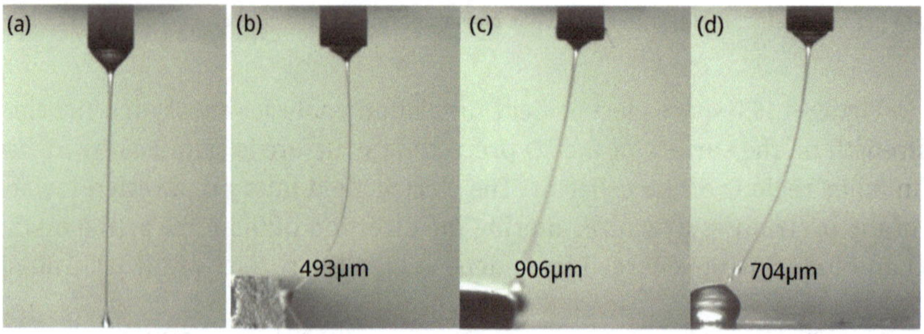

Figure 4.164: Deposition of EDW jet on substrates with different raised structures: (a) flat structure; (b) raised structure of glass sheet; (c) raised structure of iron block; and (d) raised structure of polymer droplet (solution concentration, nozzle inner diameter, spinneret-to-collector distance, solution supply rate, applied voltage, and sheath gas pressure are 10%, 0.21 mm, 3 mm, 20 μL/h, 2.6 kV, and 0 kPa, respectively).

As shown in Figure 4.165, the critical distance for the protruding structure to induce jet deflection increases with the applied voltage. This is because a higher voltage results in a stronger electric field, which enables the protruding structure to induce jet bending and deflection at a greater distance under the same conditions.

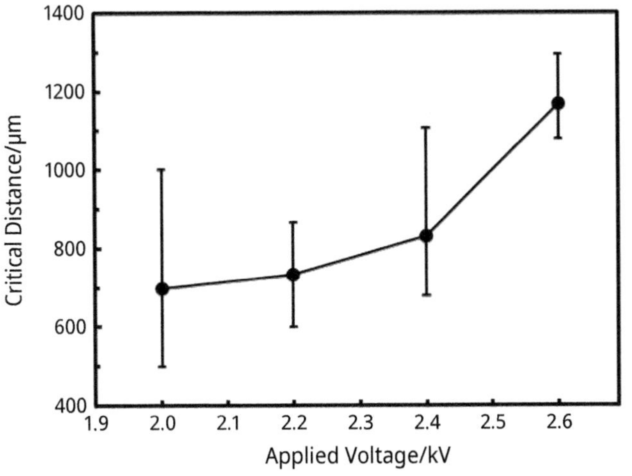

Figure 4.165: Critical distances at different applied voltages (solution concentration is 10%, needle inner diameter is 0.21 mm, spinneret-to-collector distance is 2 mm, liquid supply rate is 20 μL/h, and sheath gas pressure is 0 kPa; the raised structure is iron).

As shown in Figure 4.166, the critical distance for the protruding structure to induce jet deflection increases with the spinneret-to-collector distance. As the spinneret-to-collector distance increases, the electric field strength decreases, which theoretically requires the protruding structure to be closer to the jet to induce deflection. However, with a larger spinneret-to-collector distance, the jet becomes more prone to oscillation. Under the same jet deflection (swing angle), the critical distance becomes larger. Therefore, the critical distance increases as the spinneret-to-collector distance increases.

When applying EDW technology to the fabrication of micro-nano devices and patterned structures, it is inevitable to encounter substrates with protruding structures. Experimental results indicate that the deposition trajectory of the jet bends and deflects when deposited on such substrates, accumulating on the surface of the micro-protruding structures. To address this issue, a guided positioning deposition control technology is necessary to achieve precise positioning deposition of the jet. In sheath-gas-confined EDW technology, the sheath gas constrains the jet, mitigates the inductive effect of the protruding structures, reduces external interference, and enhances the stability of jet ejection.

To better demonstrate the confinement effect of sheath gas, the gas flow field distribution at the outlet of the EDW spinneret is simulated and analyzed, as illustrated in Figure 4.167. The results reveal a uniform gas flow field at the spinneret outlet, with the gas flowing vertically downward. This flow exerts a vertical "guiding force" on the charged jet, effectively mitigating the influence of protruding structures. Consequently, the jet's deflection is constrained, allowing it to deposit stably in a direction perpendicular

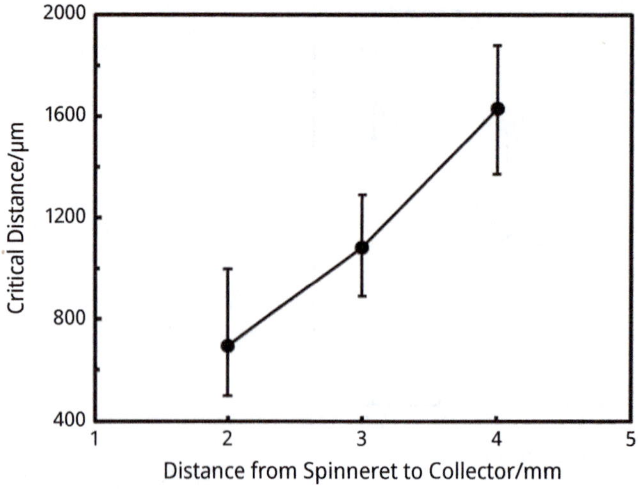

Figure 4.166: Critical distances at different spinneret-to-collector distances (solution concentration is 10%, needle inner diameter is 0.21 mm, spinneret-to-collector distance is 2 mm, liquid supply rate is 20 μL/h, and sheath gas pressure is 0 kPa, raised structure is iron).

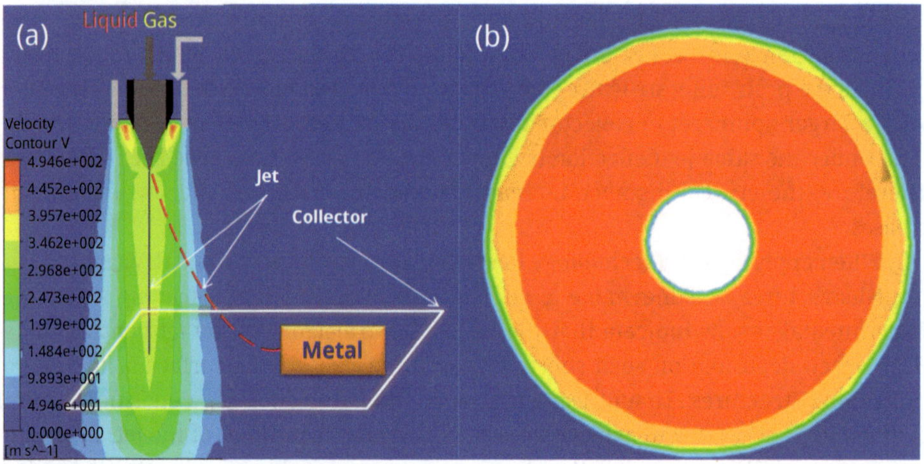

Figure 4.167: Simulation of sheath gas flow field at the spinneret outlet: (a) axial gas velocity distribution cloud map at the spinneret outlet and (b) radial gas velocity distribution cloud image at the spinneret outlet.

to the collector without interference from the protruding structures. These findings indicate that the sheath gas enhances the jet's stability and resistance to external disturbances.

To further illustrate the influence of sheath gas on the charged jet, we compared the EDW jet and the fiber deposition patterns on a collector with a protruding structure at sheath gas pressure of 0 and 25 kPa, as shown in Figures 4.168 and 4.169. In Figure 4.168, when the sheath gas pressure is 0 kPa, the jet exhibits obvious deflection and deposits on the surface of the raise due to the non-uniform space electric field distribution. After introducing the sheath gas at a pressure of 25 kPa, the jet deposits vertically downward on the substrates featuring various protrusions (glass, iron, polymer droplet), demonstrating the stabilizing effect of the sheath gas.

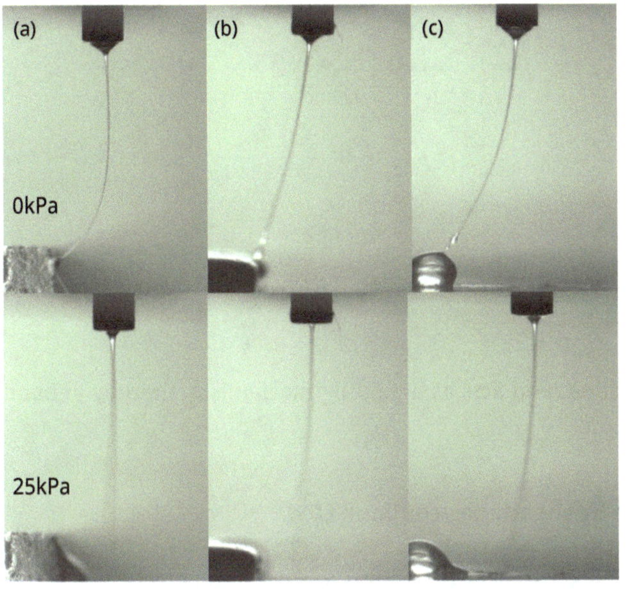

Figure 4.168: Inducing effect of sheath gas on the jet flow with raises of (a) glass sheet; (b) iron block; and (c) polymer droplet (solution concentration is 10%, nozzle inner diameter is 0.21 mm, distance from spinneret to collector is 3 mm, solution supply rate 20 is μL/h, and applied voltage is 2.6 kV).

Further investigation is conducted on silicon substrates with a 250-μm diameter acupuncture needle serving as the protruding structure. The results are shown in Figure 4.169. In Figure 4.169(a), when the sheath gas pressure is 0 kPa, the jet deflects significantly and deposits on the surface of the acupuncture needle at an interference distance of $d = 255$ μm. However, with the introduction of sheath gas, the jet no longer deflects or deposits on the protruding structure, even at interference distances $d = 129$ μm (less than 255 μm) or $d = 597$ μm (more than 255 μm), effectively eliminating the influence of the protrusion.

Based on the results shown in Figures 4.168 and 4.169, it is evident that the introduction of sheath gas in electrohydrodynamic direct-writing provides a vertically downward force to the jet. This force shields the jet from the influence of protruding

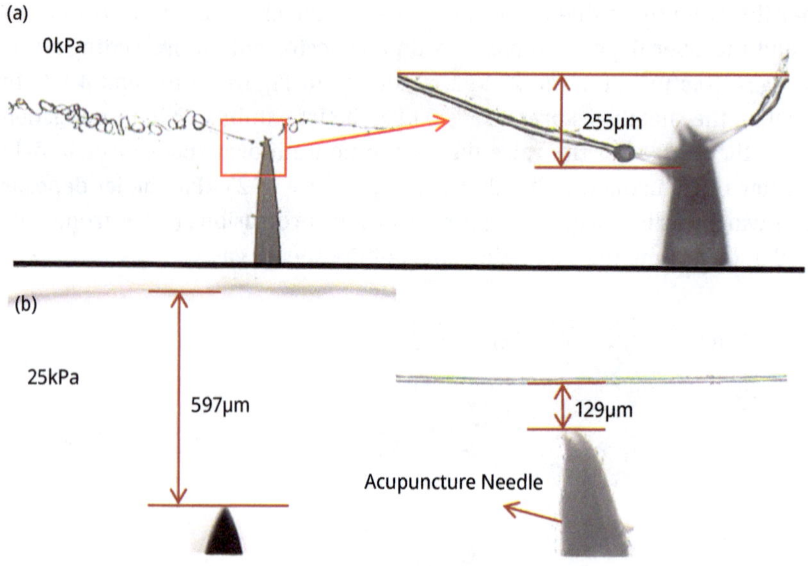

Figure 4.169: Jet deposition on substrate with raised structures via (a) EDW without sheath gas constraint and (b) EDW with sheath gas constraint.

structures, prevents jet deflection, and acts as a guiding mechanism, thereby enhancing the stability of jet deposition.

4.3.4.5 Patterned deposition by sheath-gas-confined EDW

4.3.4.5.1 Direct-writing of fiber structures by sheath-gas-confined EDW

Using a silicon substrate as the collector, the deposition morphology of electrohydrodynamic direct-written fibers is investigated. When the distance between the spinneret and the collector is too short, the solvent cannot fully evaporate before reaching the collector. Similarly, when the applied voltage is too low, the tensile force on the jet is insufficient, preventing the jet from being adequately stretched during deposition. These conditions result in the formation of bead-like structures on the fibers. In addition, the concentration of solution also significantly influences the formation of these bead structures. For solution of low concentration, the solvent fails to fully volatilize during the process of jet solidification, causing the jet to deposit as droplets on the collector and resulting in numerous bead structures, as shown in Figure 4.170 (a)–(c).

With the introduction of sheath gas confinement, the additional tensile force of sheath gas can not only deform the pendant droplet at the spinneret, forming a stable Taylor cone ejection mode with a larger cone angle, but also accelerate the evaporation of the solvent in the jet. This promotes fiber formation and prevents the generation of liquid films, as shown in Figure 4.170(d). The sheath-gas-confined EDW can effectively eliminate the formation of bead-structured fibers.

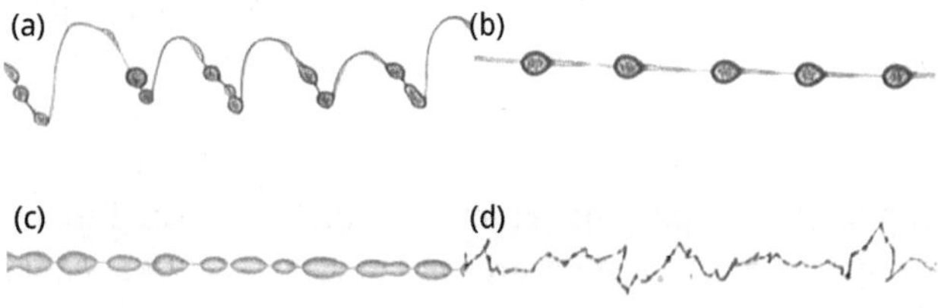

Figure 4.170: Fiber deposition morphology by sheath-gas-confined EDW with (a) solution concentration of 8%, collector speed of 10 mm/s; (b) solution concentration of 8%, collector speed of 50 mm/s; (c) solution concentration of 10%, collector speed of 10 mm/s; and (d) solution concentration of 10%, collector speed of 10 mm/s. The sheath gas pressure is 25 kPa.

Furthermore, sheath gas accelerates the volatilization of solvent in the jet, which generates an impact force on the surface of the deposited fibers. Figure 4.171 shows the SEM images of the fibers deposited on the collector. Without sheath gas confinement, the fibers exhibit a smooth and uniform surface. However, with the introduction of sheath gas, the deposited fibers become rough and the specific surface area increases significantly.

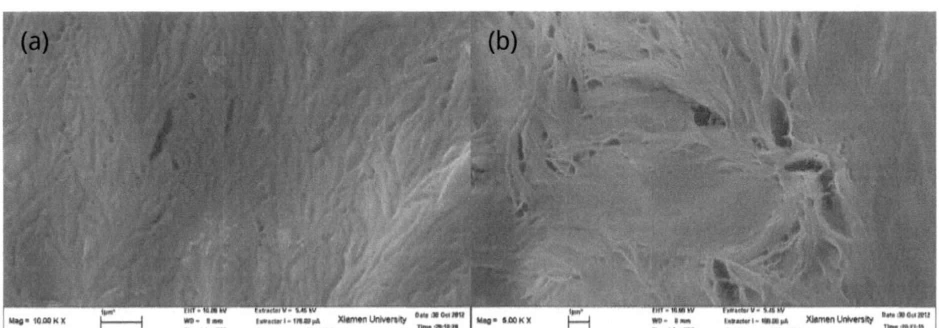

Figure 4.171: SEM images of the surface morphology of the fibers deposited by (a) EDW without sheath gas confinement and (b) sheath-gas-confined EDW (solution concentration, inner diameter of nozzle, spinneret-to-collector distance, solution feeding rate, applied voltage, and sheath gas pressure are 12%, 0.21 mm, 2 mm, 100 μL/h, 2.0 kV, and 25 kPa, respectively).

Although the sheath gas causes surface roughness, it increases the surface-to-volume ratio of the fibers, which is advantageous for applications in sensing and detection.

Fibers electrospun by sheath-gas-confined EDW exhibit increased effective contact area of the sensitive material, improving sensor performance [7], thus enabling the fabrication of high-performance sensing and detection devices.

The influence of applied voltage, sheath gas pressure, and collector moving speed on fiber diameter during sheath-gas-confined EDW is also investigated. The results, presented in Figures 4.172–4.174, demonstrate that the fiber diameter decreases with the increasing applied voltage, sheath gas pressure, and collector moving speed.

The stretching and thinning of electrospinning jet are primarily governed by the electric field strength. A higher applied voltage generates greater electric force acting on the polymer solution droplet, leading to a higher surface charge density on the jet and, consequently, greater charge repulsion. Additionally, a stronger electric field accelerates the jet to a higher velocity. These combined effects enhance the tensile force on the jet, resulting in smaller fiber diameters deposited on the collector. As shown in Figure 4.172, the average fiber diameter decreases from 4.56 to 2.23 μm as the voltage increases from 2.0 to 2.4 kV.

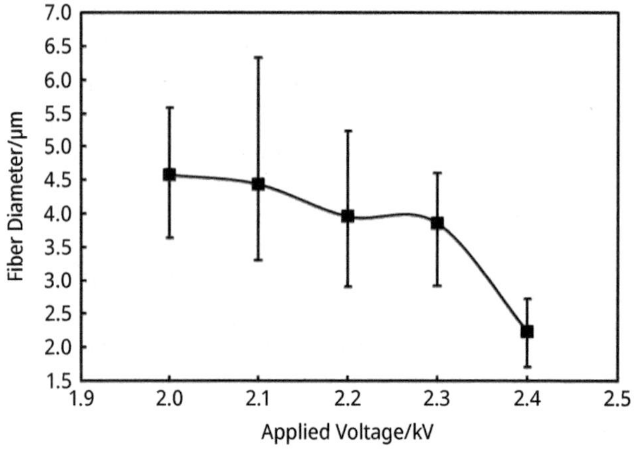

Figure 4.172: Diameters of fibers prepared by sheath-gas-confined EDW at different applied voltages (the solution concentration is 12%, the inner diameter of the nozzle is 0.21 mm, the spinneret-to-collector distance is 2 mm, the liquid feeding speed is 100 μL/h, the collector speed is 100 mm/s, and the sheath gas pressure is 25 kPa).

When sheath gas is introduced into EDW, it exerts a confinement effect on the Taylor cone and the charged jet. As shown in Figure 4.173, the average diameters of the fibers are 4.46 μm, 1.83 μm, 1.01 μm, 0.99 μm, 1.12 μm, and 845 nm at the sheath pressure of 0, 5, 10, 15, and 25 kPa, respectively. The results demonstrate that the average diameter of the fiber decreases obviously with the introduction of sheath gas. As the sheath gas pressure increases, the fiber diameter initially decreases and then stabilizes. This indicates that the sheath gas ensures the continuity of the jet and limit the growth of jet diameter. It provides an additional tensile force, and, together with the electric field force, further elongates the jet and reduces the fiber diameter. However, the reduction in the fiber diameter is limited, as the primary influence on fiber diameter remains the stretching and thinning effect of the electric field force.

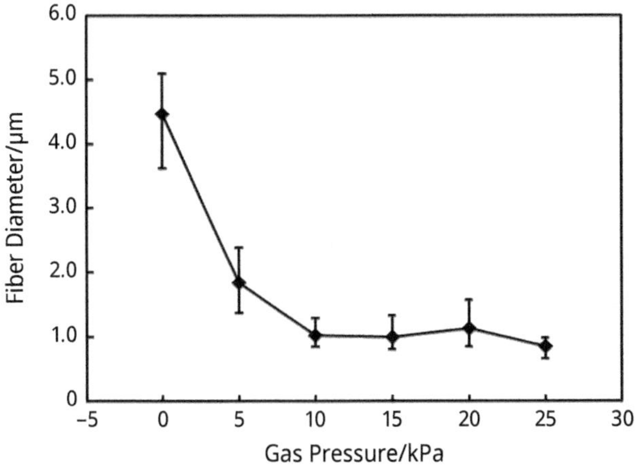

Figure 4.173: Fiber diameters prepared by sheath-gas-confined EDW at different sheath gas pressures (the solution concentration is 10%, the inside diameter of the needle is 0.21 mm, the spinneret-to-collector distance 2 mm, the solution feeding rate is 100 μL/h, the collector speed is 100 mm/s, and the voltage is 2.3 kV).

An increase in collector moving speed provides a greater mechanical tensile force to the jet. The combined action of the electric field force and mechanical tensile force, analogous to the "kite effect" [8, 9], facilitates the reduction of fiber diameter. As shown in Figure 4.174, the fiber diameters are 873, 845, 743, and 677 nm at collector velocities of 100, 200, 300, and 400 mm/s, respectively. This demonstrates that the fiber diameter decreases with increasing collector moving speed. Under the same experimental conditions, the motion of the collector applies a mechanical tensile force to the deposited fibers, and this force increases with collector moving speed.

Patterned fabrication and accurate positioning deposition of micro-nanostructure are critical for applying EDW to the integrated manufacturing of micro-nanosystems. In the following section, we focus on the direct-writing of patterned micro-nanostructures through sheath-gas-confined EDW based on a precision motion platform.

Increasing the collector movement speed to match the jet deposition speed, the influence of jet spiral whipping can be eliminated, and straight nanofibers can be obtained, as shown in Figure 4.175(f). Increasing the voltage enhances the electric field force acting on the charged jet, thereby increasing its deposition speed. Enlarging the distance between the spinneret and the collector makes the jet more susceptible to interference, leading to spiral whipping. Consequently, the matching speed increases with increasing applied voltage and spinneret-to-collector distance.

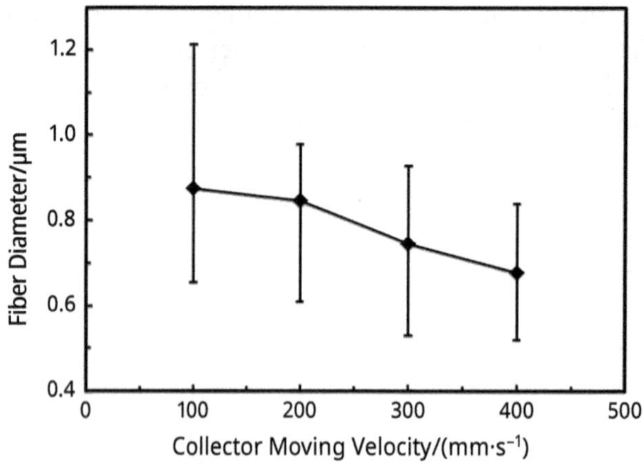

Figure 4.174: Fiber diameter produced by sheath-gas-confined EDW at different collector movement velocities (solution concentration: 10%, nozzle inner diameter: 0.21 mm, nozzle-to-collector distance: 2 mm, flow rate: 100 μL/h, voltage: 2.3 kV, and sheath gas pressure: 25 kPa).

After the jet front is deposited on the collector, it undergoes bending and spiraling due to the influence of electric field force, charge repulsion force, and its own viscoelastic force. Although the spinneret-to-collector distance is reduced in EDW, the residual charge on the collector has a more pronounced impact on the deposition behavior of the charged jet. Under the combined effect of repulsive force from residual charge on the collector and viscoelastic force of the jet, the jet exhibits spiral whipping. As the collector moving speed increases, the effect of jet whipping gradually diminishes. Figure 4.175 shows the fibers prepared at different collector moving speeds. At a collector moving speed of 20 mm/s and a solution concentration of 10%, the influence of jet whipping is significant, resulting in mutually overlapped spiral fibers as shown in Figure 4.175(a). As the collector moving speed increases, the tensile force on the deposited fibers increases, inhibiting the spiral deposition of fibers. The degree of spiral stacking between fiber structures decreases progressively with higher collector moving speeds, as shown in Figure 4.175(b)–(d). Increasing the collector moving speed to 100 mm/s to match the jet deposition speed, straight nanofibers can be obtained, as shown in Figure 4.175(f).

As the collector moving speed increases, the degree of spiral stacking between fiber structures gradually decreases, leading to variations in the width of the fiber-deposited band on the collector. Therefore, precise control of the width of the fiber-deposited band at different collector moving speeds is critical for achieving patterned and localized deposition of electrospun nanofibers via sheath-gas-confined EDW.

As shown in Figures 4.176 and 4.177, when the collector moving speed increases from 5 to 15 mm/s, the average width of the fiber-deposited band increases from 99.59 to 161.91 μm. However, as the collector moving speed continues to increase, the average width of the fiber-deposited band decreases instead, reaching 115.98 μm at 25 mm/s. At

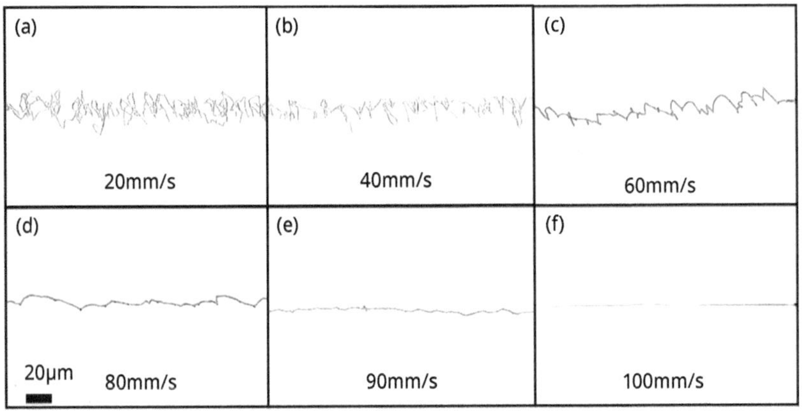

Figure 4.175: Fibers deposited by sheath-gas-confined EDW at different collector speeds (nozzle inner diameter is 0.21 mm, spinneret-to-collector distance is 2 mm, solution supply rate is 200 μL/h, applied voltage is 2.3 kV, and sheath gas pressure is 25 kPa).

100 mm/s, straight fibers are obtained, and the average width equals the fiber diameter (873.00 nm). Figuratively, the width of the fiber-deposited band is influenced by both the degree of jet spiral whipping and the collector moving speed. At lower collector moving speeds (below 15 mm/s), jet spiral whipping dominates, causing the average band width to increase with collector moving speed. At higher collector moving speeds (above 15 mm/s), the collector's motion provides greater mechanical tensile force, which suppresses jet spiral whipping. As a result, the average fiber band width decreases with increasing collector moving speed, eventually forming a linear distribution.

Parallel lines with 400 μm line spacing are designed to investigate the positioning accuracy of deposited fibers which are at different collector moving speeds. As illustrated in Figures 4.178 and 4.179, at collector moving speed of 100, 150, 200, 300, 500, 700, and 1,000 mm/s, the average line spacing errors of fibers deposited through sheath-gas-confined EDW are 36.09, 48.75, 37.41, 30.18, 17.39, 21.69, and 4.35 μm, respectively. With the increase of collector moving speed, the average error of fiber line spacing shows a decreased trend. It is influenced by the joint effect of the helical whipping of the jet and the mechanical tensile force imparted by the moving collector. The helical whipping shows uncertainty as it is closely associated to factors such as the viscoelastic force within the jet, the charge density on the jet surface, and the charge repulsion forces during the deposition process. Therefore, the average error of fiber line spacing can occasionally increase with the collector moving speed.

As the charged jet deposits on the collector, the residual charge density on the collector increases, which subsequently enhances the charge repulsion force to the subsequent charged jet. This aggravates the whipping behavior during the jet deposition process. As the fibers deposited on the collector increase, the residual charge density becomes larger, which affects the positioning deposition of subsequent fibers.

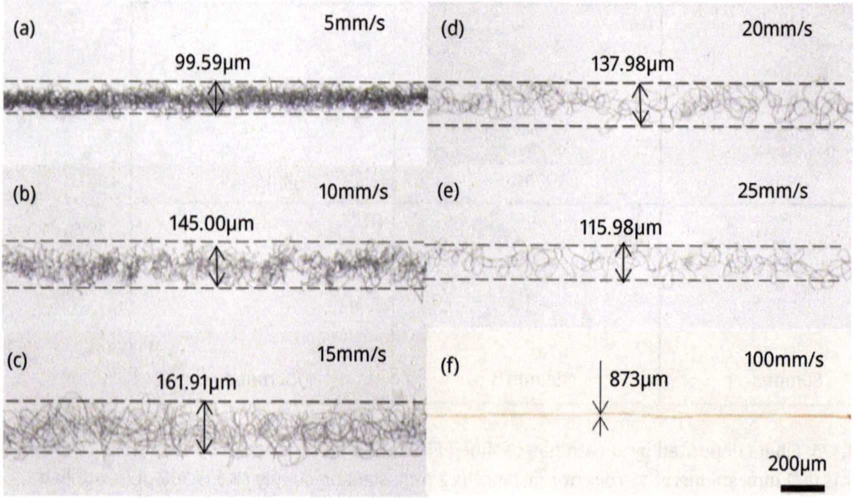

Figure 4.176: Fiber distribution width at different collector speeds (solution concentration is 10%, nozzle diameter is 0.21 mm, spinneret-to-collector distance is 2 mm, liquid supply rate is 100 μL/h, applied voltage is 2.3 kV, and sheath gas pressure is 25 kPa).

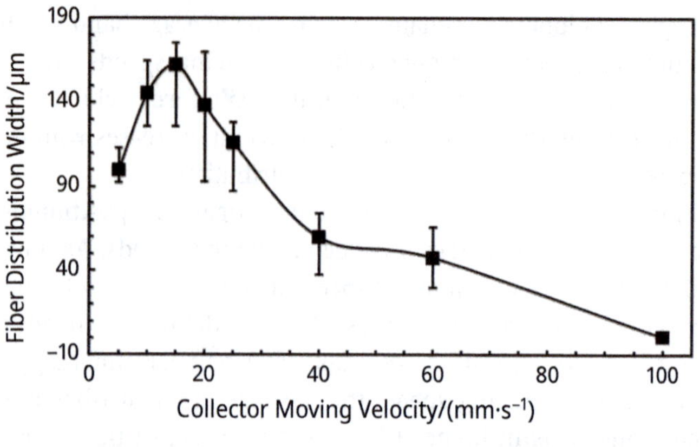

Figure 4.177: Fiber distribution width at different collector speeds (solution concentration is 10%, needle diameter is 0.21 mm, spinneret-to-collector distance is 2 mm, liquid supply speed is 100 μL/h, applied voltage is 2.3 kV, and the sheath gas pressure is 25 kPa).

We also investigate the influence of fiber deposition sequence on line spacing at varying collector moving speeds. At collector moving speeds of 100, 150, and 200 mm/s, five parallel lines with a target spacing of 400 μm each are deposited. The first deposited straight fiber is designated as the base line, with subsequent fibers numbered 1–4. Figure 4.180 shows five parallel fibers deposited at different collector moving speeds and the comparison between the experimentally fitted lines based on actual

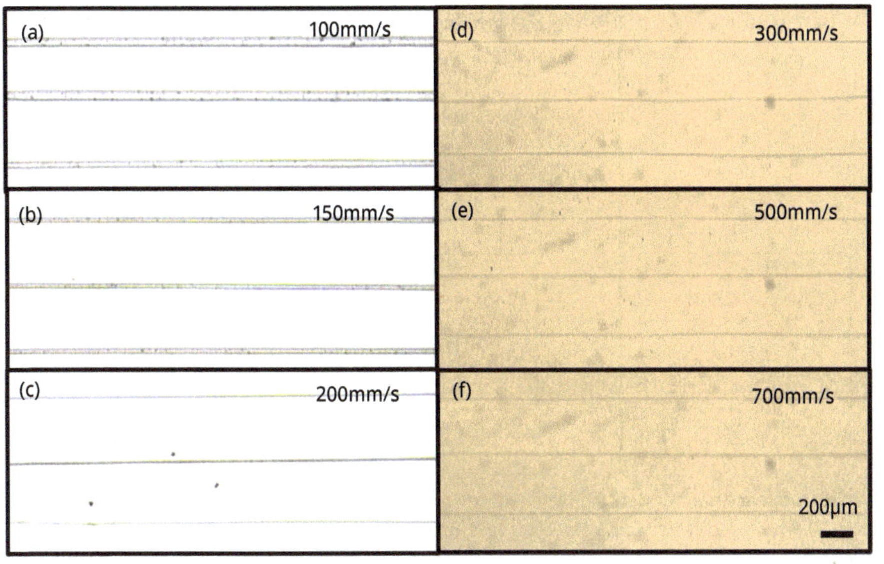

Figure 4.178: Line spacing errors at different collector speeds (solution concentration is 10%, nozzle diameter is 0.21 mm, spinneret-to-collector distance is 2 mm, liquid supply rate is 100 μL/h, applied voltage is 2.3 kV, and sheath gas pressure is 25 kPa).

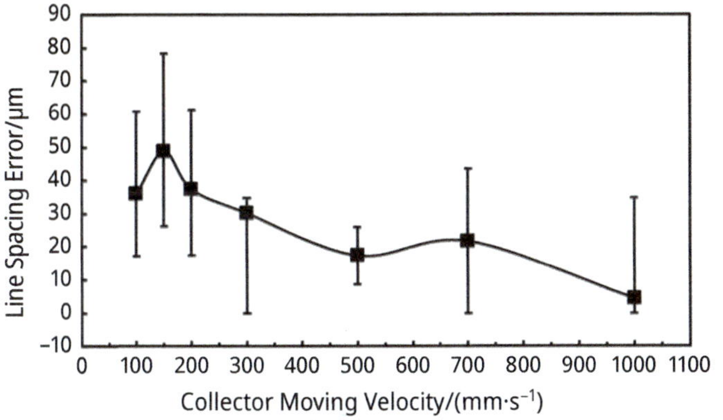

Figure 4.179: Line spacing errors at different collector speeds (solution concentration is 10%, nozzle diameter is 0.21 mm, spinneret-to-collector distance is 2 mm, liquid supply rate is 100 μL/h, applied voltage is 2.3 kV, and sheath gas pressure is 25 kPa).

measurements and the set lines (line spacing of 400 μm). At a collector moving speed of 100 mm/s, the average errors in line spacing for the fibers numbered 1, 2, 3, and 4 are 36.09, 46.17, 54.19, and 124.67 μm, respectively. As indicated in Figures 4.180 and 4.181, the average error in line spacing increases with the sequence of fiber deposition progresses at all the collector moving speeds.

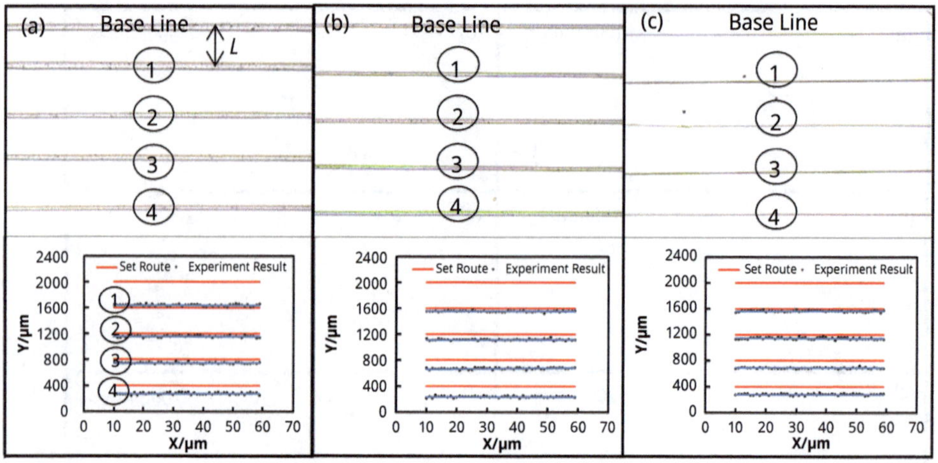

Figure 4.180: Influence of fiber deposition sequence on line spacing. The collector speeds are (a) 100; (b) 150; and (c) 200 mm/s (solution concentration is 10%, inner diameter of nozzle is 0.21 mm, spinneret-to-collector distance is 2 mm, liquid supply rate is 100 μL/h, applied voltage is 2.3 kV, and sheath gas pressure is 25 kPa).

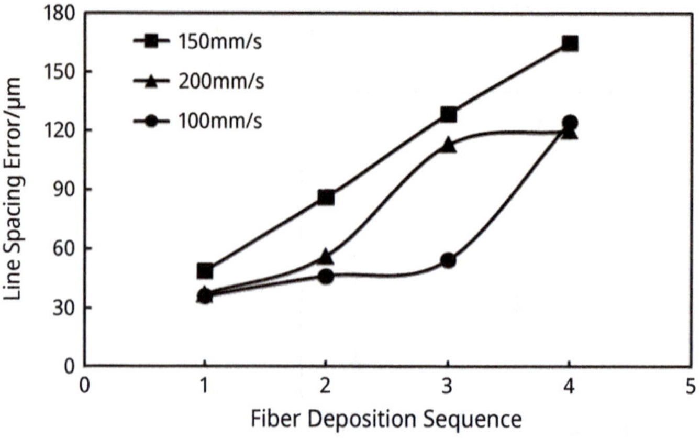

Figure 4.181: Influence of fiber deposition sequence on line spacing error under different collector speeds (solution concentration is 10%, inner diameter of nozzle is 0.21 mm, spinneret-to-collector distance is 2 mm, liquid supply rate is 100 μL/h, applied voltage is 2.3 kV, and sheath gas pressure is 25 kPa).

4.3.4.5.2 Printing of patterned structures by sheath-gas-confined EDW

The continuous jet ejection and patterned fabrication of single nanofibers on a substrate using sheath-gas-constrained EDW is critical for ensuring the compatibility of EDW technology and other micro-nano manufacturing processes as well as for enabling micro-nanosystem integration. By controlling the motion trajectory of the collector, which is mounted on a precision motion platform, via an industrial computer, the sheath-gas-constrained EDW jet is guided to deposit fibers along the predefined trajectory, enabling precise patterning.

As previously analyzed, the degree of fiber spiral stacking gradually decreases with increasing collector moving speed. Single straight nanofibers are obtained only when the collector moving speed matches the jet deposition speed. If the collector moving speed is lower than the jet deposition speed, spiral fiber deposition occurs, as shown in Figure 4.182.

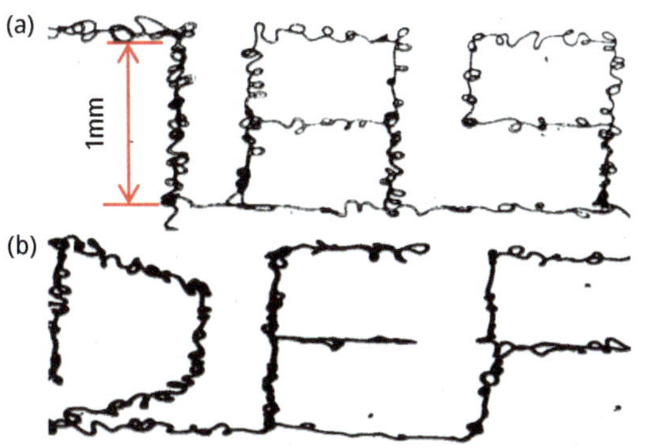

Figure 4.182: Patterns produced by sheath-gas-confined EDW: (a) digits "7," "8," and "9," and (b) letters "D," "E," and "F" (collector speed is 50 mm/s, solution concentration is 10%, inner diameter of nozzle is 0.21 mm, spinneret-to-collector distance is 2 mm, the liquid supply rate is 100 µL/h, applied voltage is 2.3 kV, and sheath gas pressure is 25 kPa).

Figure 4.183(a) illustrates a designed pattern consisting of parallel lines. To ensure the formation of straight fibers, the linear travel distance of the motion platform is increased to 20 mm. This adjustment allows the platform to reach the set speed promptly, ensuring that the collector moving speed matches the jet deposition speed. As a result, uniform straight fibers and overall well-preserved micro-nano patterns are successfully achieved. Figure 4.183(b) to (e) shows the patterns deposited on a silicon substrate. By setting the line spacing L of the designed patterns to 400, 300, and 200 µm and a nonuniform line spacing of 200 or 400 µm, the corresponding patterns are successfully deposited on the collector.

Based on the designed structures shown in Figure 4.183(a), parallel micro-nanostructures can be fabricated. By altering the motion trajectory of the collector and the line spacings, a grid pattern, as depicted in Figure 4.184, can be deposited. With sufficient acceleration distance, the collector moving speed matches the jet deposition speed, ensuring that the fiber deposition trajectory closely follows the motion platform's programmed path.

However, due to the residual like charges on adjacent fibers, there is a charge repulsion force between the already deposited fibers and the subsequently deposited fiber. This

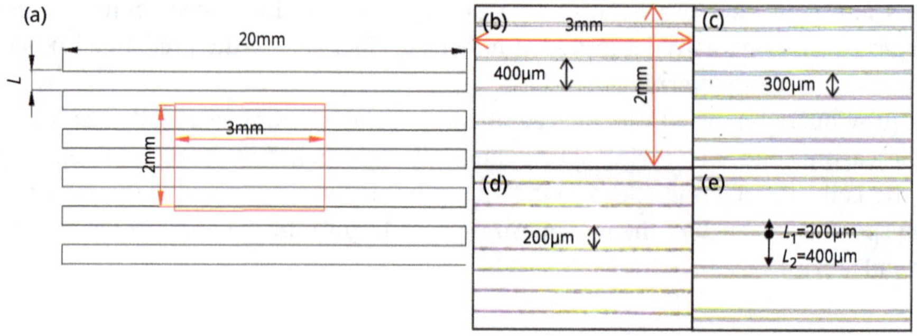

Figure 4.183: Direct-writing of parallel line structures through sheath-gas confined EDW: (a) designed pattern; (b)–(e) direct-writing parallel line patterns at different collector speeds; (b) 100 mm/s and line spacing $L = 400$ μm; (c) 200 mm/s, line spacing $L = 300$ μm; (d) 200 mm/s and line spacing $L = 200$ μm; and (e) 200 mm/s, line spacing $L = 200$ μm/400 μm (the solution concentration is 10%, the inner diameter of needle is 0.21 mm, the spinneret-to-collector distance is 2 mm, the liquid supply speed is 100 μL/h, the applied voltage is 2.3 kV, and the sheath gas pressure is 25 kPa).

causes the deposition trajectory to deviate from the pre-programmed positions, resulting in positional errors in the grid. As shown in Figure 4.184(a), while the pre-programmed side length of the grid is set to 400 μm, the measured average side length is 476.35 μm.

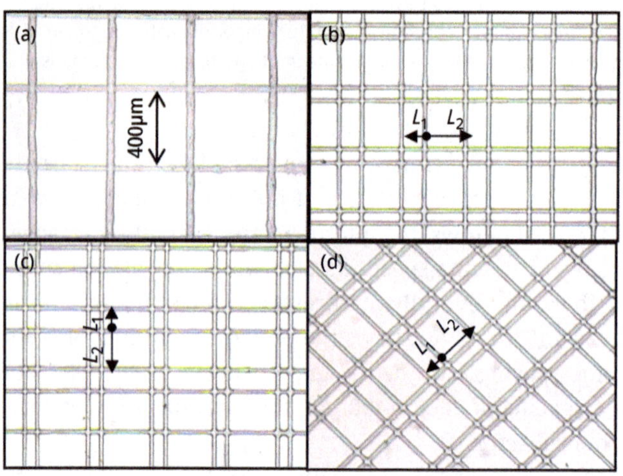

Figure 4.184: Direct-written grids through sheath-gas-confined EDW at different collector speeds: (a) 100 mm/s; (b)–(d) 200 mm/s, $L_1 = 200$ μm, $L_2 = 400$ μm (the solution concentration, inner diameter of the needle, spinneret-to-collector distance, solution supply rate, applied voltage, and sheath gas pressure are 10%, 0.21 mm, 2 mm, 100 μL/h, 2.3 kV, and 25 kPa, respectively).

References

[1] Ding Z, Salim A, Ziaie B. Selective nanofiber deposition through field-enhanced electrospinning. Langmuir 2009, 25(17), 9648–9652.

[2] Kim S Y, Kim Y, Park J, Hwang J. Design and evaluation of single nozzle with a non-conductive tip for reducing applied voltage and pattern width in electrohydrodynamic jet printing (EHDP). J Micromech Microeng 2010, 20(5), 055009.

[3] Ganan-Calvo A M. Generation of steady liquid microthreads and micron-sized monodisperse sprays in gas streams. Phys Rev Lett 1998, 80(2), 285–288.

[4] Ganan-Calvo A M. Electro-flow focusing: The high-conductivity low-viscosity limit. Phys Rev Lett, 2007, 98(13), 134503.

[5] Li F, Ganan-Calvo, LÓPEZ-HERRERA J M. Absolute-convective instability transition of low permittivity, low conductivity charged viscous liquid jets under axial electric fields. Phys Fluids. 2011, 23(9), 094108.

[6] Liu C. Formation of electrospun fibers. Mod Silk Sci Technol 2008, 23(4), 11–15.

[7] Pati S, Maity A, Banerji P, Majumder S B. Temperature dependent donor-acceptor transition of ZnO thin film gas sensor during butane detection. Sensor Actuat B-Chem 2013(183), 172–178.

[8] Bu N, Huang Y, Deng H, Yin Z. Tunable bead-on-string microstructures fabricated by mechano-electrospinning. J Phys D Appl Phys 2012, 45(40), 405301.

[9] Huang Y A, Wang X, Duan Y, Bu N, Yin Z. Controllable self-organization of colloid microarrays based on finite length effects of electrospun ribbons. Soft Matter 2012, 8(32), 8302–8311.

Chapter 5
Electrohydrodynamic direct-writing of 3D micro-nanostructures

5.1 Superposition deposition and integration behaviors of micro-nano jet

5.1.1 Superposition deposition behaviors of charged EDW jet

The three-dimensional (3D) micro-nanostructure formation process is observed and recorded using a CCD camera to study the spiral deposition behavior of charged jets under electric field focusing. The spiral whipping and layer-by-layer stacking behavior of charged jets under electric field focusing is illustrated in Figure 5.1. The time when the jet front contacts the surface of the collector is defined as 0 s. After the jet contacts the collector surface, it begins to bend and undergoes spiral deposition. At 0.4 s, the bending radius of the jet's spiral whipping is relatively large, and the formed nanofiber loop serves as the bottom layer of the 3D cylindrical structure. At 0.8 s and 1.2 s, as the jet deposits layer by layer in a spiral manner, the bending radius of the nanofiber continuously decreases with the increase in the number of stacking layers. At 2.0 s, the number of fiber stacking layers further increases, and the diameter of the nanofiber loops decreases accordingly, resulting in a frustum-shaped structure as shown in Figure 5.1(f).

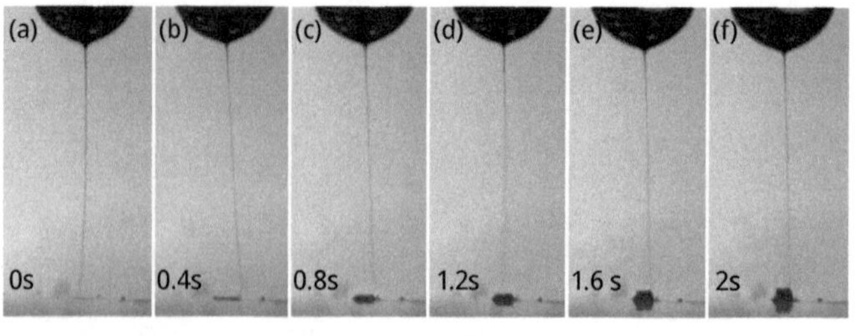

Figure 5.1: Spiral superposition deposition process of the EDW jet: (a) 0 s; (b) 0.4 s; (c) 0.8 s; (d) 1.2 s; (e) 1.6 s; and (f) 2.0 s (solution concentration is 10%, spinneret-to-collector distance is 2 mm, and applied voltage is 1.5 kV).

Based on the hydrodynamic characteristics of electrohydrodynamic direct-writing (EDW) jet, the stacking deposition behaviors of 3D nanofiber helical structure are further analyzed. The spiral whipping behavior of the jet on the collector surface is influenced by many factors, including the viscoelastic force, the electric field force, and the charge repulsion force of the jet. The bending radius of the jet's helical whipping can be given by the following formula [1]:

https://doi.org/10.1515/9783110797879-005

$$R \sim \left(\frac{Y}{\varepsilon_0 E^2}\right)^{1/3} r \tag{5.1}$$

where the electric field strength on the axis of the spinneret is [2]

$$E(x) = \frac{hC}{x(2h-x)+(h-x)r} \tag{5.2}$$

and the electric field constant is

$$C = \frac{U}{\ln 2\sqrt{\frac{h}{r}}} \tag{5.3}$$

Thus, the electric field strength on the collector can be described as

$$E = \frac{C}{h} = \frac{U}{h\ \ln 2\sqrt{\frac{h}{r}}} \tag{5.4}$$

where R is the bending radius of the jet's spiral whipping, Y is the Young's modulus, r is the radius of the spinneret, E is the electric field strength, ε_0 is the vacuum dielectric constant, h is the distance from the spinneret to the collector, x is the distance from a point on the axis to the spinneret, U is the applied voltage, and C is the electric field constant.

The electric field strength is the most important process factor that affects the bending radius of the jet's spiral whipping. According to formula (5.1), the bending radius is inversely proportional to the electric field strength. This is because the increase of the applied voltage raises the space electric field strength, and the vertical downward electric field force on the charged jet increases, enhancing the confinement effect to the jet and reducing the whipping amplitude, thereby decreasing the bending radius of the jet's spiral whipping. Formula (5.2) shows that the bending radius is positively correlated with the distance from the spinneret to the collector. As the distance increases, the electric field intensity decreases, reducing the confinement ability of the focused electric field on the jet, and consequently, the whipping amplitude and the bending radius of the spiral whipping also decrease. However, the free movement distance of the charged jet increases with the spinneret-to-collector distance, leading to an increase in the bending radius of the jet's spiral whipping.

5.1.2 Electrohydrodynamic direct-writing 3D micro-nanohelical structures

5.1.2.1 Electrohydrodynamic direct-writing self-spiraling 3D micro-nano structures

The jet spiral whipping guides nanofibers to stack on a glass substrate, resulting in frustum-shaped spiral fiber structures shown in Figure 5.2. The focused electric field can effectively restrain the jet motion and deposition, guiding the jet to stack layer by layer and avoiding the disordered deposition of the fiber. The sidewall of the frustum-shaped structure is composed of multiple layers of fibers, which are stacked evenly and show a hollow shape, as shown in Figure 5.2(a). Constrained by the stress of the deposited fibers, the diameter of the nanofiber loop-like structure decreases with the increase of the number of the deposition layers, with a bending radius of about 58.13 μm for the bottom layer, and 32.75 μm for the top layer. 3D spiral structures are formed by the layer-by-layer stacking of nanofibers, with ordered nanofibers uniformly arranged on the side of the frustum-shaped structure, as shown in Figure 5.2(b). The average diameter of spiral fibers in the conical structure is about 1.23 μm.

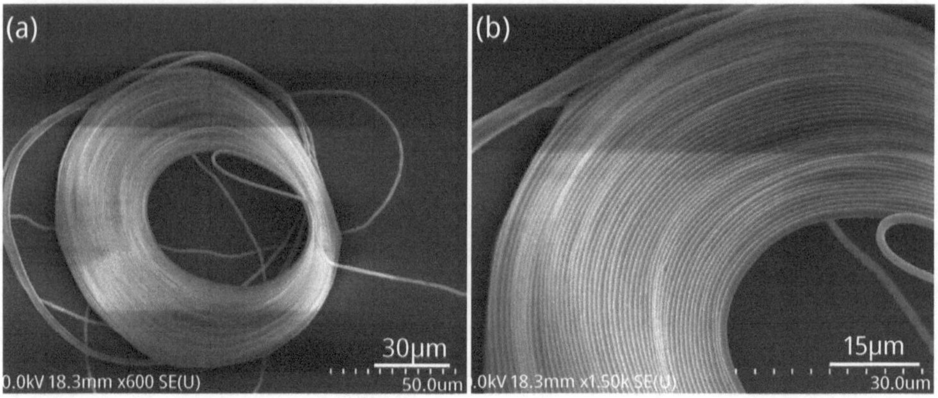

Figure 5.2: Electrohydrodynamic direct-writing spiral nanofiber conical structure: (a) 3D hollow frustum-shaped structure; and (b) local magnification of 3D fiber structure (solution concentration is 10%, spinneret-to-collector distance is 2 mm, and applied voltage is 1.5 kV).

EDW jet is deposited layer by layer under the focusing effect of the electric field. The distance between the jet deposition point and the spinneret decreases as the spiral fibers are stacked. According to Eq. (5.2), the electric field strength at the deposition point increases as the distance to the spinneret decreases. Under the confinement of higher electric field strength, the bending radius of the jet's helical whipping also decreases with the increase in the number of fiber layers. By adjusting the process parameters, the deposition behavior of charged jet can be controlled, allowing manipulation of the diameter and height of the 3D spiral structure. This provides a promising technical approach for the controlled formation of 3D micro-nanostructures. The influence of the applied voltage and the spinneret-to-collector distance on the diameter of the 3D helical structure is analyzed.

The relationship between the bending radius of the spiral fiber on the collector and the applied voltage is demonstrated in Figure 5.3. When the voltage increases from 1.3 to 1.7 kV, the bending radius decreases slightly, from 61.54 to 57.87 μm. An increase in the applied voltage raises the surface charge on the jet. This, in turn, increases the charge repulsive force on the jet, reducing its stability. As a result, there is a significant deviation in the deposition position, and the fiber deposition range becomes larger. As the voltage increases from 1.7 to 1.9 kV, the bending radius increases significantly, from 57.87 μm to 70.64 μm. Although the bending radius of the fiber can be changed to a certain extent by controlling the voltage, the change is relatively slow, and the increase of the charge repulsion force will also negatively affect the deposition range and accuracy of the jet. Experimental results show that high voltage is detrimental to the precise fabrication of 3D helical fiber structures.

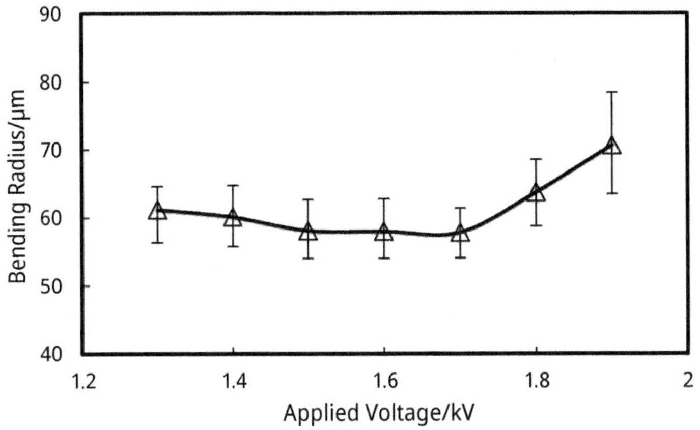

Figure 5.3: The plot of spiral fiber bending radius versus the applied voltage (the solution concentration is 10%, and the spinneret-to-collector distance is 2 mm).

The distance between the spinneret and the collector is a critical parameter influencing the jet ejection and deposition behavior in EDW. The relationship between the bending radius of the spiral fiber on the collector and the distance between the spinneret and the collector is illustrated in Figure 5.4. At a voltage of 1.5 kV, as the spinneret-to-collector distance increases, the free movement distance of the charged jet also increases, leading to a corresponding increase in the bending radius of the fibers deposited on the collector surface. When the spinneret-to-collector distance is less than 1.90 mm, the change in the fiber bending radius is relatively gentle. However, when the spinneret-to-collector distance exceeds 1.90 mm, the bending radius increases sharply.

An equation can be derived through curve fitting:

$$y = 3388.9x^3 - 18144x^2 + 32376x - 19228 \tag{5.5}$$

5.1.2.2 Controllable fabrication of 3D spiral micro-nanostructure via electrohydrodynamic direct-writing

Using Eq. (5.5), the bending radius of fibers can be calculated for different spinneret-to-collector distances at a voltage of 1.5 kV. By adjusting the spinneret-to-collector dis-

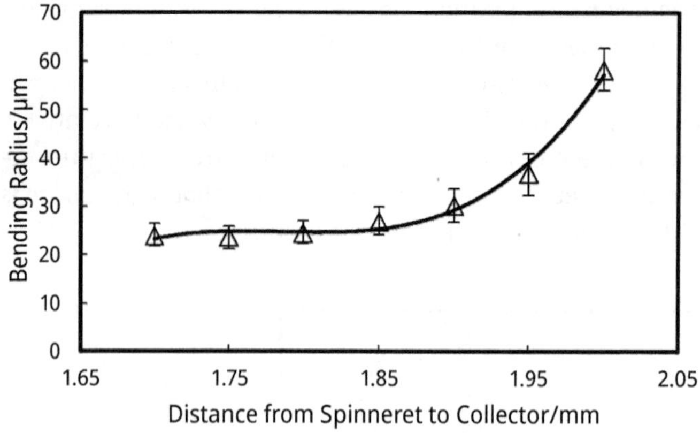

Figure 5.4: Relationship of the bending radius and the spinneret-to-collector distance (solution concentration is 10%, and applied voltage is 1.5 kV).

tance, 3D columnar fiber structure of various shapes can be fabricated. Figure 5.5 illustrates the dynamic evolution of the 3D columnar fiber structure over deposition time, achieved by varying the spinneret-to-collector distance, with the initial distance set to 2 mm. The layer-by-layer stacking of fibers increases the height of the 3D spiral structure while reducing the distance between the spinneret and the jet deposition point. Consequently, during the initial stage of jet deposition, when the spinneret-to-collector distance remains constant, a frustum-shaped structure is formed, as shown in Figure 5.5(a). At this stage, the height of the frustum-shaped structure is approximately 280.36 μm, corresponding to a distance of 1.72 mm between the spinneret and the jet deposition point. And the bending radius of the fiber is 25.75 μm according to Eq. (5.5). As the distance between the spinneret and the deposition point further decreases, the fibers no longer stack along the frustum-shaped structure, but stacks with varying bending radius. By programming the spinneret-to-collector distance to follow a "stop-go-stop-go" cyclic motion, a bead-on-string 3D micro-nanofiber structure can be formed, as shown in Figure 5.5(c). As the spinneret-to-collector distance continues to increase, the fabricated 3D structure grows taller. However, when the top layer accumulates excessive fibers and a fiber layer with a smaller bending radius exists within the columnar structure, the structure tends to break at the layer with the smaller bending radius, as shown in Figure 5.5(n). At this point, the 3D fiber structure ceases to grow further in height.

Figure 5.6 illustrates the dynamic layer-by-layer deposition process of the 3D spiral fiber structure over a period of 2 s. Measurements show that the height of the 3D spiral fiber structure increases from 325.24 μm to 375.67 μm within 2 s, resulting in a total height increase of approximately 50.43 μm. This corresponds to an average stacking rate of 25.22 μm per second, indicating a fiber stacking speed of about 25.22 μm/s.

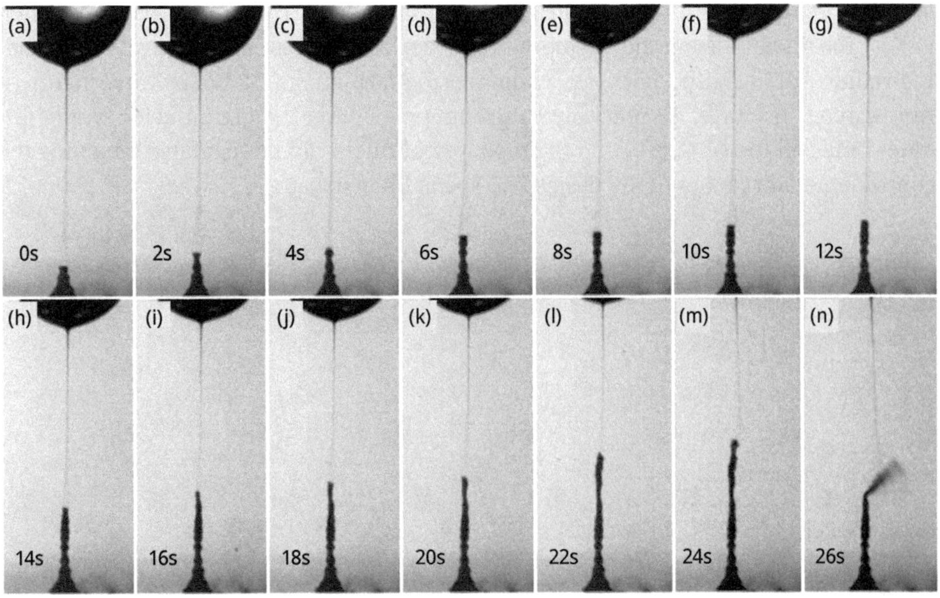

Figure 5.5: Formation of the 3D columnar fiber structure (solution concentration is 10%, initial spinneret-to-collector distance is 2 mm, and applied voltage is 1.5 kV).

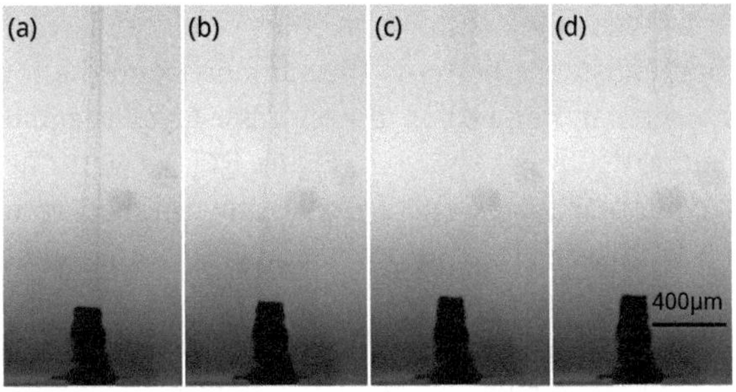

Figure 5.6: Dynamic layer-by-layer deposition of spiral fibers (solution concentration is 10%, distance from spinneret to collector is 2 mm, and applied voltage is 1.5 kV).

During the fiber spiral stacking process, the spinneret motion velocity v_n serves as a controllable variable. Together with the fiber stacking speed v_f, v_n determines the deposition morphology of the spiral fiber stacking. Figure 5.7 illustrates the relationship between the spirally stacked fiber morphology and the spinneret motion velocity. When $v_n = v_f$, the distance from the spinneret to the deposition point remains constant, and the bending radius of the fiber stays unchanged, resulting in the formation of a hollow columnar fiber structure. When $v_n < v_f$, the distance from the spinneret to the deposition point continuously decreases, and the fiber's bending radius decreases accordingly, forming a hollow upright frustum-shaped structure. Conversely, when $v_n > v_f$, the distance from the spinneret to the deposition point increases and the bending radius of fiber also increases, leading to the formation of a hollow inverted frustum-shaped structure. By precisely controlling the spinneret motion velocity, the distance between the spinneret and the top layer of fibers can be regulated, enabling the controllable fabrication of the desired 3D spiral fiber structures.

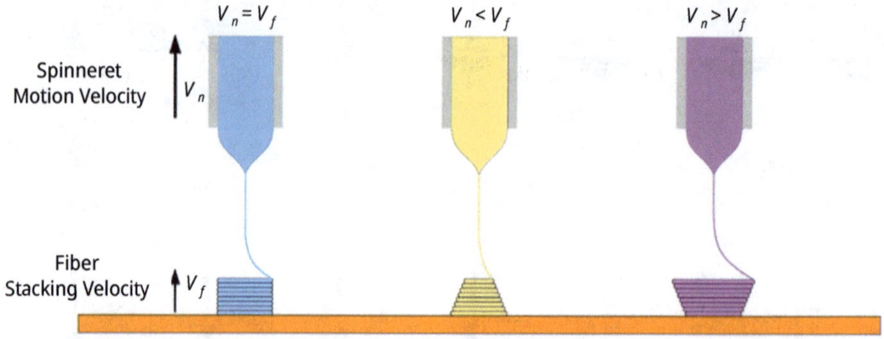

Figure 5.7: Schematic diagram of the relationship between fiber spiral stacking morphology and spinneret motion velocity.

Figure 5.8 illustrates the deposition morphology of 3D spiral fiber structures under different spinneret motion velocities. When the spinneret motion velocity matches the fiber stacking rate of 25.22 μm/s, the distance between the spinneret and the top layer of fibers remains constant, and the bending radius of each fiber layer is equal, resulting in a 3D columnar spiral fiber structure, as shown in Figure 5.8(a). Over time, as the fibers continue to stack, a 3D columnar fiber structure with a high aspect ratio is formed, as shown in Figure 5.8(b), with a height of approximately 1.68 mm. When the spinneret motion velocity exceeds 25.22 μm/s, the distance between the spinneret and the top layer of fibers continuously increases, and the fiber's bending radius also increases, forming a bowl-shaped 3D fiber structure, as shown in Figure 5.8(c). Conversely, when the spinneret motion velocity is less than 25.22 μm /s, the fiber's bending radius continuously decreases, resulting in a conical 3D fiber structure, as shown in Figure 5.8(d). When the spinneret motion velocity varies, initially exceeding 25.22 μm/s and then dropping below 25.22 μm/s, an oval-shaped 3D fiber structure is formed, as shown in Figure 5.8(e) and (f). By controlling the spinneret motion velocity to match the EDW process, the direct-writing and fabrication of 3D structures with various shapes can be achieved

5.1.2.3 Electrohydrodynamic direct-writing of three-dimensional spiral fiber arrays

Figure 5.9 shows the spiral fiber structure after a deposition time of 3 s, with a small number of scattered fibers hanging from the top and the sidewalls of the spiral struc-

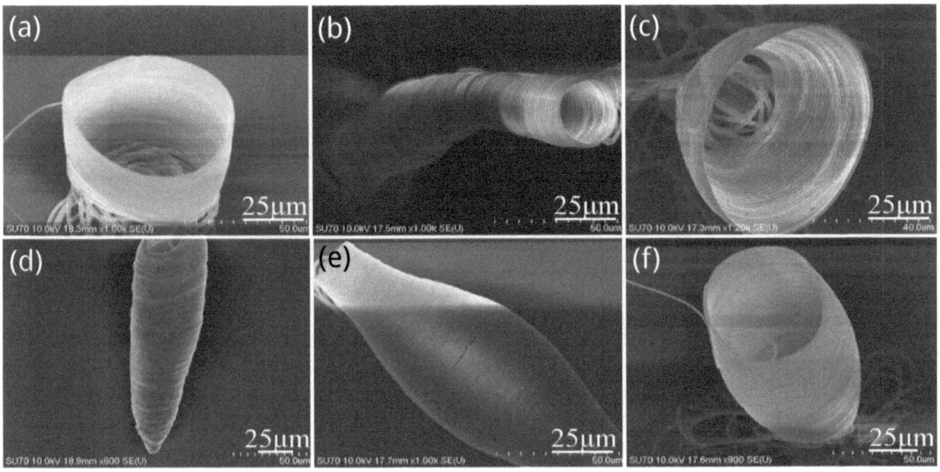

Figure 5.8: Three-dimensional columnar fiber structures under different spinneret motion velocities: (a) cylindrical structure prepared at a spinneret motion velocity of 25.22 μm/s; (b) cylindrical structure with a high aspect ratio, prepared at a spinneret motion speed of 25.22 μm/s; (c) bowl-shaped structure, prepared at a spinneret motion velocity of 40 μm/s; (d) conical structure, prepared at a spinneret motion velocity of 20 μm/s; and (e), (f) oval structure, prepared at a spinneret motion velocity of 40 μm/s to 20 μm/s (solution concentration is 10%, applied voltage is 1.5 kV, and initial spinneret-to-collector distance is 2 mm).

ture, extending down to the collector. Through these scattered fibers, it is evident that the sidewalls of the cylindrical spiral structure are composed of layers of fibers stacked on top of each other, as shown in Figure 5.10(a). The boundaries between individual fibers are clearly distinguishable, indicating that the jet solidifies into independent nanofibers before reaching the collector and stacking into the helical structure under internal tension.

To further investigate the morphology of the spiral structure, we analyze a spiral structure that is fractured by external forces. Figure 5.10(b) shows a 3D spiral structure that broke at the 23rd fiber layer (each fiber layer is labeled in the figure). The fracture exhibits a circular shape, confirming that the spiral structure is hollow rather than formed by irregular stacking of oscillating fibers.

By examining the height of the spiral structure at different deposition times, the jet volume can be calculated. Figures 5.11 and 5.12 show the spiral structure formed for deposition times ranging from 1 to 5 s.

Patterns composed of multiple spiral structures and nanofibers fabricated via EDW with electric field focusing are shown in Figure 5.13. Single nanofibers are prepared using the straight segment of the jet, and a triangle pattern is created by controlling the collector motion, with pauses of 1 s, 2 s, and 3 s at the three vertices of the right triangle, respectively. The vertices are labeled as #1, #2, and #3, respectively, according to the deposition sequence. The number of fiber layers in the spiral strucures

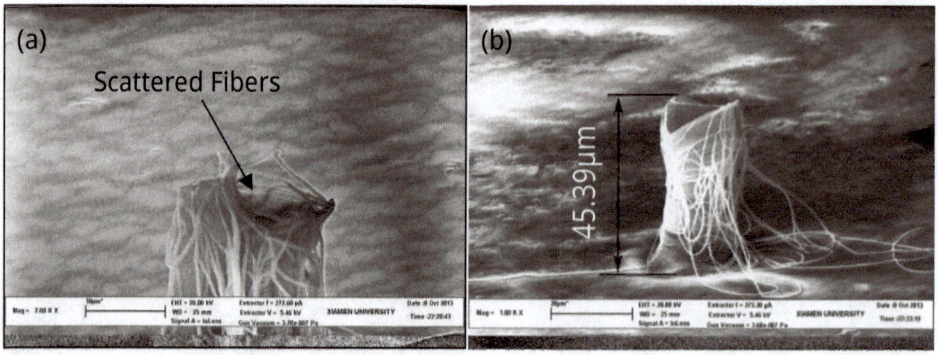

Figure 5.9: Magnified view of the spiral fiber structure. The spiral structure is 37 μm in height with a diameter of 20 μm, and the diameter of a single fiber is 600 nm (solution concentration is 10%, applied voltage is 1 kV, deposition time is 3 s, and spinneret-to-collector distance is 1 mm).

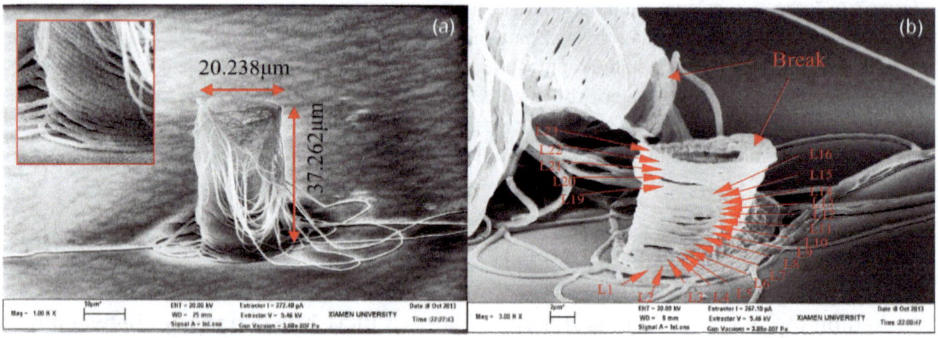

Figure 5.10: Stacked fibers on the spiral structure: (a) sidewalls of spiral structures; and (b) spiral structure split into halves (solution concentration is 10%, applied voltage is 1 kV, deposition time is 3 s, and spinneret-to-collector distance is 1 mm).

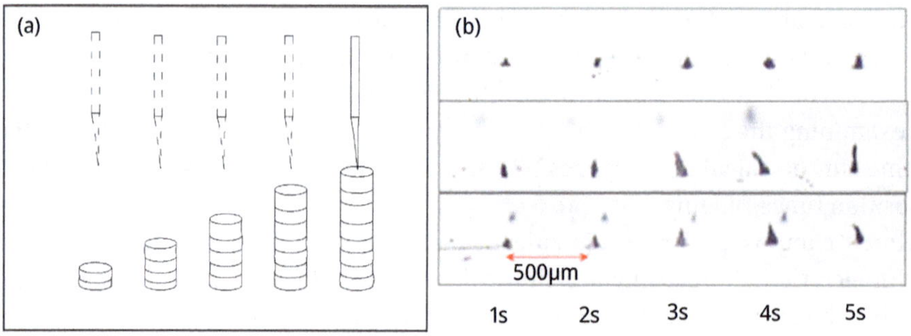

Figure 5.11: Stacked spiral structure formed at deposition times ranging from 1 to 5 s, with a 500 μm interval between adjacent spiral structures (applied voltage is 1 kV, spinneret-to-collector distance is 1 mm, 10% solution concentration).

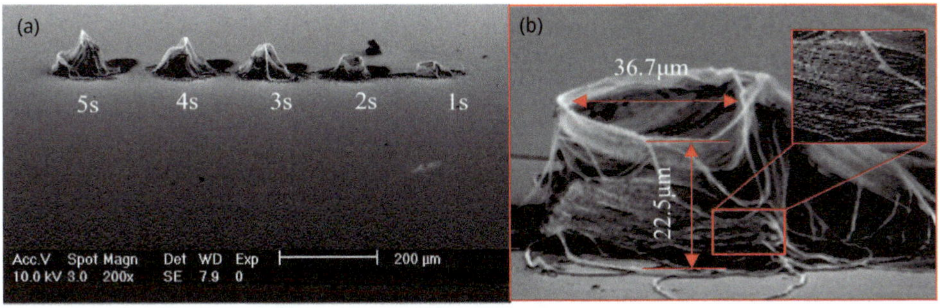

Figure 5.12: SEM images of spiral structures formed at deposition times ranging from 1 to 5 s (applied voltage is 1 kV, spinneret-to-collector distance is 1 mm, 10% solution concentration).

at each vertex can be calculated after local magnification. Figure 5.14(a) shows the SEM images of the vertices with deposition time of 3 s, 2 s, and 1 s, respectively. The corresponding number of fiber layers stacking at each vertex are calculated to be 32, 17, and 13 layers.

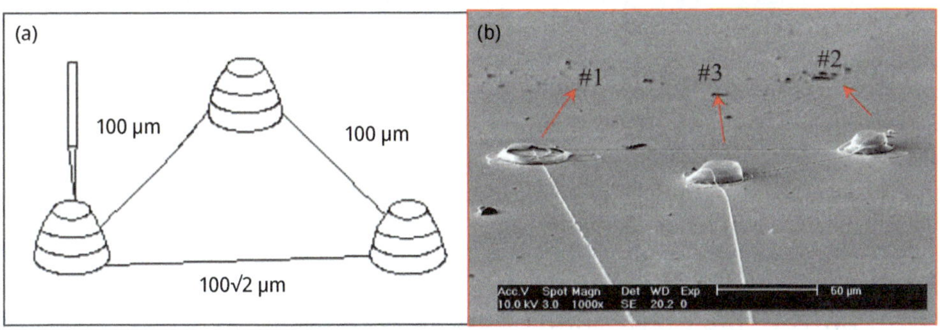

Figure 5.13: Direct-writing of isosceles right triangle with a side leg of 100 µm: (a) schematic diagram of the triangular pattern composed of spiral structures and nanofibers; and (b) SEM image of the direct-written triangle (applied voltage is 1 kV, spinneret-to-collector distance is 1 mm, pause times at the vertices are 1 s at #1, 2 s for #2, 3 s for #3, solution concentration is 12%, and spinneret inner diameter is 60 µm).

As the number of stacking layers increases, the whipping amplitude of the jet decreases and the diameter of the spiral structure reduces. As shown in Figure 5.14(a), the diameter of the spiral structure decreases from an initial value of 34.15 µm to 17.5 µm at deposition time of 3 s, representing a reduction of nearly a half.

During the deposition process of spiral stacking, the viscoelastic force plays a dominant role, causing a decrease in the bending radius of the fibers. Simultaneously, as the jet solidifies during its motion, the viscoelastic force increases, significantly slowing down the jet's moving speed. Additionally, the spiral fibers already deposited

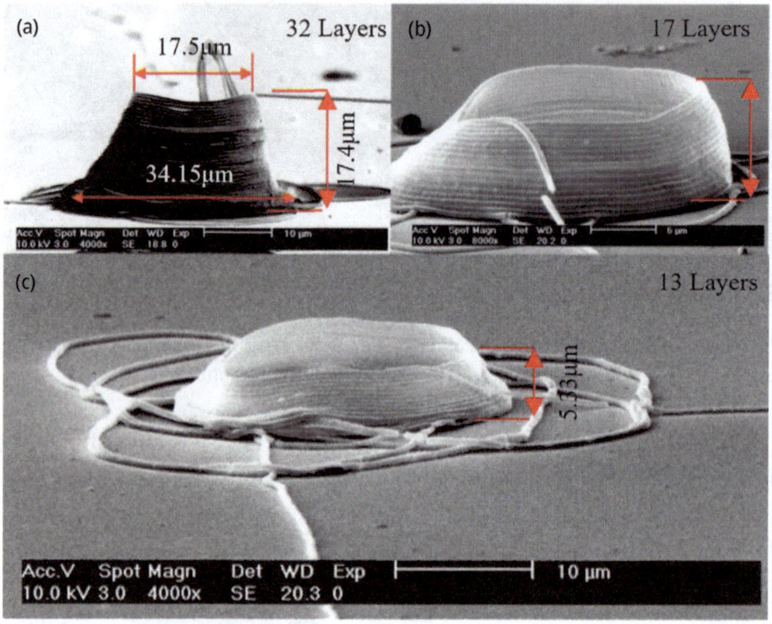

Figure 5.14: SEM images of the spiral structure at the vertices of the right triangle. The deposition times are: (a) 3 s; (b) 2 s; and (c) 1 s (applied voltage is 1 kV, spinneret-to-collector distance is 1 mm, solution concentration is 12%, and spinneret inner diameter is 60 μm).

on the collector generate an inward charge repulsion force on subsequent fibers, resulting in a gradual reduction in the range of the fiber's spiral whipping.

5.2 Controllable fabrication of 3D patterned micro-nanostructures

5.2.1 Superposition deposition behavior of EDW fiber

5.2.1.1 Influence of deposited fibers on superposition deposition

ANSYS is employed to construct a space electric field simulation model, aiming to analyze the influence of deposited fibers on the space electric field. The simulation results for two modes, namely, with and without electric charges on the deposited fibers, are compared. The parameter settings of the simulation model are as follows: the inner diameter of the spinneret is 0.11 mm; the distance between the spinneret and the underlying fibers is 2 mm; the relative permittivity of the simulation region is 1; a voltage of 1 kV is applied to the spinneret, and the lower electrodes are all grounded, with a voltage of 0 kV. When there are fibers deposited on the collector, the simulation results of the electric field distribution are shown in Figure 5.15.

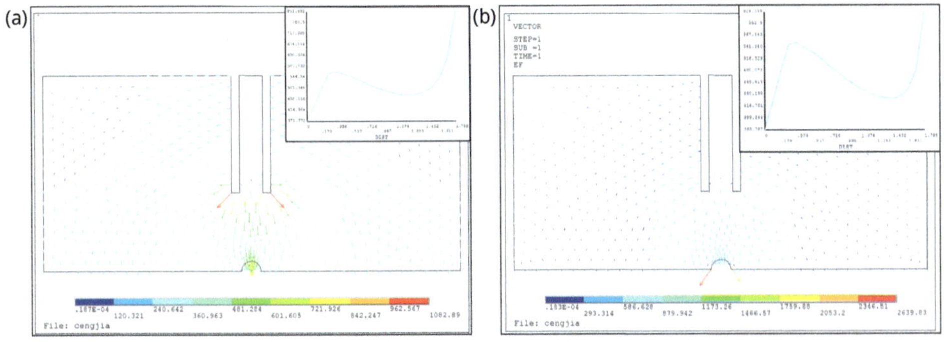

Figure 5.15: Simulation of the space electric field above the collector with deposited nanofibers: (a) deposited fibers with no residual charge; and (b) deposited fibers with residual charge.

As shown in Figure 5.15, when the deposited fibers carry no residual charge, the electric field strength above the fibers is higher, and the electric field lines near the fibers point toward the fiber surface. The maximum electric field strength on the fiber surface is 803.69 V/mm, occurring at the top of the fiber, while on both sides of the fibers, the electric field strength is significantly lower at 281.56 V/mm. This creates a focused electric field directed toward the fiber, causing the charged jet to move toward the surface of the deposited fiber. In contrast, when the deposited fibers carry residual charges, most of the electric field lines near the fiber point toward the collector surface (bypassing the fiber). The electric field strength is the highest at the edge of the fiber (2,639.83 V/mm) and significantly lower on the fiber surface (610.36 V/mm), causing the charged jet to move toward the sides of the deposited fiber.

To further analyze the impact of residual charges on the deposited fibers, the deposition morphology of multiple fibers electrospun along the same straight-line path is investigated under two conditions: with and without residual charges. To eliminate residual charges on the fibers, the power supply is turned off for 5 s after depositing each fiber, allowing the charges to completely dissipate before re-applying the voltage to deposit the next fiber. This process is repeated to deposit the third, fourth, and fifth fibers, as illustrated in Figure 5.16. The results are shown in Figure 5.17.

When no residual charge is present on the deposited fibers, the multilayer stacked fibers can completely coincide. The diameter of a single fiber is 5.84 μm, and the next fiber tends to move towards the already deposited one. After stacking five layers, the overall distribution width of the fiber structures is approximately 12.84 μm. Moreover, when the third to fifth fibers are stacked, the change in fiber distribution width is relatively small, as shown in Figure 5.18. In contrast, when there is residual charge, the charges carried by the nanofibers cannot be conducted away in a timely manner. The direct-written fibers with the same path (with a fiber diameter of approximately 6.24 μm) are difficult to stack together. As ahown in Figure 5.5 17(b), the distribution of the fibers is relatively disordered. The residual charges on the al-

Figure 5.16: Schematic diagram of EDW fiber superposition.

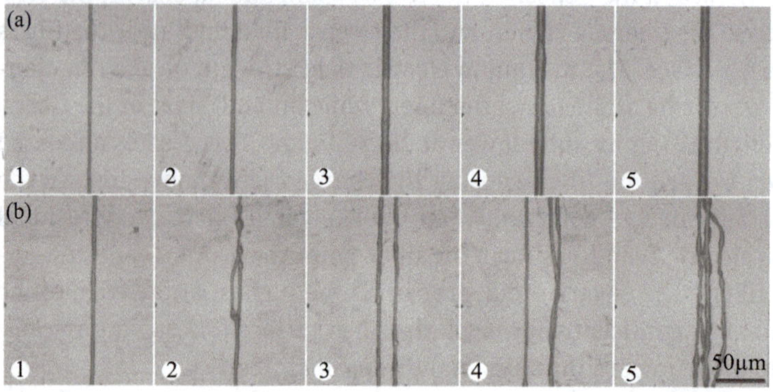

Figure 5.17: Deposition morphology of EDW fiber superposition: (a) Nanofibers with no residual charge; (b) nanofibers with residual charges (solution concentration is 10%, spinneret-to-collector distance is 2 mm, applied voltage is 1.5 kV, collector speed is 30 mm/s, silicon dioxide substrate).

ready deposited fibers cannot be released quickly, generating a relatively large charge repulsion force on the subsequent fiber deposition. As a result, even if the movement path of the collector is the same during the EDW process, the second fiber is also very difficult to stably and accurately stack on the first fiber, and the gap between the next fiber and the already deposited fiber is relatively large and irregular. When five fibers are deposited, the positioning error is as high as 56.12 μm.

Under the influence of the electric field, the deposited fibers carrying residual charges exert a repulsive force on subsequent fibers, making it difficult to achieve stable and precise stacking and hindering the formation of 3D structures. The residual charges on the deposited fibers have crucial impact on the distribution of the electric field during the direct-writing of 3D structures. To achieve more stable and uniform

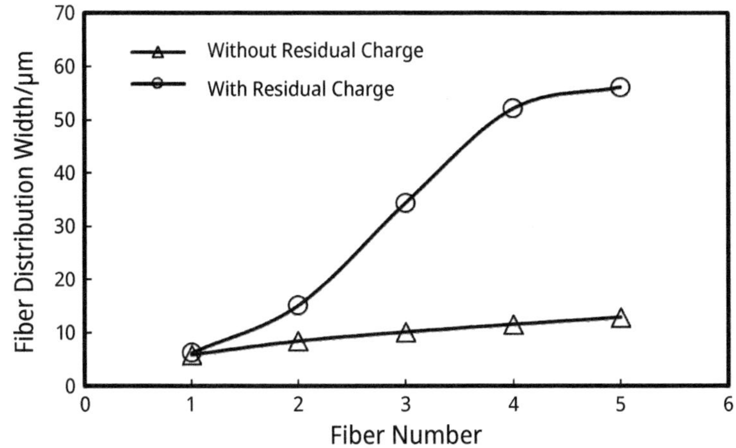

Figure 5.18: Relationship between fiber superposition deposition width and number of stacked fibers (solution concentration is 10%, spinneret-to-collector distance is 2 mm, applied voltage is 1.5 kV, and collector speed is 30 mm/s, silicon dioxide substrate).

fiber stacking, it is essential to eliminate the residual charges carried by the deposited fibers. By using an auxiliary needle tip to induce the rapid charge transfer from deposited fibers, precise and controllable fabrication of 3D structures can be achieved.

5.2.1.2 Multilayer fiber stacking and molding via electrohydrodynamic direct-writing

Figure 5.19 shows the stacking morphology of fibers during EDW process at different collector speeds. The results for stacking of first to fifth fibers are displayed from left to right. As the the number of fiber layers increases, the stacking height also increases. (In the SEM images, higher stacking heights correspond to darker shadow areas.)

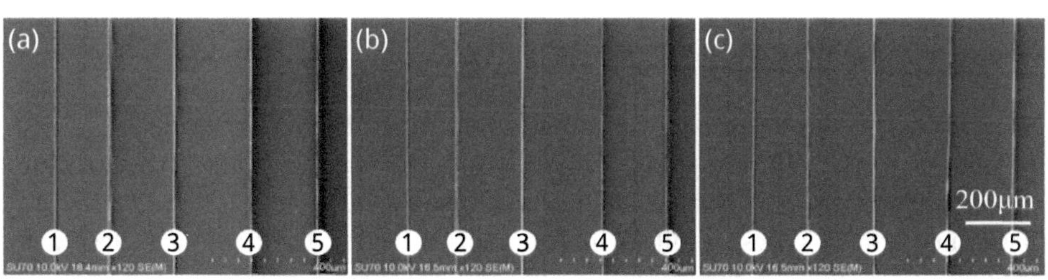

Figure 5.19: SEM images of the deposited structures with first to fifth stacked fibers at various collector speeds: (a) 15 mm/s; (b) 20 mm/s; and (c) 25 mm/s (magnification: 120×, 10% solution concentration, spinneret-to-collector distance is 2 mm, and applied voltage is 1.5 kV).

To further investigate the morphology of the stacked fibers, the total width of fiber stacking at different collector speeds is studied. The fiber morphology and the curves of the total width of stacked fibers are shown in Figures 5.20 and 5.21, respectively. At a collector speed of 15 mm/s, the width of the first fiber is 10.54 μm; after stacking the second fiber, the total width of fibers increases to 11.34 μm. With the addition of the third, fourth, and fifth layers, the total width grows to 13.24 μm, 14.21 μm, and 14.82 μm, respectively. The total width increases by 4.28 μm after the fifth layer, with an average width increase of 1.07 μm per layer. The first fiber is deposited on a flat surface, and the evaporation of the jet solvent takes time, causing the originally cylindrical jet to deform into an elliptical shape. The subsequent fibers deposit on the previous layer of fibers, with some solvent permeating into the previous layer, resulting in minimal deformation. The average width of the subsequent fibers is about 7.68 μm. When the collector speed is increased to 20 mm/s, the total width of the first to fifth layers of fibers measures 8.94 μm, 9.87 μm, 9.47 μm, 10.14 μm, and 10.12 μm, respectively. The width of a single fiber is about 5.68 μm, with each layer contributing an average increase of 0.30 μm. When the collector speed is further increased to 25 mm/s, the total width of the first to fifth layers measures 8.11 μm, 9.75 μm, 9.67 μm, 10.02 μm, and 9.86 μm, respectively. The average width increase per layer is about 0.48 um, and the width of single fiber is about 5.38 μm.

As the collector speed increases, both the width of a single fiber and the total width of the stacked fibers decrease. The enhanced stretching effect of the collector on the jet helps eliminate the interference of charge repulsion, maintaining jet stability and promoting accurate layer-by-layer stacking, and precise deposition. Consequently, the total width of the stacked fibers decreases with increasing collector speed.

Additionally, an increase in the number of nanofiber deposition layers enhances the electric field strength on the surface of the 3D fiber structure, improving the induction effect on the charged jet and facilitating precise fiber deposition. As shown in Figure 5.20, the total width of the stacked fibers remains almost unchanged from the second to the fifth layer, enabling accurate 3D stacking of the fibers.

5.2.2 Electrohydrodynamic direct-writing of patterned 3D micro-nanostructure

5.2.2.1 Superposition deposition of high-aspect-ratio fiber wall structures

High-aspect-ratio 3D structures with multilayer fiber stacking can be fabricated using EDW with electric field focusing, as shown in Figure 5.22. The fibers are direct-written back and forth 60 times, resulting in a total of 60 layers. The fibers are closely and neatly stacked, with most layers aligned in the same vertical plane. The total height of the fiber wall is approximately 102.15 μm, with an average height increase of 1.70 μm per layer. Measurements indicate that the height increase per layer is about 1.84 μm.

The influence of the collector moving speed, applied voltage and spinneret-to-collector distance on the total width of the fiber wall are shown in Figures 5.23–5.25,

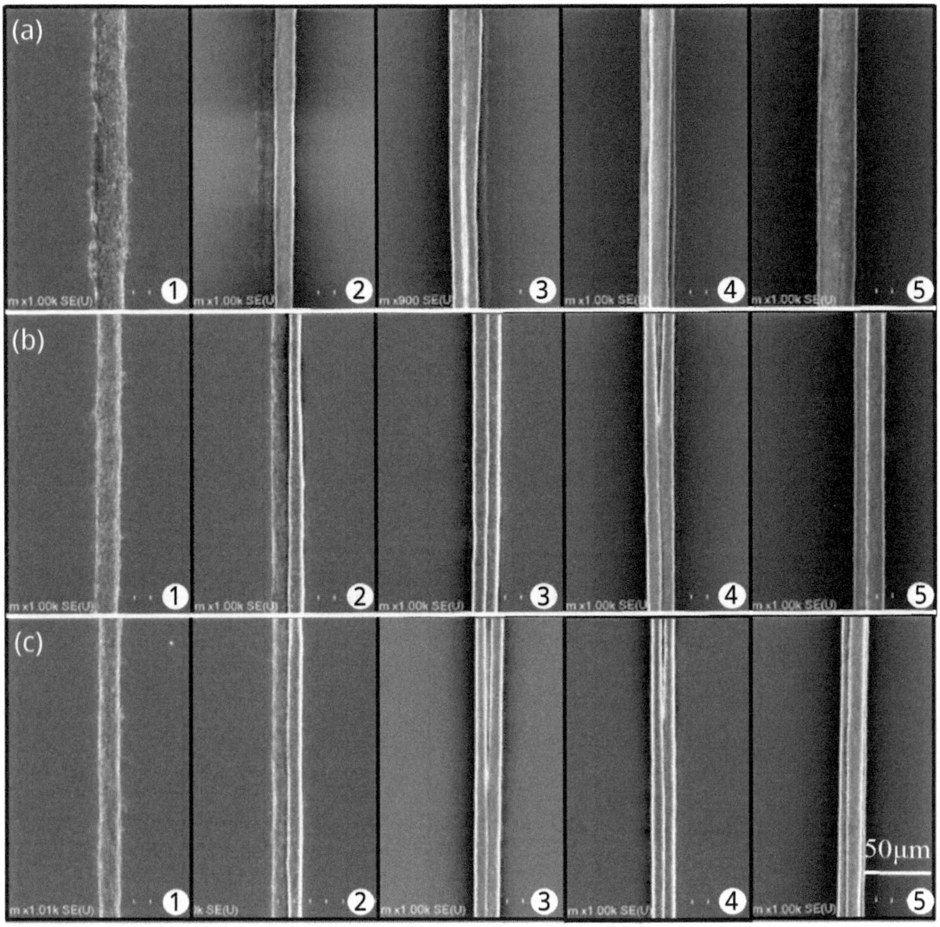

Figure 5.20: SEM images of fiber stacking structures at different collector speeds: (a) 15 mm/s; (b) 20 mm/s; and (c) 25 mm/s (magnification: 1,000×, solution concentration is 10%, spinneret-to-collector distance is 2 mm, and applied voltage is 1.5 kV).

respectively. Increasing the collector speed enhances the stretching force on the fibers, thus reducing the overall width. Higher applied voltages accelerate the jet ejection speed, which also decreases the fiber overall width while increasing the instability of the deposition position as the charge repulsion force on the jet grows. The spinneret-to-collector distance affects the jet's travel distance; increasing this distance reduces the fiber width but extends the fiber whipping time. Among these factors, collector speed has the most significant impact on fiber width. If the collector speed is too high, the jet deposition speed may fall below the collector speed, resulting in large positioning errors. Matching the collector speed with the jet deposition speed is essential for maintaining consistent 3D fiber morphology.

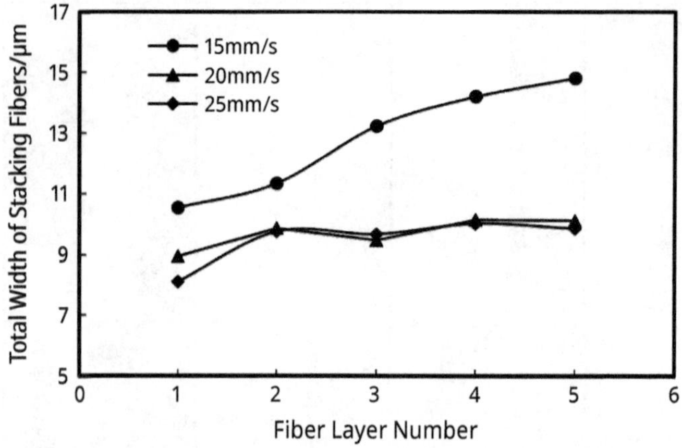

Figure 5.21: Relationships between the fiber stacking width and the number of fiber layers (solution concentration is 10%, spinneret-to-collector distance is 2 mm, and applied voltage is 1.5 kV).

Figure 5.22: SEM images of 60 layers of stacked fibers: (a) 100× magnification; (b), (c) 300× magnification; (d) 1,000× magnification; (e) 1,500× magnification; and (f) 4,000× magnification (solution concentration is 10%, spinneret-to-collector distance is 2 mm, applied voltage is 1.5 kV, and collector speed is 25 mm/s).

5.2.2.2 Formation of patterned 3D micro-nanostructures

Figure 5.26 shows a schematic diagram of a 3D box structure formed by stacking fibers. The direct-written pattern is a square with a side length of 500 μm. During the electrospinning process, no pause time is left at the corners of the square.

The 3D electrospun box structure is shown in Figure 5.27. The designed box in Figure 5.27(a) has a side length of 500 μm. Measured side lengths of the fabricated box are 556.55 μm and 514.89 μm, with errors of approximately 11.0% and 2.8%, respectively.

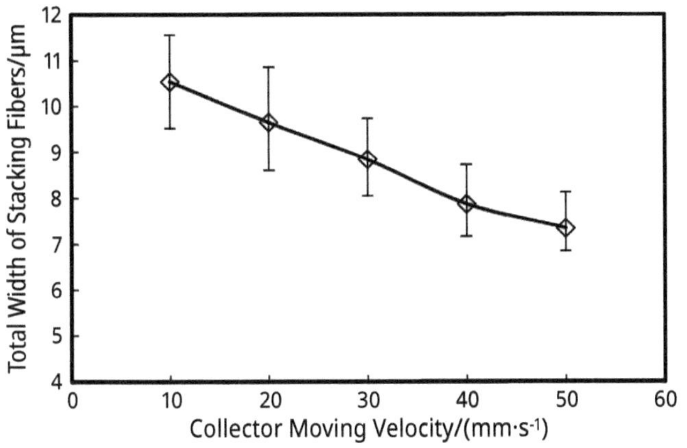

Figure 5.23: Total fiber width versus collector moving speed (10% solution concentration, spinneret-to-collector distance is 2 mm, and applied voltage is 1.5 kV).

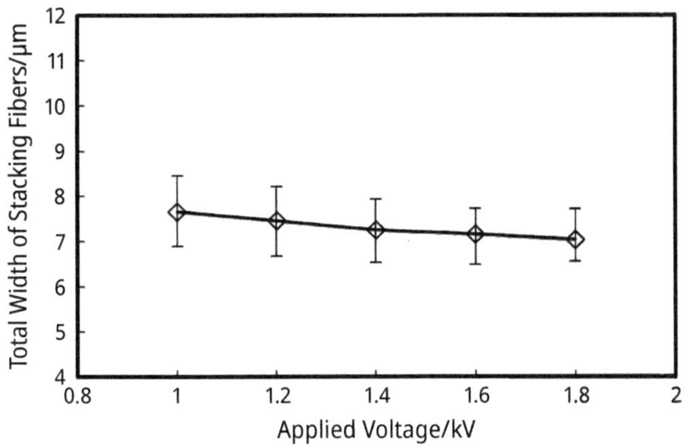

Figure 5.24: Total fiber width versus applied voltage (spinneret-to-collector is 2 mm, and collector speed is 25 mm/s).

The details of the box structure are further observed through SEM images. As shown in Figure 5.27(b), the box has a designed side length of 300 μm. During the initial deposition stage, the fibers exhibit vigorous whipping. However, as the number of fiber layers increases, the deposited fibers aggregate, and subsequent fibers deposit on these aggregated fibers. The side length of the prepared box is approximately 273 μm, smaller than the designed value, indicating that the jet gradually deposits inward. The stacked structure is located on the inner side of the fiber wave band, with a stacked width of about 17 μm and an inner side length of 239 μm.

The sidewall morphology of the box structures with different numbers of deposition layers is shown Figure 5.28. As the number of layers increases, more fibers are

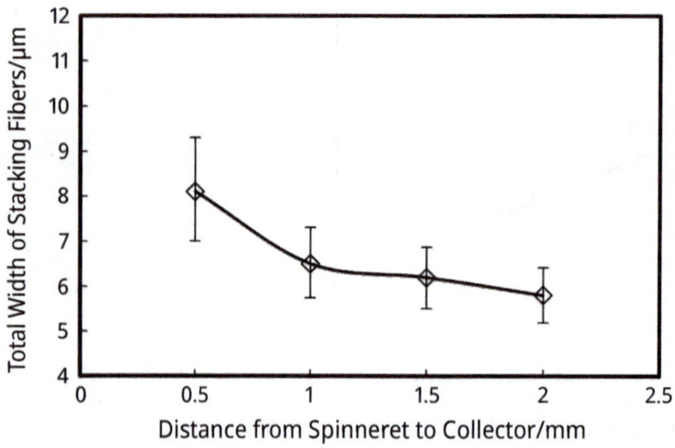

Figure 5.25: Total fiber width versus spinneret-to-collector distance (solution concentration is 10%, applied voltage is 1.5 kV, and collector speed is 25 mm/s).

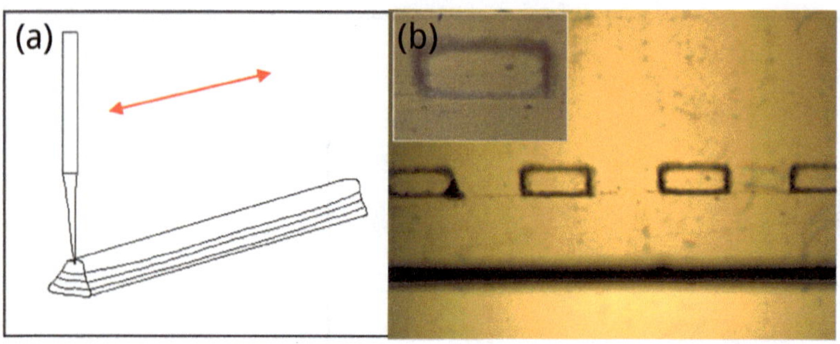

Figure 5.26: Schematic diagram and optical microscope diagram of the electrohydrodynamic direct-written 3D box structure: (a) schematic diagram of layer-by-layer fiber stacking; and (b) array of fiber-stacked box (10% solution concentration, collector speed is 5 mm/s, and spinneret-to-collector distance is 1 mm).

stacked, and the sidewall height increases. However, the number of disordered fibers also increases. As shown in Figure 5.28(e), although the fibers are tightly arranged and stacked layer-by-layer, they extend toward the inner side of the box, forming an adhesive arc transition at the inner corners.

When the number of deposition layers exceeds 50, the box corners exhibit an arc transition. Increasing the number of layers to 70 results in box morphology as shown in Figure 5.29, with an outer side length of 300 μm and an inner circular diameter of 170.5 μm. Since the designed pattern is square, it can be inferred that the fibers initially deposit from the outer contour of the pattern and gradually extend inward as the number of layers increases. After reaching a certain height, the square corners gradually become blurred, forming adhesive arc transitions. Rather than tight allignment at the be-

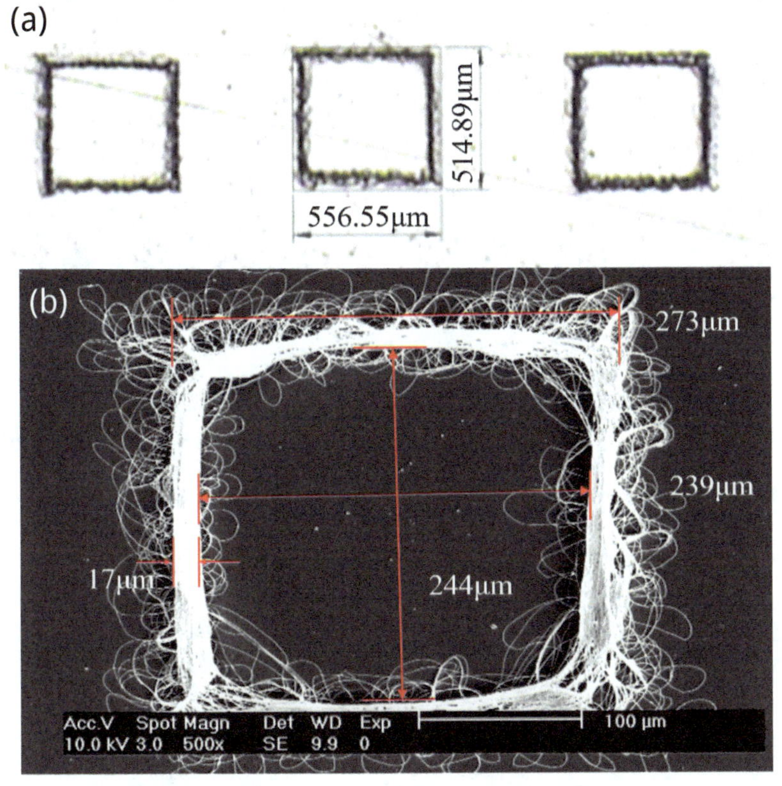

Figure 5.27: Electrohydrodynamic direct-written 3D box structure: (a) box with 55 stacked layers (designed size: 0.5 mm × 0.5 mm); and (b) box with 70 stacked layers (designed size: 300 μm × 300 μm) (solution concentration is 10%, applied voltage is 1 kV, collector speed is 5 mm/s, and spinneret-to-collector distance is 1 mm).

ginning of the deposition, gaps are observed among the arc-shaped fibers, indicating that the fibers experience significant tensile forces as they pass through the corners.

To investigate the size error of the square deposition affected by the number of stacked layers, the relationship between the diagonal dimensional error and the number of stacked layers is plotted, as shown in Figure 5.30. The dimensional error reflects the overall shape distortion. As the number of stacked layers increases, the diagonal dimensional error fluctuates slightly, with an average error of 40–50 μm for 10–90 layers. The designed diagonal length is 424 μm, corresponding to an error of 9–11%.

Precise fabrication of 3D grid structure is further investigated. Figure 5.31 illustrates a three-layer 3D grid structure. During the fiber stacking process, the collector speed is set to 20 mm/s in the process of fiber stacking to ensure that the jet deposition speed matches the collector speed. Under the stretching effect of the moving collector, the EDW jet deposits fibers along the preset trajectory, layer by layer, to form a 3D grid

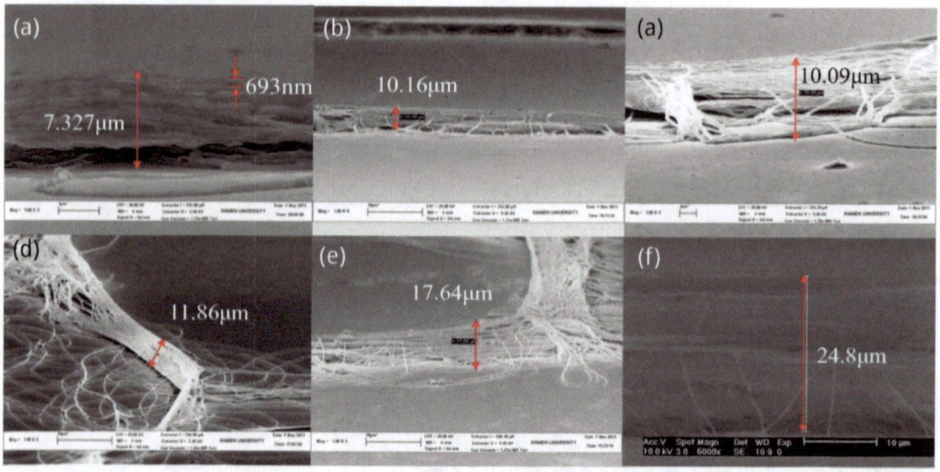

Figure 5.28: SEM images of box structures with different numbers of layers: (a) 15 layers; (b) 25 layers; (c) 35 layers; (d) 45 layers; (e) 55 layers; and (f) 70 layers (solution concentration is 10%, applied voltage is 1 kV, collector speed is 5 mm/s, and spinneret-to-collector distance is 1 mm).

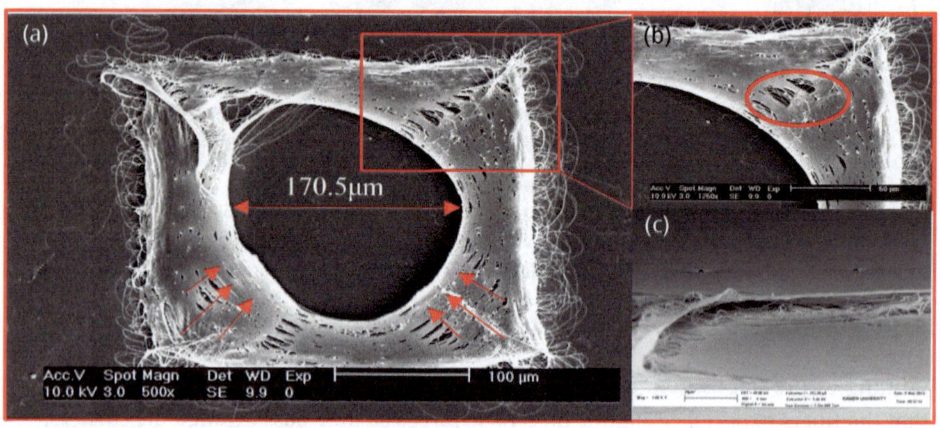

Figure 5.29: Circularly distributed fibers inside the box: (a) circularly distributed fibers formed inside the box, with a diameter of 170.5 μm; (b) enlarged view of the arc transition fibers formed at the corner; and (c) a hollow arc corner (10% solution concentration, applied voltage is 1 kV, collector speed is 5 mm/s, and spinneret-to-collector distance is 1 mm).

structure. The actual dimensions of the as-prepared fiber grid structure in Figure 5.31 are approximately (101.12 ± 0.25) μm × (100.57 ± 0.27) μm, which is basically consistent with the designed size of 100 μm × 100 μm with, an error of about 0.85%. The total width of the stacked fibers in the grid structure is about 10.24 μm. At the intersections of the grid pattern, the fibers stack stably, with the fiber width at the intersections slightly larger than that of the stacked fibers, measuring about 13.54 μm.

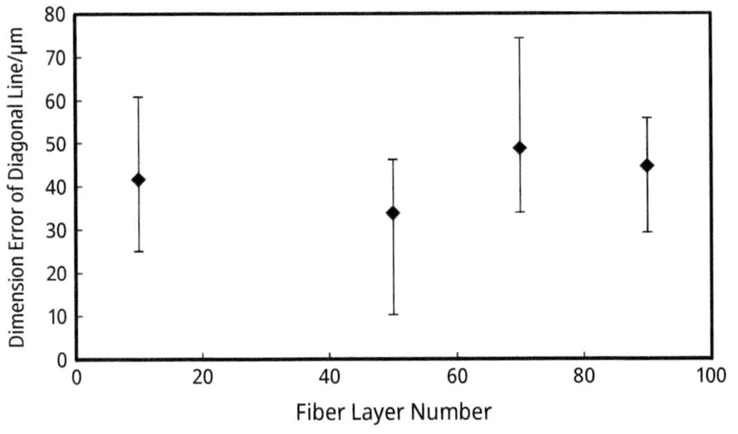

Figure 5.30: Relationship between diagonal dimensional error and number of stacked layers (solution concentration is 10%, applied voltage is 1 kV, collector speed is 5 mm/s, and spinneret-to-collector distance is 1 mm).

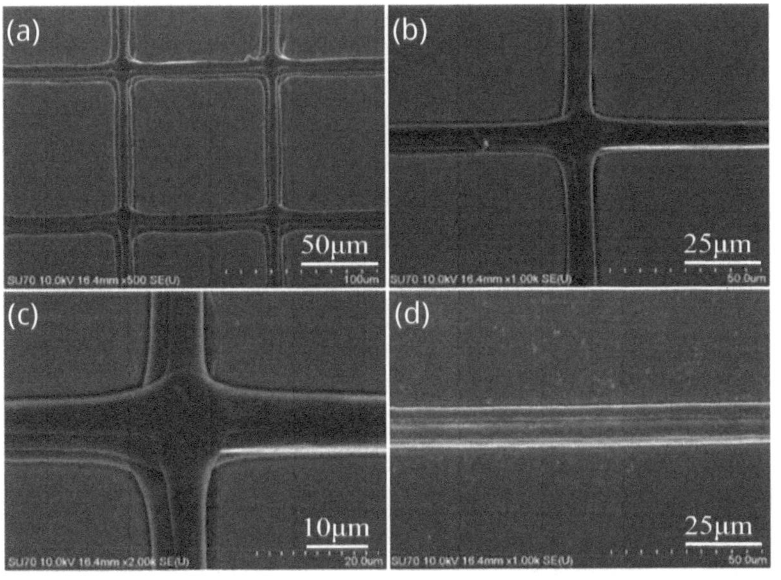

Figure 5.31: Three-layer 3D grid structure: (a) 100 μm × 100 μm fiber grid; (b) and (c) fiber grid intersections; and (d) horizontal fibers (solution concentration: 10%, spinneret-to-collector distance: 2 mm, applied voltage: 1.5 kV, collector speed: 25 mm/s).

Figure 5.32 shows 3D fiber grid structures with 10 and 20 stacked layers. As the number of direct-writing repetitions increases, the height of the 3D grid structure also increases. Additionally, with more layers, the enhanced local electric field strength improves the guidance of fiber deposition, leading to greater stability in jet deposition and higher precision in the direct-written patterns. The fibers in the lower layer are flattened and deformed due to the weight of the upper layers, with the degree of deformation decreasing in higher layers, as shown in Figure 5.33. Specifically, the fibers in the bottom four layers are significantly flattened, with the deformation decreasing in the order of the first layer > second layer > third layer > fourth layer. The fifth layer shows minimal deformation, and the deformation in subsequent layers remains consistent. In the 20-layer structure, all fibers except those in the bottom layer have equal widths, and the surface of the stacked fibers is smooth, which ensures the accurate direct-writing of the 3D fiber structures.

Figure 5.32: 3D grid structures with 10 and 20 layers of stacked fiber (a)–(c) 10 layers; (d)–(f) 20 layers (solution concentration is 10%, spinneret-to-collector distance is 2 mm, applied voltage is 1.5 kV, and collector speed is 25 mm/s).

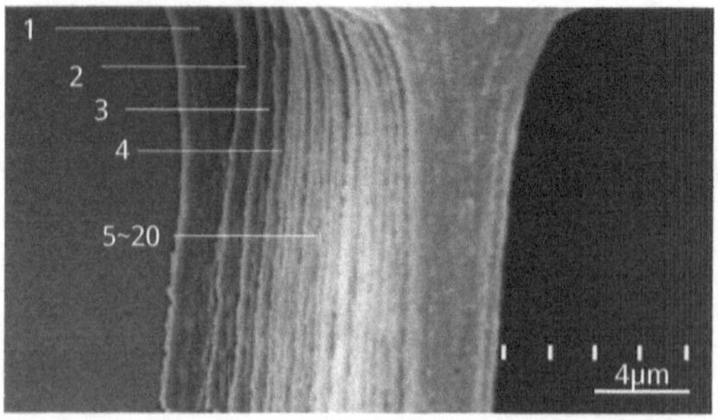

Figure 5.33: Deformation of stacked fibers in grid structures (solution concentration is 10%, spinneret-to-collector distance is 2 mm, applied voltage is 1.5 kV, and collector speed is 25 mm/s).

To further investigate the behavior of fibers at the intersection, a stacked grid intersection was analyzed, as shown in Figure 5.34. The width of the transverse fibers before and after passing through the intersection is approximately 5.12 μm and 5.07 μm, respectively. Similarly, the widths of the longitudinal fibers are 5.21 μm and 5.13 μm, respectively. The presence of the intersection does not affect the width of the continuous fibers before and after the intersection. However, the number of stacked layers at the intersection is twice that of the straight sections, resulting in a higher stacking height and the formation of small protrusions at the intersections. Due to the increased number of fiber layers at the intersections, the deformation of the fibers is more pronounced than in the straight sections. Fibers closer to the center of the intersection experience greater deformation, ultimately forming a cross-shaped intersection.

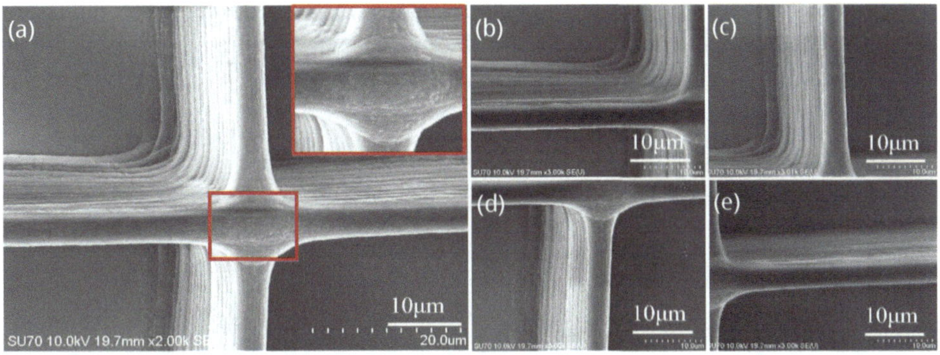

Figure 5.34: Fiber structures at grid intersections: (a) 20 layers of cross-stacked deposition fibers; (b) stacked fibers on the left side of the intersection; (c) stacked fibers on the upper side of the intersection; (d) stacked fibers fibers on the under side of the intersection; and (e) stacked fibers on the right side of the intersection (solution concentration is 10%, spinneret-to-collector distance is 2 mm, applied voltage is 1.5 kV, and collector speed is 25 mm/s).

Based on the study of the fundamental process control principles described above, we further investigate the printing and molding of more complex 3D micro-nanostructures. Leveraging the controllability of electric field-focused EDW jets, along with precise control of the collector's motion trajectory, fibers are guided to deposit along the preset trajectory, enabling the direct-writing of various complex 3D structures.

Figure 5.35 presents a 3D pentagram structure formed by orderly fiber stacking. The motion path of the collector is composed of straight lines, with a total of 15 layers. In the designed pentagram path, each side length is 1.5 mm, and each angle of the pentagram is 36°. The final deposited pentagram exhibits side lengths of 1.493 mm, 1.536 mm, 1.542 mm, 1.491 mm, and 1.506 mm, respectively, with an average of 1.514 mm, and an error of 0.91%. The morphology of the pentagram's apex is shown in Figure 5.35(c), where straight fibers continuously intersect and overlap at a fixed angle. Measurements reveal that the angle formed by the bottom-layer fibers is 36.3°, while that by the top-layer fibers is 36.2°, indicating that the angle of subsequent fibers remains consistent during the stacking process, which is close to the designed angle. The five angles in the 3D pentagram pattern measure 34.5°, 36.4°, 35.7°, 35.3°, and 36.5°, respectively, with an average of 35.7° and an average error of 0.89%. Both the side lengths and the angles exhibit minimal errors, allowing the pentagram structure to be fully and accurately reproduced.

A regular pentagon pattern is formed. The average side length of the pentagon is measured to be 0.356 mm, which is close to the designed value of 0.354 mm. The fiber deposition and stacking morphology at the intersection point are shown in Figure 5.35 (d) and (e). The fibers cross and stack uniformly and stably, with a fiber width of approximately 4.16 μm and a height of approximately 2.56 μm per layer, as shown in Figure 5.35(f). The height of the whole 3D pentagram structure is about 38.4 μm. During the EDW process, the collector moving speed should match the jet deposition speed to ensure stable jet ejection, eliminating spiral whipping interference, and achieve precise fiber positioning and 3D shaping.

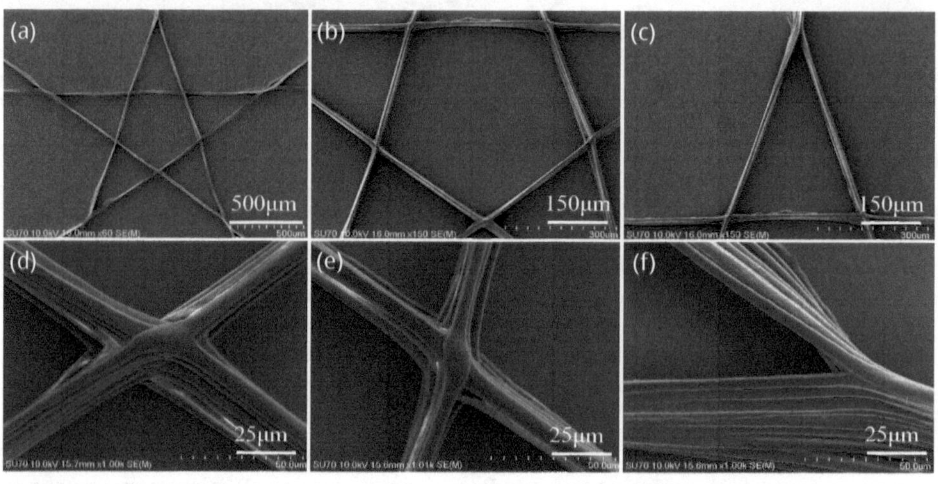

Figure 5.35: Electrohydro direct-writing of 3D pentagram fiber structure: (a) complete pentagram; (b) pentagon formed by intersecting fibers; (c) morphology of fiber stacking at the pentagram corner; (d) and (e) deposition morphology at intersection point; and (f) sidewall fibers (solution concentration of 10%, spinneret-to-collector distance of 2 mm, applied voltage of 1.5 kV, collector speed of 25 mm/s).

5.2.2.3 Fabrication of 3D hollow cylinder structures

In addition to stacking straight lines to form 3D structures, curve 3D structures are also essential in practical applications. When the collector follows an arc-shaped trajectory, a centripetal force is generated:

$$F_n = mv^2/r = m\omega^2 r \tag{5.6}$$

where m is the mass of the jet, v is the velocity of the collector, ω is the angular velocity, and r is the curvature radius of the motion trajectory.

Due to the presence of the centripetal force, the jet is stretched and pulled, causing the fibers to move inward along the trajectory during the arc-shaped deposition. Figure 5.36(a) shows a 3D elliptical structure with a major axis of 600 μm and a minor axis of 500 μm. The curvature radii at the ends of the major and minor axes are 720 μm and 416 μm, respectively. The moving speed of the collector is 30 mm/s. Due to the influence of the centripetal force on the jet, the deposition point of each fiber layer is slightly offset, resulting in uniform fiber arrangement. As shown in Figure 5.36(b), the fibers eventually tilt and deposit, forming a hollow, tilted elliptical 3D structure. After stacking 20 layers, the fiber at the vertex of the major axis is offset by 57.12 μm, and the deposition point of each layer is offset by an average of 2.86 μm. Similarly, the fiber offset at the vertex of the minor axis is 48.23 μm, with an average offset of 2.41 μm per layer. The offset of the deposition point is affected by the centripetal force. Increasing the collector speed or reducing the curvature radius of the motion trajectory can increase the fiber offset, and vice versa. Figure 5.36(c) shows the deposition morphology of a 3D structure with an elliptical trajectory where the major and minor axes are 300 μm and 250 μm, respectively. The collector speed remains constant at 20 mm/s, and the curvature radii at the vertices of the major and minor axes are 360 μm and 208 μm, respectively. As the centripetal force increases, the fibers are evenly wound side by side on the collector, as shown in Figure 5.36(d). Although the curvature radii at the vertices of the major and minor axes differ; the deposition results are similar. Due to the continuous ejection of the jet along the arc-shaped trajectory, the jet is subjected to internal stress, and the offset of the deposition point will not exceed the diameter of the fiber. This indicates that at curvature radii of 360 μm and 208 μm, the fibers will not stack into a 3D structure. As the number of fiber layers increases, the jet offset increases, and the centripetal force acting on the jet decreases. When the centripetal force decreases to match the jet's internal stress, the fibers begin to deposit and stack, forming a hollow cylinder as shown in Figure 5.36(c).

The forces acting on the jet during its superposition along an arc-shaped trajectory are illustrated in Figure 5.37. When printing the bottom layer of the fiber structure, the jet is primarily subjected to two forces: the downward electric field force F_E and the upward viscoelastic force F_c. The resultant force is directed vertically downward, enabling the jet to eject stably in that direction. When direct-writing the second layer of fibers, the solvent in the jet rapidly penetrates into the first layer, creating a strong bond between the two layers. At this stage, the new jet experiences a frictional force f_1, which generates a centripetal force F_a directed towards the center of the arc. Under the combined action of the centripetal force, the electric field force and the viscoelastic force, the jet deposition deviates, and the viscoelastic force F_c changes along the jet direction. As the jet's deviation distance increases, its actual motion radius r decreases. According to Eq. (5.6), when the angular velocity ω remains unchanged, the centripetal force decreases with the radius. When the centripetal force and horizontal

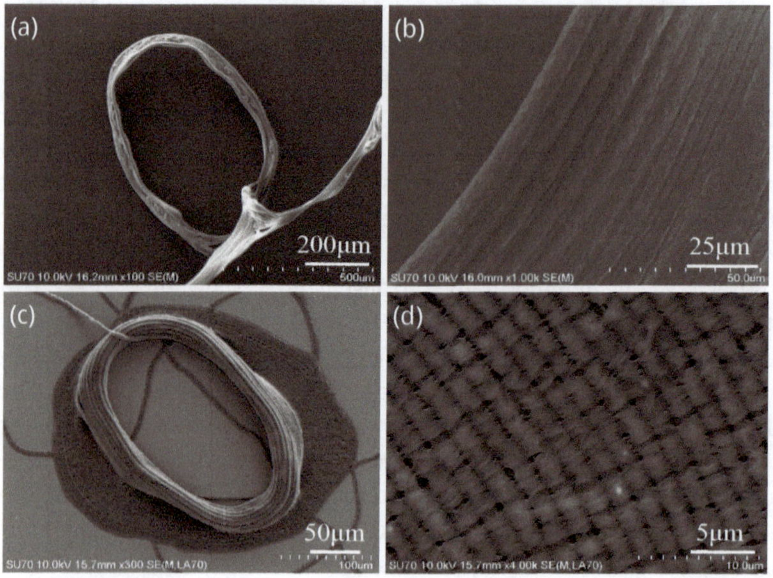

Figure 5.36: Electrohydrodyanmic direct-writing 3D elliptical structure: (a) elliptical fiber structure with a major axis of 600 μm and a minor axis of 500 μm; (b) inclined stacked fibers; (c) elliptical fiber structure with the major axis of 300 μm and the minor axis of 250 μm; and (d) side-by-side winding fibers (solution concentration is 10%, spinneret-to-colletor distance is 2 mm, applied voltage is 1.5 kV, and collector speed is 30 mm/s).

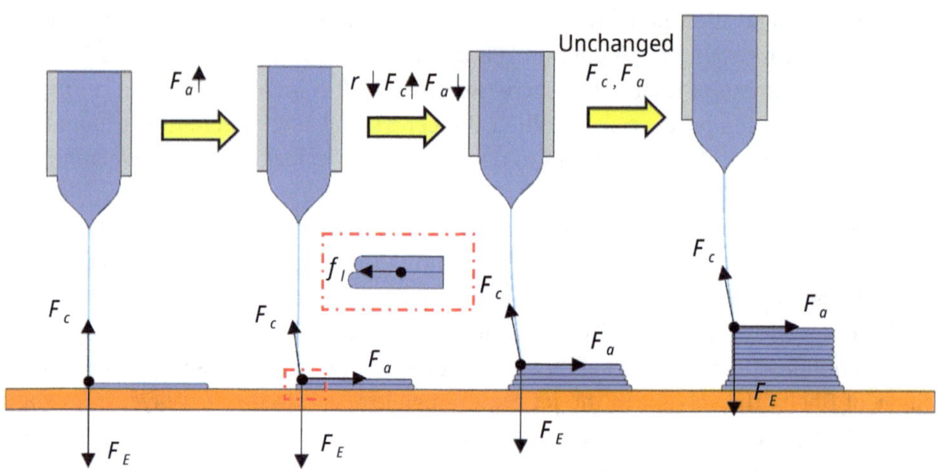

Figure 5.37: Forces acting on the jet during fiber deposition along an arc-shaped trajectory.

component of the viscoelastic force reach equilibrium, the jet ceases to deviate and the fibers will continue to stack vertically upward.

Figure 5.38(a) shows a 3D hollow cylinder structure, the lowest layer of which is a circle with a diameter of 300 μm. The structure was fabricated by repeating the direct-writing process 400 times at a collector speed of 20 mm/s. During fiber stacking, the height of the spinneret is adjusted at a speed of 20 μm/s to maintain a constant distance between the spinneret and the deposition point. Initially, the jet experiences a strong centripetal force, resulting in a significant deposition point offset and inclined fiber deposition. After stacking around 200 layers, the deposition point deviates by approximately 60 μm, averaging 0.30 μm per layer. At this stage, the centripetal force acting on the jet decreases, and the offset begins to decrease. Finally, as the number of layers increases to 400, the diameter of the circular fiber structure measures about 160.15 μm, with an offset of about 9.93 μm over the last 200 layers, averaging 0.05 μm per layer. The fibers stacked vertically upward, maintaining a tight and orderly arrangement along the curved edge without significant misalignment as the number of layers increases, as shown in Figure 5.38(b) and (c). The height of a single fiber stack is about 2.64 μm, as shown in Figure 5.38(d). The total height of the hollow cylinder structure is about 1.056 mm.

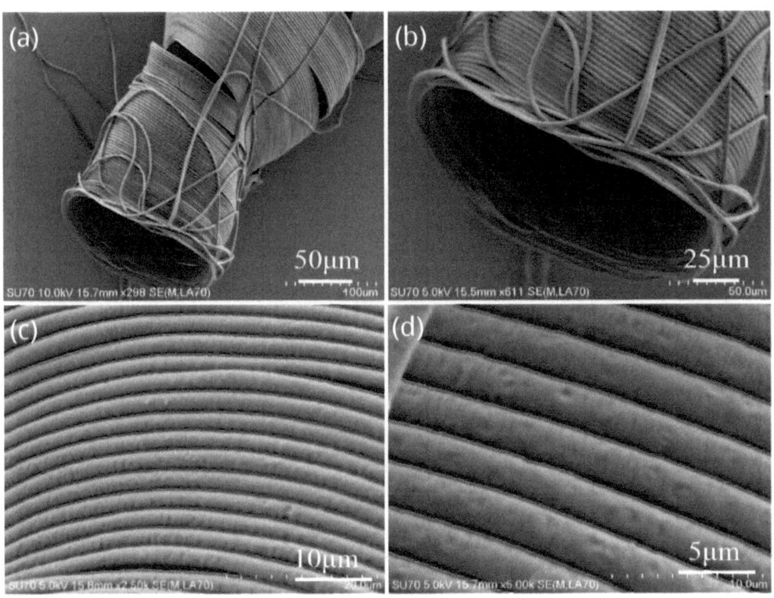

Figure 5.38: Electrohydrodynamic direct-writing 3D hollow cylinder structures: (a) overall 3D fiber structure; (b) fibers at the top; and (c) and (d) fibers at the sidewall (solution concentration of 10%, spinneret motion speed of 20 μm/s, initial spinneret-to-collector distance of 2 mm, applied voltage of 1.5 kV, collector speed of 20 mm/s).

References

[1] Kim H-Y, Lee M, Park K J, Kim S, Mahadevan L. Nanopottery: Coiling of electrospun polymer nanofibers [J]. Nano Lett 2010, 10(6), 2138–2140.
[2] Coelho R, Debeau J. Properties of the tip-plane configuration [J]. J Phys D: Appl Phys 1971, 4(9), 1266–1280.

Chapter 6
Electrohydrodynamic direct-writing functional micro-nanostructure and its applications

6.1 Micro-nanochannels

6.1.1 Fabrication of micro-nanochannels

In fabrication of micro-nanochannel, the sacrificial layer nanofibers are produced by electrohydrodynamic direct-writing (EDW) [1]. The printing ink is made of 18% PEO in a mixed solvent of deionized water and alcohol (3:2, v/v). The spinneret used here is a solid metal probe with a tip diameter of 0.04 mm to produce finer nanofibers. The probe handle is hydrophobically treated before the experiment, while the probe tip remains its hydrophilicity.

Figure 6.1(a) shows the EDW experimental setup. The distance between the spinneret and the collector is 1.5 mm. The process flow for preparing micro-nanochannels is shown in Figure 6.1(b), including:

(1) Substrate cleaning: A silicon wafer substrate is ultrasonically cleaned in acetone.
(2) Nanofiber EDW: Nanofibers are directly electrospun on the substrate as the sacrificial layers for micro-nanochannel preparation.
(3) Gold sputtering: A gold film about 30 nm in thickness is used as the channel material and is sputtered above the nanofibers.
(4) Sacrificial layer (nanofiber) removal: Removal of sacrificial nanofiber and the gold layer covering is done by rinsing the gold-coated substrate in chloroform solution, accompanied by ultrasonic agitation. The micro-nanochannel is then fabricated. In this process, ultrasonic cleaning power is set to 80 W, cleaning frequency of 40 kHz, for a duration of 15 min.

Smooth-edged micro-nanochannels are well fabricated through gold sputtering and sacrificial layer removal. Figure 6.2(a) shows a residual nanofiber of insufficient ultrasonic cleaning (ultrasonic time: 10 min). It is observed that the ultrasonically cleaned nanofibers are dissolved, split, and separated from the silicon substrate and nearby metal layers. The resulted micro-nanochannels have basically the same width as that of nanofibers. It is clear that the width of the micro-nanochannels can be effectively controlled by the diameter of the nanofiber. To confirm the complete removal of sacrificial layer and the adherent gold film, energy-dispersive X-ray analysis is used to examine the energy spectrum on gold film surface at the bottom and outer regions of micro-nanochannels. Graphs in Figure 6.2(b) show that only the elements silicon and oxygen are present at the bottom of micro-nanochannels, indicating that polymer nanofibers and the gold film have been completely removed. The comparison results

https://doi.org/10.1515/9783110797879-006

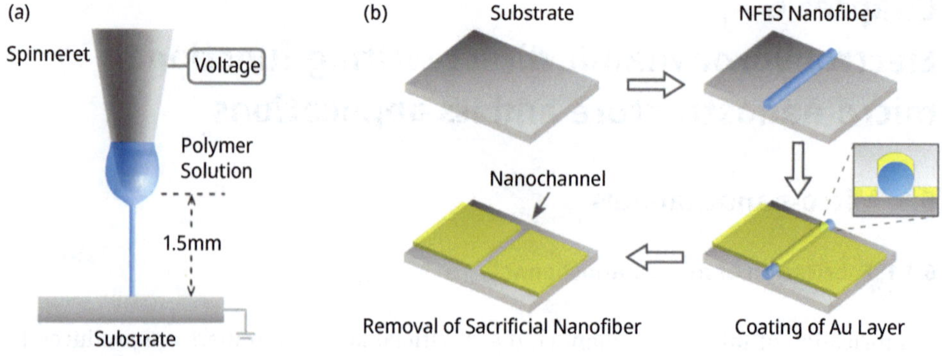

Figure 6.1: Experimental details of nanochannel fabricated via NFES. (a) The experimental setup of NFES. (b) The schematic diagram of nanochannel fabrication process [1].

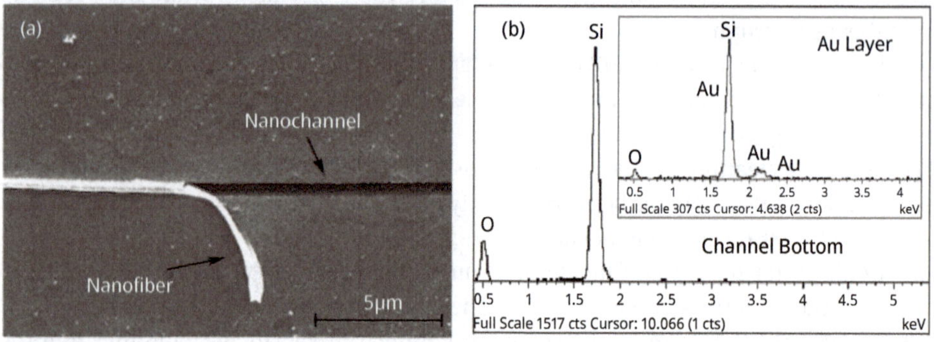

Figure 6.2: Micro-nanochannel and its energy spectrum analysis. (a) A residual nanofiber on the substrate under an insufficient ultrasonic cleaning time (10 min). The channel width is almost equal to the nanofiber diameter. (b) EDX spectra of nanochannel bottom and Au layer beside the channel [1].

of the non-micro-nanochannel region shown in the inset reveal the presence of Au along with Si and O, confirming that the gold film remained intact.

6.1.2 Fabrication of micro-nanochannels by EDW

As mentioned in earlier section, the nanofiber diameter determines the width of the micro-nanochannel. Therefore, one can fabricate micro-nanochannels of specific dimensions by controlling the EDW process parameters. The micro-nanochannel width is positively related to the applied voltage as shown in Figure 6.3. The corresponding average width of the micro-nanochannels increases from 133 nm to 13.54 μm when the applied voltage rises from 1.2 kV to 2.5 kV.

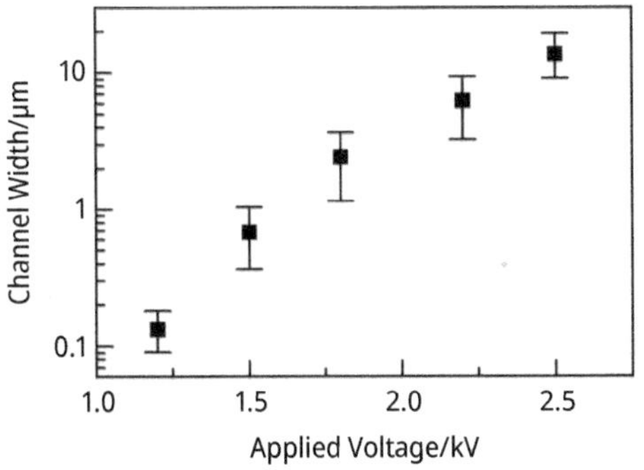

Figure 6.3: Relationship between channel width and applied voltage [1].

In addition, during the EDW process, by changing the motion trajectory of the collector, nanofiber patterns of different structures can be obtained. This further enables the control of the shape of the micro-nanochannels, resulting in the different micro-nanochannel structures shown in Figure 6.4.

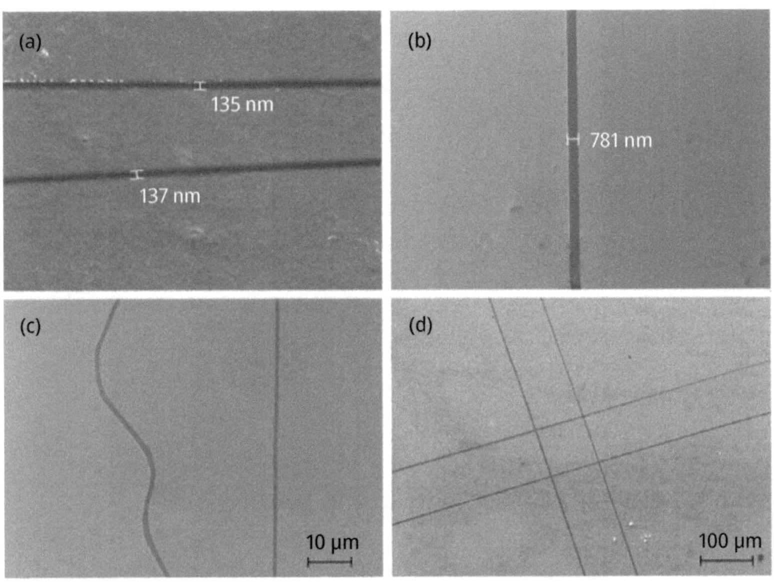

Figure 6.4: Different micro-nanochannel structures fabricated by EDW. (a) Two aligned nanochannels, with widths of 135 nm and 137 nm, applied voltage 1.2 kV. (b) Nanochannel with a width of 781 nm, applied voltage 1.5 kV. (c) Straight and curve micro-nanochannels on single substrate, with widths of 1.5 μm and 750 nm, applied voltages of 1.8 kV and 1.5 kV, respectively. (d) Nanochannel grid, with channel widths of 800 nm and spacing of 100 μm, applied voltage of 1.5 kV [1].

Technology of micro-nanochannel fabrication is further developed by combining EDW with the traditional microelectromechanical system (MEMS) process. First, the required patterned substrate is fabricated by the traditional MEMS process to define the shape of the required electrode (Figure 6.5) on the silicon substrate. Thereafter, micro-nanochannels are prepared by EDW as part of the circuit. The nanofibers are electrospun directly on the patterned substrate (Figure 6.6(a)). The deposition morphology of the nanofibers on the substrate can be controlled by changing the trajectory of the collector. Successive steps, such as gold film sputtering, photoresist spin coating and nanofiber removal, are carried out. The gold film covering the photoresist and nanofiber is removed, while the remaining gold film retains on the silicon substrate, forming an electrode with a micro-nanochannel structure (Figure 6.6(b)). This composite technology is also applied to produce multielectrode structures as shown in Figure 6.6(c).

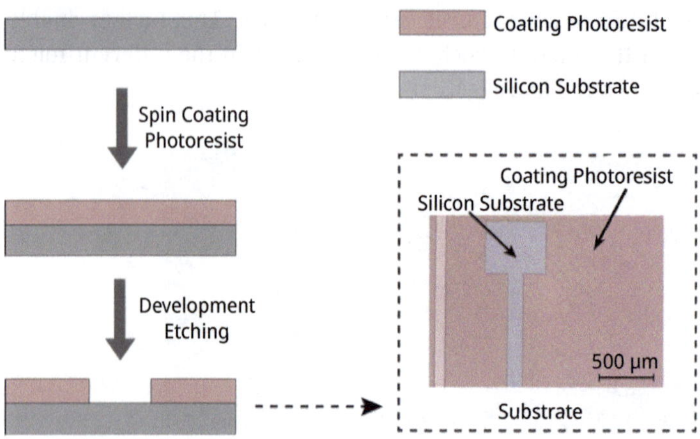

Figure 6.5: Fabrication process of patterned silicon substrate.

6.2 Typical circuit components

6.2.1 Fabrication of silver-based conductive structures by EDW

The silver-based conductive structures are EDW fabricated with commercial silver ink (TEC-IJ-040, InkTec). The silver ink, as a soluble organometallic compound, decomposes to silver monomer precipitate by sintering and curing to form conductive structures. This is achieved by using EDW with a spinneret of 0.06 mm inner diameter, followed by curing the silver ink at 250 °C on a hot plate for 20 min.

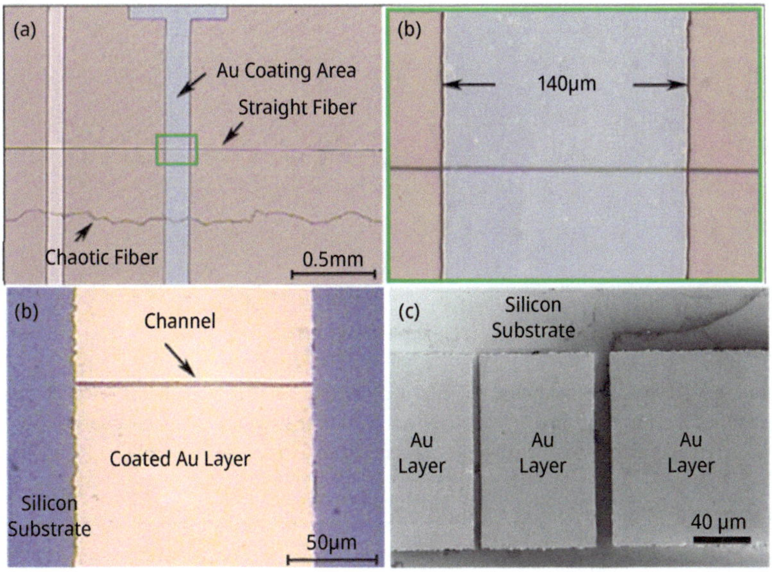

Figure 6.6: Optical images of electrospun nanofibers, micro-nanochannels, and electrodes: (a) Linear and curved nanofibers on a silicon substrate, collected at speeds of 150 mm/s and 100 mm/s, respectively; (b) micro-nanochannel transferred from linear nanofiber; and (c) multiple electrodes separated by micro-nanochannels [1].

6.2.1.1 Electrohydrodynamic direct-writing of silver structures with DC voltage

Figure 6.7 presents the silver structures that are electrospun at different voltages applied between the spinneret and collector [2]. The silver ink starts to eject at a critical voltage of 1.0 kV, having the spinneret-to-collector distance at 0.25 mm. The diameter of silver structures increases with the applied voltage, while damaging the uniformity of the structure. Numerous satellite droplets are observed around the silver structure at applied voltages of 1.4 and 1.5 kV. If the spinneret-to-collector distance increases to 0.75 mm, the required voltage also increases. The ejection starts at approximately 1.5 kV. The surrounding satellite droplets and its poor uniformity remain unresolved (Figure 6.8). Moreover, at this distance of 0.75 mm, the voltage has little effects on the feature size of silver structures.

Because of the low viscosity of the silver ink used in the experiment, atomization is prone to occur during the EDW process. With the increase in the applied voltage, the amount of charge increases during the silver ink ejection, so the atomization phenomenon becomes more significant, and the diameter of the silver structure increases accordingly. When the distance between the spinneret and the collector increases to 0.75 mm, the atomization phenomenon becomes more severe, so the relationship between the diameter of the silver structure and the applied voltage is not significant.

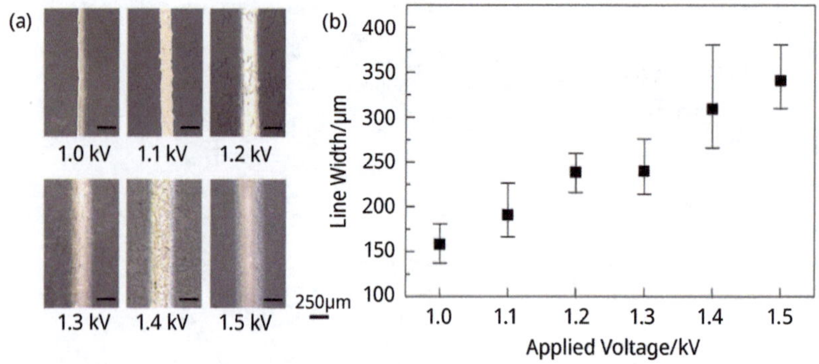

Figure 6.7: Control of applied voltage on dimensions of EDW silver structures at a spinneret-to-collector distance of 0.25 mm: (a) EDW silver structure morphology; and (b) silver structure diameter dependence on applied voltage (constant collector speed at 10 mm/s, silver ink feeding rate at 20 μL/h) [2].

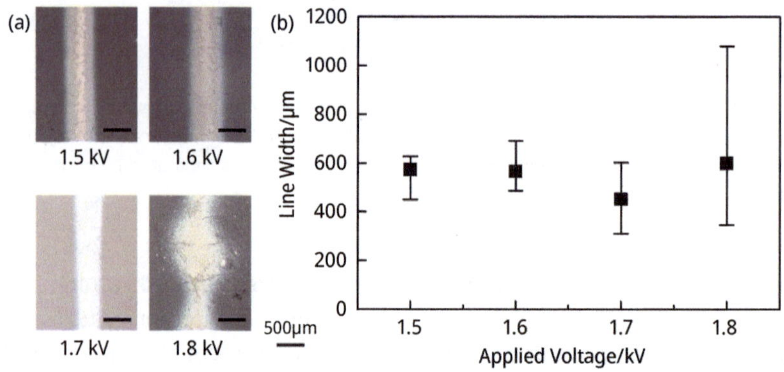

Figure 6.8: Control of applied voltage on dimensions of EDW silver structures at the spinneret-to-collector distance of 0.75 mm. (a) EDW silver structure morphology. (b) Silver structure diameter dependence on applied voltage (constant collector speed at 10 mm/s and silver ink feeding rate at 20 μL/h).

Figure 6.9 presents the silver structures that are electrospun at different ink feeding rates. The silver structure diameter increases with the feeding rate at the spinneret-to-collector distance of 0.25 mm and the applied voltage of 1.1 kV. When the parameters are increased to 0.75 mm and 1.5 kV, changing the feeding rate does not reduce the spray of numerous satellite droplets surrounding the printed silver structure, and the uniformity of the prepared silver diameter remains poor (Figure 6.10).

Additionally, increasing the collector speed can reduce the ink volume accumulated at one particular point along the writing path, and thus reduces the diameter of the direct-written silver structure. Figure 6.11 shows the silver structures with different collector speeds (at the spinneret-to-collector distance of 0.25 mm). Diameter of the silver structures is compressed from 0.278 mm to 0.128 mm when the collector speed is increased from 5 mm/s to 20 mm/s.

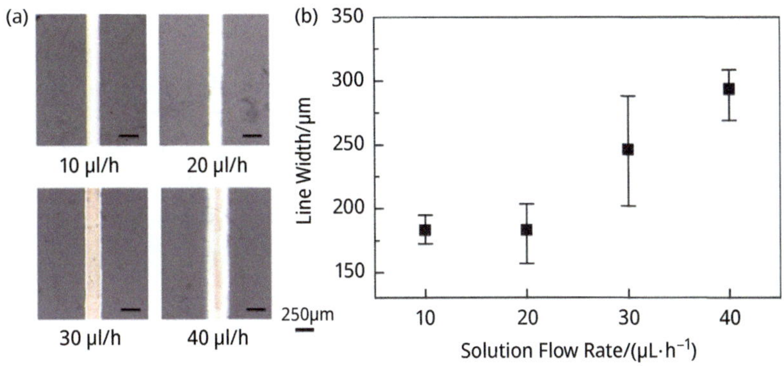

Figure 6.9: Impact of silver ink feeding rate on EDW structure diameter at the spinneret-to-collector distance of 0.25 mm. (a) Morphology of the silver structures. (b) Relationship between silver structure diameter and ink feeding rate (collector speed at 10 mm/s, applied voltage at 1.1 kV) [2].

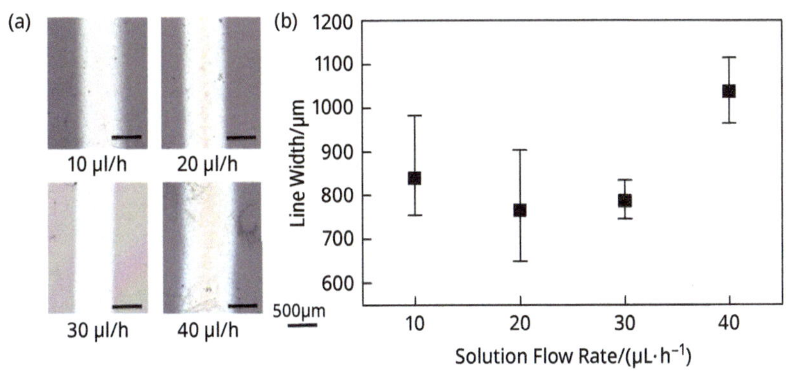

Figure 6.10: Impact of silver ink feeding rate on EDW structure diameter at the spinneret-to-collector distance of 0.75 mm. (a) Morphology of the silver structure. (b) Relationship between silver structure diameter and ink feeding rate (collector speed is 10 mm/s, and applied voltage is 1.5 kV).

It must be noticed that the ink atomization remains significant when the spinneret-to-collector distance is set to 0.75 mm (Figure 6.12). In the aforementioned process, the silver ink has low viscosity, and it is easy to atomize. Another factor is that the silver ink jet is continuously ejected at a high speed under the DC voltage, so the silver structures of larger feature size are obtained.

6.2.1.2 Electrohydrodynamic direct-writing of silver structures under pulse voltage

When DC voltage is applied in patterned EDW, the obtained silver structure morphology is in a continuous linear form, which limits the application of EDW under DC voltage to a certain extent. Therefore, we also apply pulse voltage to print silver structure on demand. In the process of pulse ejection, the jet mainly deposits in the form of

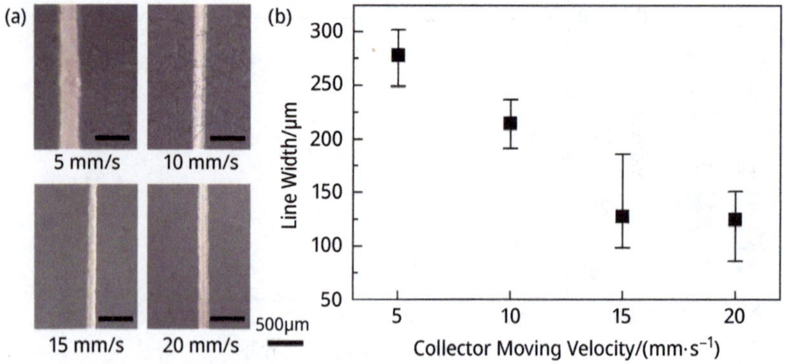

Figure 6.11: Impact of collector speed on EDW silver structure at the spinneret-to-collector distance of 0.25 mm. (a) Morphology of the silver structure. (b) Relationship between silver structure diameter and collector speed (ink feeding rate 20 μL/h, applied voltage 1.1 kV).

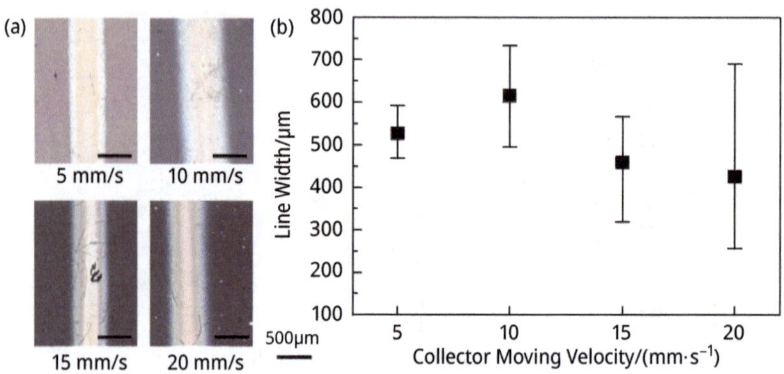

Figure 6.12: Collector speed has less effect on EDW structure diameter at the spinneret-to-collector distance of 0.75 mm. (a) Morphology of the silver structure. (b) Results of silver structure diameter with collector speed (silver ink feeding rate: 20 μL/h, applied voltage: 1.5kV).

droplets, and various continuous structures can be formed through the overlap of droplets.

The DC voltage leads to stable jet ejection, while pulse voltage causes intermittent jet ejection [3]. Figure 6.13 shows the ejection of silver ink within a single pulse cycle. At 0 s, when the output is at a low level (0 kV), due to the continuous ink supply and the absence of an electric field force, the solution forms a free-hanging droplet at the spinneret. Starting from 8.5 ms, when the output is at a high level (1.2 kV), the electric force stretches the surface of the solution into a cone. Between 9.0 and 9.5 ms, before the ejection stabilizes, the solution shows a pulsation retraction process. This phenomenon mainly results from the unstable motion of charges and ink during the formation of the cone jet. After self-regulation, a continuous stable jet is formed from

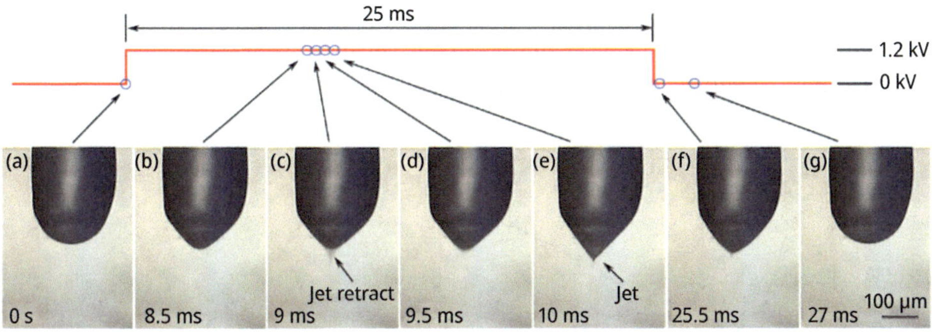

Figure 6.13: EDW jet ejection behavior in a single pulse cycle (spinneret-to-collector distance is 0.4 mm, the inner diameter of the spinneret is 0.11 mm, the ink feeding rate is 5 μL/h, the pulse voltage frequency is 20 Hz and the duty cycle is 50%) [3].

10.0 ms and persists until the high-level output ends. The ejection process terminates when the electric field disappears (25.5–27.0 ms).

The silver ink printing is performed at a repetition rate of 20 Hz, producing both droplet-like and line structures. Droplet-like silver structures with diameters ranging from 45 to 60 μm and intervals of 125 μm are obtained at a collector speed of 2.5 mm/s (Figure 6.14(a)), while a line structure with a width of 60 μm is produced through droplet stacking at a collector speed of 0.4 mm/s (Figure 6.14(b)). At high magnification, the line exhibits a clear edge contour and minimal satellite droplets in Figure 6.14(c).

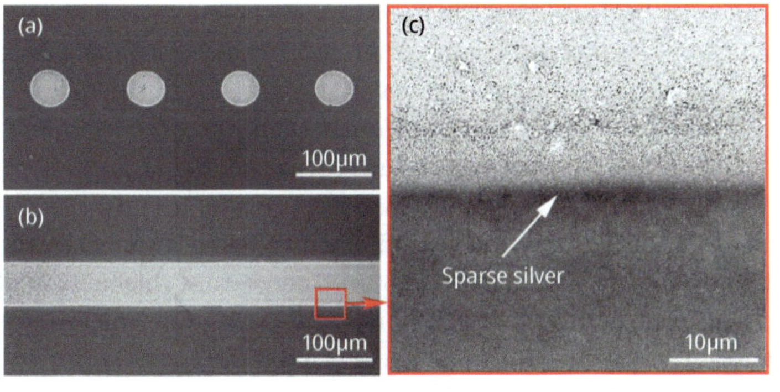

Figure 6.14: Silver structure printed on the collector under pulse voltage. (a) Discrete round droplets. (b) A line structure formed by the stacked droplets. (c) Profile of the linear structure edge (spinneret-to-collector distance 0.4 mm, pulse voltage frequency 20 Hz, duty cycle 50%, voltage amplitude 1.2 kV, and collector speed 2.5 m/s) [3].

The pulse voltage supply has an important impact on the morphology of printed silver droplet patterns. Average diameters of the ejected droplets increase approximately from 50 to 200 μm with increase in the pulse voltage from 1.2 to 2.0 kV, as shown in

Figure 6.15. Meanwhile, larger voltage induces greater charge repulsion force inside the solution, resulting in the jet atomization and formation of excessive satellite droplets around the main structures, roughening the structure edge.

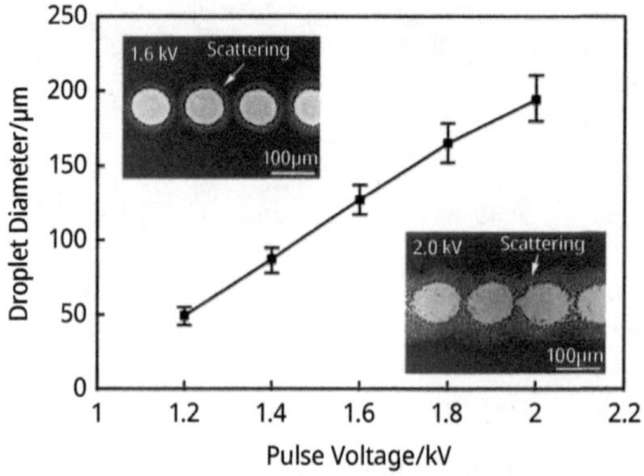

Figure 6.15: Effect of pulse voltage amplitude on the morphology of the silver droplets. The insets show ink scattering round deposited dots under applied voltages of 1.6 kV (top-left) and 2.0 kV (bottom-right) (spinneret-to-collector distance is 0.4 mm, the pulse voltage frequency is 20 Hz, the duty cycle is 50%, and the collector speed is 2.5 m/s) [3].

6.2.2 Electrical properties of conducting silver structure

Silver ink has low conductivity before curing. Silver monomers precipitate and join to produce excellent conductivity by thermal curing. Therefore, curing conditions have significant impacts on the electrical properties of silver structures. The resistivity of direct-written silver structures produced under different conditions are examined. The silver wires of length $L = 10$ mm (Figure 6.16(a)) are cured at 130, 150, and 200 °C, respectively, and the curing periods are set to 5–60 min. We use a multimeter to measure the terminal resistance R of the wire and a profilometer (Dektak3 Series, Veeco) for the cross-sectional profile of the wire (Figure 6.16(b)). The average thickness H of silver structures is about 300 nm, and the average diameter D is 0.25 mm. The resistivity is calculated by the formula $\rho = RHD/L$, and the effects of curing temperature and time on changes in resistivity are therefore obtained, as shown in Figure 6.17.

The experimental results show that the resistivity of direct-written silver structure decreases with higher curing temperature and longer curing time, and it eventually stabilizes. The wire resistivity reaches its minimum value of about 9 μΩ · cm after curing at 130 °C for 30 min. As the curing temperature increases, the resistivity decreases, and the required curing time also shortens accordingly. After curing at 200 °C for 10 min, the resistivity stabilizes at around 3.3 μΩ · cm, which is about 2.1 times that of pure silver. We use scanning electron microscopy (SEM) to analyze the surface

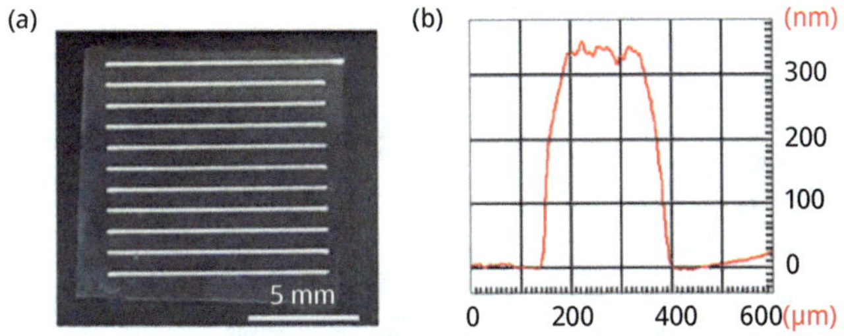

Figure 6.16: Electrohydrodynamic direct-written silver structure and its profile: (a) the EDW silver wire structures and (b) cross-sectional profile of the wire [3].

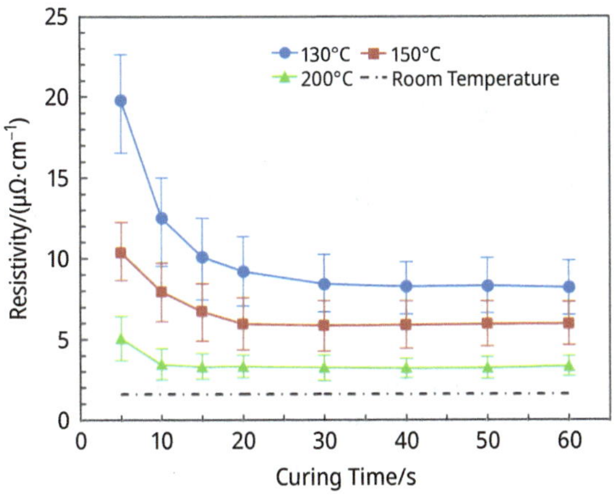

Figure 6.17: Impact of curing condition on the resistivity of silver structure [3].

morphology of silver structures after curing at different temperatures for 60 min, as shown in Figure 6.18. At room temperature, silver monomers precipitate and agglomerate. It shows a high resistivity without thermal curing. When the curing temperature rises, the silver particles melt and link to form a conductive layer. High temperature makes the silver particles bind tightly together and thus successfully decreases the resistivity of conductive layer. It can still be seen that there are a lot of pores in the cured silver structure, which is why the resistivity remains higher than that of pure silver.

6.2.3 Fabrication of typical circuit components vias EDW

6.2.3.1 Typical passive components

Based on previous work, we use the EDW technology to print three typical passive components – resistors , capacitors, and inductors. Figure 6.19(a) shows a polymer re-

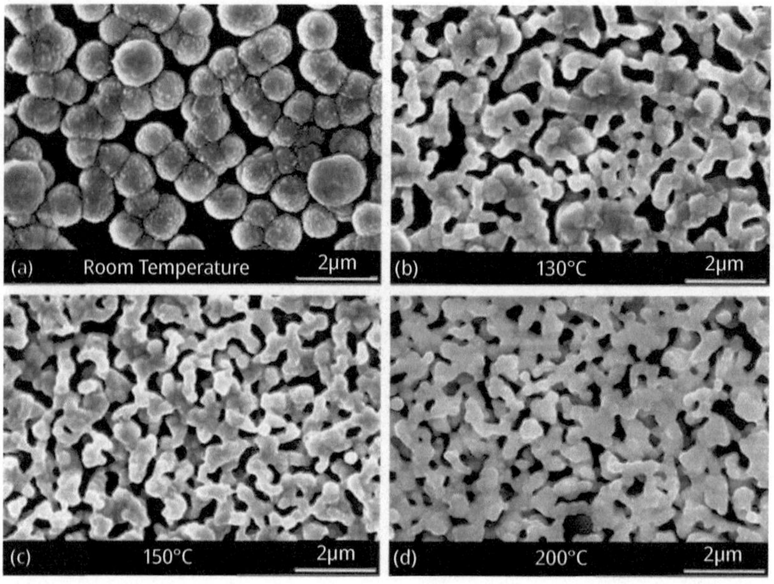

Figure 6.18: Surface morphology of silver structures after curing at different temperatures for 60 min: (a) room temperature; (b) 130 °C; (c) 150 °C; and (d) 200 °C [3].

sistor device using EDW-printed silver structure as the conductor. The thermoplastic polymer PEDOT:PSS (PEDOT: poly3, 4-vinyl dioxythiophene monomer; PSS: sodium polystyrene sulfonate)/PVP(polyvinylpyrrolidone) composes the resistance layer. The content ratio of the compositions directly affects the resistance value. Figure 6.19(b) shows a 30-tooth comb capacitor having a tooth length of 9 mm, a line width of 60–70 μm, tooth spacing of about 80 μm, and a measured capacitance value of about 6 pF. Figure 6.19(c) shows a planar inductor coil with 12 turns. The average line width and interval are 0.22 mm and 0.1 mm, respectively. The inductance value is about 0.6 μH.

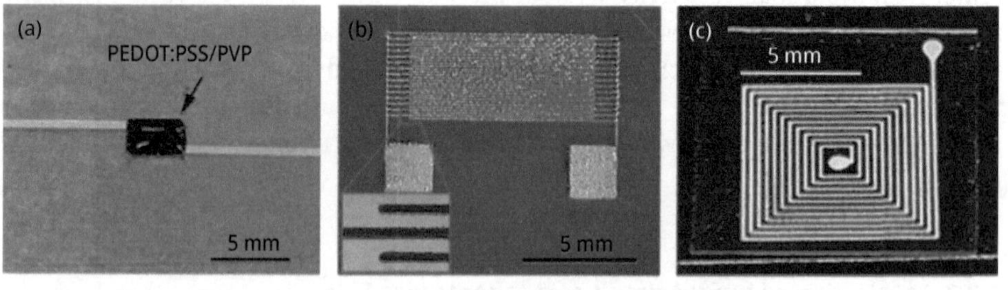

Figure 6.19: Passive components printed by EDW technology: (a) resistance; (b) capacitance; and (c) inductor [3].

6.2.3.2 Multilayer flexible interconnect circuit

We attempted to perform EDW of multilayer composite structures on flexible substrate (Figures 6.20 and 6.21) [2,4]. The interconnection circuit use flexible polyimide film as substrate, EDW silver wire as structures I and II, and printed PVP insulator as the middle layer (Figure 6.20). PVP middle layer shows the outstanding insulation property, and thus the electrical test detects almost zero current across the I–II composite structure. This work provides a solid research foundation for the low cost, rapid preparation of flexible circuits through EDW.

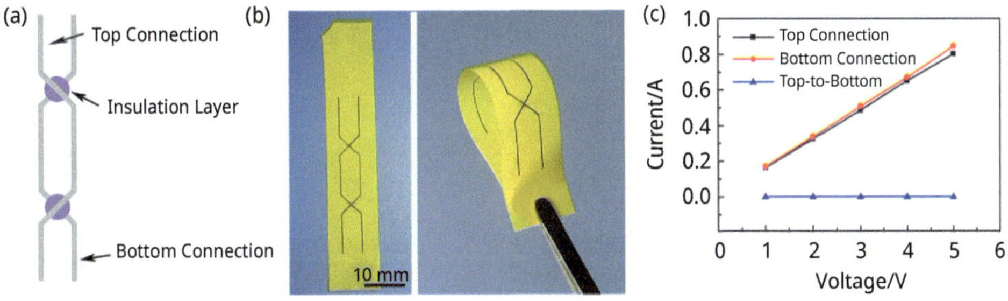

Figure 6.20: Multilayer composite flexible circuit: (a) schematic diagram; (b) EDW flexible circuit; and (c) electrical test results [4].

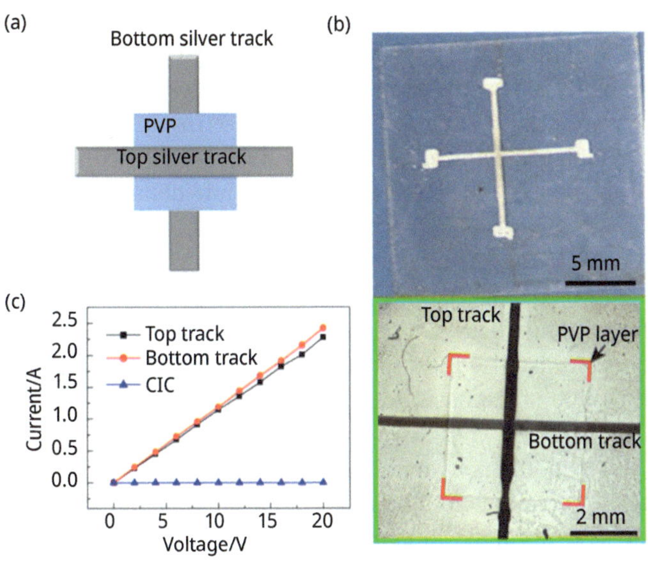

Figure 6.21: Multilayer flexible interconnects circuit: (a) schematic diagram; (b) printed flexible circuit; and (c) electrical characteristics test [2].

6.3 Microsemiconductor sensor components

6.3.1 Fabrication of ZnO nanofibers by EDW

ZnO is a common semiconductor functional material with a wide forbidden band, high exciton binding energy, good electromechanical coupling performance, environmental friendliness, and good biocompatibility, making it a promising material for applications in the fields of optoelectronics and sensing. Using EDW technology, the preparation of ZnO nanofibers mainly involves three steps, as shown in Figure 6.22 [5].

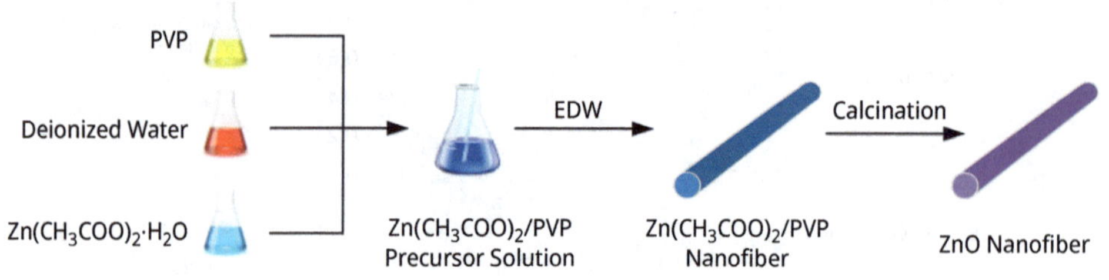

Figure 6.22: Flow diagram for preparing ZnO nanofibers by EDW.

The fabrication process of ZnO nanofibers is as follows:

(1) Preparation of precursor solution: Zinc acetate [$Zn(CH_3COO)_2 \cdot H_2O$] is used as the precursor to prepare ZnO nanofibers. The low viscosity of the pure inorganic salt solution is not suitable for EDW process of nanofibers. Therefore, an inorganic salt/polymer mixture having a larger viscosity is valid for electrospinning. Firstly, 0.8 g of PVP is stirred in 5 mL of deionized water for 4 h until PVP is completely dissolved; then 0.3 g zinc acetate is added and stirred for 12 h to obtain a clear $Zn(CH_3COO)_2$ precursor solution.

(2) EDW $Zn(CH_3COO)_2$/PVP nanofibers: The spinneret is a stainless steel hollow needle with an inner diameter 0.16 mm. The liquid feeding rate is set to 10 μL/h. A flat glass or silicon wafer with microstructures on its surface is used as the collector. The spinneret-to-collector distance is set to 2 mm, and the applied voltage is 2 kV. Patterned $Zn(CH_3COO)_2$/PVP nanofibers are then produced by controlling the collector movement.

(3) Sintering $Zn(CH_3COO)_2$/PVP nanofibers in an air atmosphere: The precursor nanofibers are heated up and sintered at 500 °C in an air atmosphere on hot plate for 1 h. This process promotes the decomposition of $Zn(CH_3COO)_2$/PVP and PVP to the final product ZnO.

The SEM image in Figure 6.23(a) presents the EDW $Zn(CH_3COO)_2$/PVP nanofibers, with a smooth surface and a diameter distribution ranging from 500 to 1,500 nm. $Zn(CH_3COO)_2$/PVP decomposes by sintering and is oxidized to form ZnO nanofibers. The nanofiber maintained a continuous fiber shape with diameter reduced to several hundred nanometers (Figure 6.23(b)). There are small ZnO crystallites adhering on the nanofibers making the surface rough. Figure 6.23(c) shows the X-ray diffraction (XRD) pattern of the ZnO nanofibers obtained after sintering the precursor nanofibers. The characteristics peaks identify that composition of the nanofibers is ZnO and it is the polycrystalline Wurtzite structure (JCPDS 36-1451).

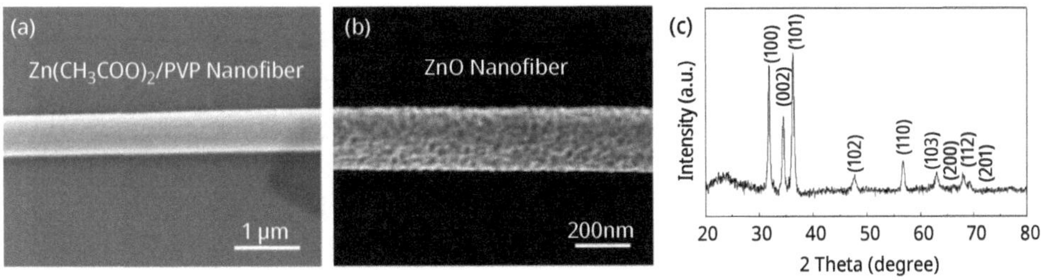

Figure 6.23: Electrospun $Zn(CH_3COO)_2$/PVP nanofibers and prepared ZnO nanofibers: (a) SEM image of electrospun $Zn(CH_3COO)_2$/PVP nanofibers; (b) ZnO nanofibers prepared by calcining the precursor; and (c) XRD pattern of prepared ZnO nanofibers [5].

6.3.2 Electrohydrodynamic direct-writing and positioning deposition of ZnO nanofiber

During the EDW process, the collector speed directly affects the deposition of nanofibers [5]. We investigated the morphology of ZnO nanofibers obtained at different collector speeds. Figure 6.24 shows the morphology of deposited ZnO nanofibers produced when the collector moves in a straight line at 20 mm/s, 40 mm/s, 80 mm/s, 100 mm/s, 150 mm/s, and 200 mm/s. When the collector moves slowly (20 mm/s), the nanofibers deposit in a spiral pattern on the collector, forming rather disordered fiber bands. As the collector moves faster, the whipping-deposition range of the nanofibers gradually decreases, and the width of the fiber band also becomes smaller. When the collector speed is further increased to match the jet deposition speed, a single straight nanofiber could be produced on the collector. At this matched collector speed, precise positioning and patterned deposition of nanofibers can be achieved. In the subsequent experiments, we fixed the collector speed at 200 mm/s for the production of ordered ZnO nanofibers.

We design the grid pattern with intervals of 15 and 30 μm (Figure 6.25(a)). The EDW nanofibers are written along the pattern trajectory, forming ordered ZnO nanofiber grid array on the collector (Figure 6.25(b)). SEM images reveal that the ZnO nanofibers integrated to a whole grid with good physical contact after sintering, given that the precursor nanofibers are written interlaced (Figure 6.25(c)). Figure 6.26(a) and (b) shows the spiral-shaped and curved ZnO nanofibers, respectively.

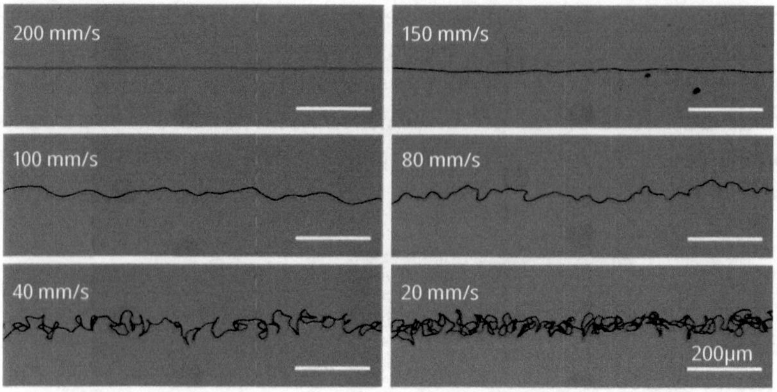

Figure 6.24: ZnO nanofibers produced at different collector speeds (applied voltage, 2 kV; spinneret-to-collector distance, 2 mm) [5].

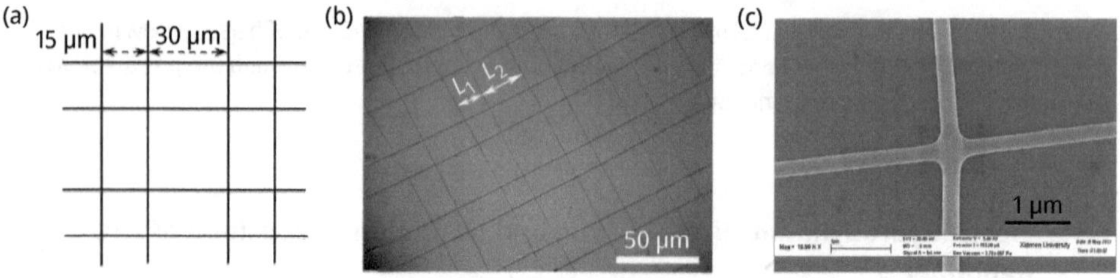

Figure 6.25: Electrospun ZnO nanofiber arrays: (a) designed pattern; (b) patterned ZnO nanofibers; and (c) interconnected ZnO nanofibers after sintering [5].

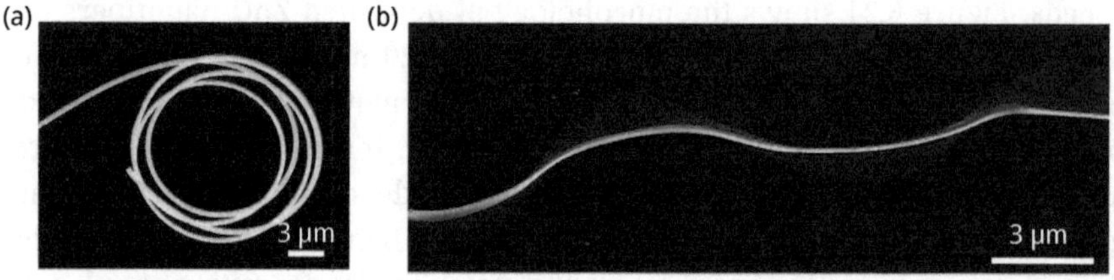

Figure 6.26: Different shapes of ZnO nanofibers: (a) helical shaped and (b) curved shape [5].

EDW technology provides a good technical approach for the integration of nanofibers and micro-nanostructures. We fabricate microstructures with 5 μm height on the silicon substrate and use it as collector to investigate the integration of EDW ZnO nanofibers and specific micro-nanostructures. The ZnO nanofibers are deposited on top of a silicon beam (Figure 6.27(a)) or suspended precisely across two micro-pillars (Figure 6.27(b)). Experimental results confirm that EDW technology is compatible with traditional silicon microfabrication processes. It becomes easy to integrate func-

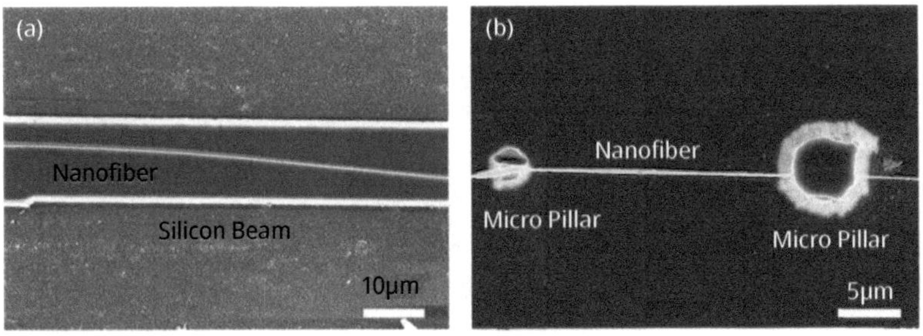

Figure 6.27: Printing and integration of ZnO nanofibers and microstructures. (a) Nanofiber on top of the silicon beam. (b) Nanofiber suspended across two micro-pillars [5].

tional nanofibers and micro-nanostructures, providing good technical conditions for the application prospects of EDW in micro-nanodevices and system integration.

6.3.3 Fabrication of microsemiconductor sensor devices by EDW

6.3.3.1 ZnO nanofiber field effect transistors (FETs)

ZnO nanofiber FET is mainly composed of ZnO nanofiber bridge, electrodes, silicon substrate and silica dielectric layer (Figure 6.28). The fabrication process includes: (1) produce silicon dioxide insulation layer and electrodes (source, drain and gate) on heavily doped (n++) silicon substrate by MEMS technology; (2) produce Zn $(CH_3COO)_2$/PVP nanofibers by EDW on the electrode; and (3) high-temperature sinter to prepare ZnO nanofibers.

Figure 6.28(a) shows the schematic structure of the ZnO nanofiber FET. The optical image of devices and the SEM image of ZnO nanofiber across the electrode are shown in Figure 6.28(b) and (c), respectively. The electrode spacing measures approximately 12 μm, and the ZnO nanofibers have a diameter of about 500 nm. Connecting the source, drain, and gate as depicted in Figure 6.28(b), the electrical characteristics of FET are performed by semiconductor parameter analyzer. Various gate bias voltages (0 V, 20 V, 40 V, 60 V, and 80 V) are applied, with a source-drain voltage range of 0–10 V. The corresponding I–V curves are depicted in Figure 6.28(d). The observed increase in Drain-Source current I_{ds} with rising gate Voltage V_g signifies the N-type behavior of the ZnO nanofiber FET, indicating ZnO nanofibers' characteristic as N-type semiconductors.

6.3.3.2 ZnO/CuO nanofiber PN junctions

To prepare PN junction devices, CuO is chosen as the p-type semiconductor material. Copper acetate $[Cu(CH_3COO)_2]$ is used as the precursor to prepare CuO nanofibers. The solution preparation method and component content, etc., are the same as $Zn(CH_3COO)_2$/PVP.

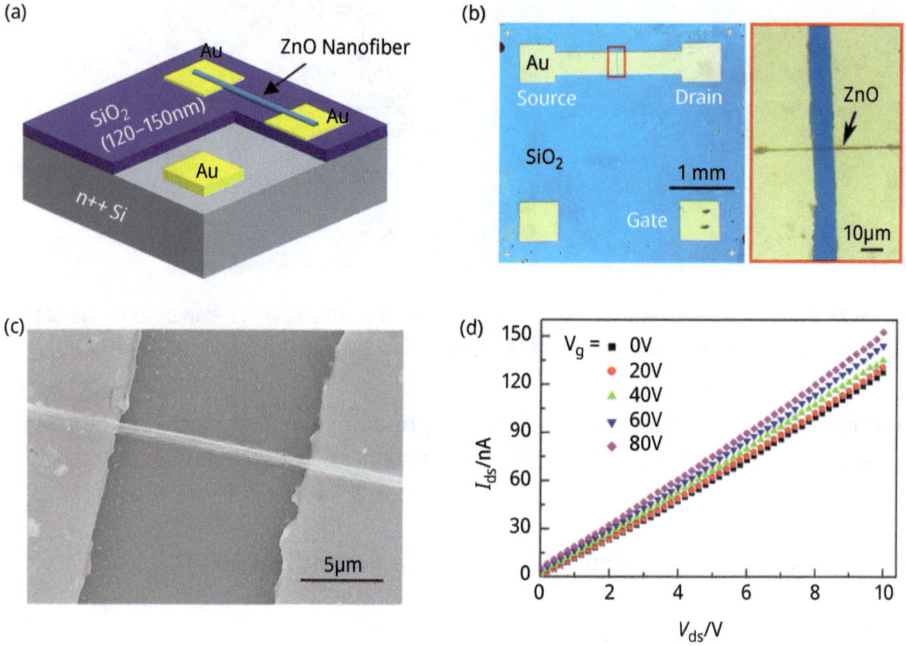

Figure 6.28: ZnO nanofiber FET. (a) Schematic diagram of the FET structure. (b) Photograph of the FET device. (c) SEM image of the ZnO nanofiber crossing two Au electrodes. (d) Gate-dependent current-voltage characteristics of the single ZnO nanofiber FET [5].

Fabrication of the designed PN junction device as shown in Figure 6.29 involves the direct-writing of $Zn(CH_3COO)_2$/PVP and $Cu(CH_3COO)_2$/PVP nanofibers on the electrodes, followed by sintering. Sintering results in good physical contact between the nanofibers, forming the PN junction at the points where ZnO nanofibers and CuO nanofibers come into contact. The electrode spacing on the collector (approximately 1 mm) is much larger than the diameter of the nanofibers, which is unfavorable for the measurement of electrical properties. Therefore, the experiment employed a repeated motion of the collector to directly write 50 strips of $Zn(CH_3COO)_2$/PVP and $Cu(CH_3COO)_2$/PVP nanofibers, respectively. The *I*–*V* testing curve of the PN junction (Figure 6.29(c)) indicates that the device exhibited effective one-way conduction.

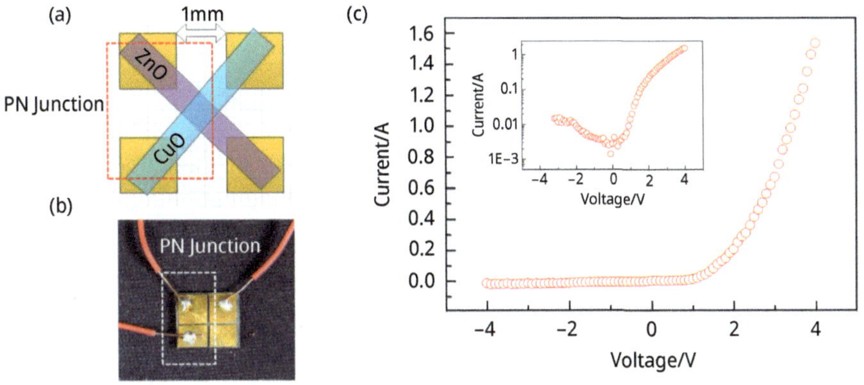

Figure 6.29: ZnO/CuO nanofiber PN junction: (a) schematic diagram of the structure; (b) physical diagram; and (c) I–V test curve on the PN junction.

References

[1] Wang X, Zheng G F, Xu L et al. Fabrication of nanochannels via near-field electrospinning. Appl Phys A-Mater, 2012, 108(4), 825–828.

[2] Zheng G F, Pei Y B, Wang X, Zheng J Y, Sun D H. Electrohydrodynamic direct-writing of conductor-insulator-conductor multi-layer interconnection. Chinese Phys B, 2014, 23(6): 066102.

[3] Wang X, Xu L, Zheng G F, Cheng W, Sun D H. Pulsed electrohydrodynamic printing of conductive silver patterns on demand. Sci China Technol Sc. 2012, 55(6), 1603–1607.

[4] Sun L L, Jiang J X, Yu Z J, Zheng G F, Sun D H. Electrohydrodynamic Direct-Write Multi-laver Micro Circuit on Flexible Substrate [C]. 5th International Conference on Flexible and Printed Electronics, ICFPE. 2014. Beijing. China:147–151.

[5] Wang X, Zheng G F, He G Q, et al. Electrohydrodynamic direct-writing ZnO nanofibers for device applications. Mater Lett. 2013, 109,58–61.

Index

https://doi.org/10.1515/9783110797879-007